AF577023

TRANSMISSION ELECTRON MICROSCOPY OF SILICON VLSI CIRCUITS AND STRUCTURES

TRANSMISSION ELECTRON MICROSCOPY OF SILICON VLSI CIRCUITS AND STRUCTURES

R. B. Marcus and T. T. Sheng

Bell Laboratories
Murray Hill, New Jersey

A Wiley-Interscience Publication

JOHN WILEY & SONS

New York Chichester Brisbane Toronto Singapore

Library of Congress Cataloging in Publication Data:

Marcus, R. B. (Robert B.)
Transmission electron microscopy of silicon VLSI circuits and structures.

"A Wiley-Interscience publication."
Includes index.
1. Integrated circuits. 2. Electron microscopy.
3. Electron microscope, Transmission. I. Sheng, T. T. (Tan Tsu) II. Title.

TK7874.M26 1983 621.381'73 83-3469
ISBN 0-471-09251-7

Printed in the United States of America

10 9 8 7 6 5 4 3 2 1

To the memory of L. O. Brockway,

R. B. Marcus

To H. H. Mollenhauer who guided me into this field,

T. T. Sheng

PREFACE

The application of transmission electron microscopy (TEM) to problems in very large scale integrated (VLSI) circuit technology has been a slow process. Until the mid-1970s, TEM studies were considered useful, but seldom essential, and were generally made to prove a point or to produce data in exploratory programs that supported processing efforts in circuits with large scale integration (LSI). Two factors coincided to bring about a change in this attitude. First, TEM sample preparation methods finally reached a stage of development where analyses could be performed quickly. Thus, information on microstructure could be generated (and interpreted) within 2–3 days. This success was largely a result of TEM sample preparation methods developed at Bell Laboratories by T. T. Sheng and the use of a special TEM test pattern on device wafers. Second, at about the same time, a critical stage was reached in the development of near-micron and submicron LSI and VLSI technology, which called for better microscope resolution than provided by optical or scanning electron microscopy (SEM). TEM was needed!

Chapter One describes the role of microscopy in silicon integrated circuit (IC) technology, and Chapter Two discusses the application of TEM to this technology, including a description of sample preparation methods. Chapters Three through Seven present electron micrographs of various classes of VLSI features, while Chapter Eight illustrates TEM techniques for analyzing device leakage and breakdown problems. Most of the micrographs in this book were chosen from studies of *n*-channel MOS structures, which were generated by X-ray lithographic processing in the Advanced LSI Development Laboratory at the Bell Laboratories in Murray Hill, New Jersey. Although some features illustrated in the micrographs pertain only to aspects of *n*-channel devices, most of the micrographs contain information that is relevant to other silicon device technologies as well.

The question often asked the electron microscopist in a VLSI technology study is, "What does a specific feature—polysilicon step coverage, edge of gate oxide, side wall slope, and so on—look like?" The micrographs in Chapters Three through Seven are typical answers to this question. Electron micrographs often reveal unexpected morphological features; these features show a lack of control over a processing step or indicate a real or potential source of device failure. Other micrographs demonstrate how well the process can control a troublesome point and show that a problem has been solved. In electron micrographs, evidence of deviations from ideal morphological configurations is relatively easy to find, but the practical importance of these deviations varies. For example, a 1000-Å penetration of the substrate during window etching may seem to indicate a serious lack of control over that processing step, and indeed, is useful information during development of a new dry etching procedure. However, in terms of the performance and reliability of the completed device, the effect of the overetch may be trivial.

Each micrograph in Chapters Three through Seven is placed on an odd-numbered page with a diagram of the micrograph plus figure caption and relevant labels on the facing page; with this format a micrograph can be viewed without the obstruction of lettering and arrows. Chapter Eight does not follow the format of the previous five chapters, primarily because other types of illustrations are needed to demonstrate clearly the application of TEM methods to device leakage and breakdown problems. The micrographs in Chapter Eight answer the question, "What unique characteristic at a specific location in the device causes leakage or breakdown?" These micrographs were made at leakage and breakdown sites found in *pn* junctions and in capacitors by electron-beam induced current (EBIC) studies using an SEM.

The successful integration of TEM into the VLSI diagnostics effort at Bell Laboratories is largely a result of the strong support and leadership given the VLSI technology program by M. P. Lepselter, Director of the Advanced LSI Development Laboratory, and H. J. Levinstein, Head of the Materials Technology Department. The micrographs in this book are from diagnostic programs that took place between 1977 and the present. Many individuals contributed to these programs, and references are given for both published and unpublished work. The interpretations of the micrographs are from numerous conversations with these individuals in corridors, over coffee, and at formal meetings. In particular, we wish to acknowledge useful and pleasant interactions with A. C. Adams, H. J. Boll, C. C. Chang, W. Fichtner, E. N. Fuls, D. B. Fraser, E. Kinsbron, R. Knoell, R. A. Kushner, R. Levin, P. S. D. Lin, W. T. Lynch, S. P. Murarka, E. I. Povilonis, A. K. Sinha, and S. Vaidya. We also wish to thank L. Doerries and W. R. Wagner for critical review of the manuscript, and R. E. Caffrey for having suggested this book.

R. B. Marcus
T. T. Sheng

Murray Hill, New Jersey
August 1983

CONTENTS

TRANSMISSION ELECTRON MICROSCOPY OF SILICON VLSI CIRCUITS AND STRUCTURES

CHAPTER ONE

VLSI TECHNOLOGY AND MICROSCOPY

1.1 INTRODUCTION TO IC TECHNOLOGY

The technology of silicon integrated circuits (ICs) has evolved from the single-point-contact transistor invented by Bardeen and Brattain in 1947 to today's ICs with over 100,000 components on a quarter-inch square chip. Progression from the point-contact transistor, to the junction transistor proposed by Shockley, to the planar process in the late 1950s, to the first silicon IC chip in 1959[1] required an increasingly enormous expenditure of effort to understand, develop, and control new materials, devices, and processing technologies. By 1961 silicon bipolar IC chips were being produced in earnest by Fairchild and Texas Instruments. Development of the junction field-effect transistor in 1953[2] led to the introduction of metal-oxide-semiconductor (MOS) technology in the early 1960s followed by the development of memory and logic circuits.

The density of components packed into a chip has increased rapidly since the 1960s. Figure 1 shows the trend in the number of components per IC chip from 1962 to 1982.[3] Since the chip area scales directly with the number of components, the chip size would become enormous if the component size were to remain the same. A larger chip has a greater statistical probability of containing defects than a smaller chip has, and the avoidance of defects has been one driving force toward the formation of smaller components. Other considerations, such as speed, power loss, and production cost also make smaller components more attractive to the device engineer, and the 10–100-μm linewidth of

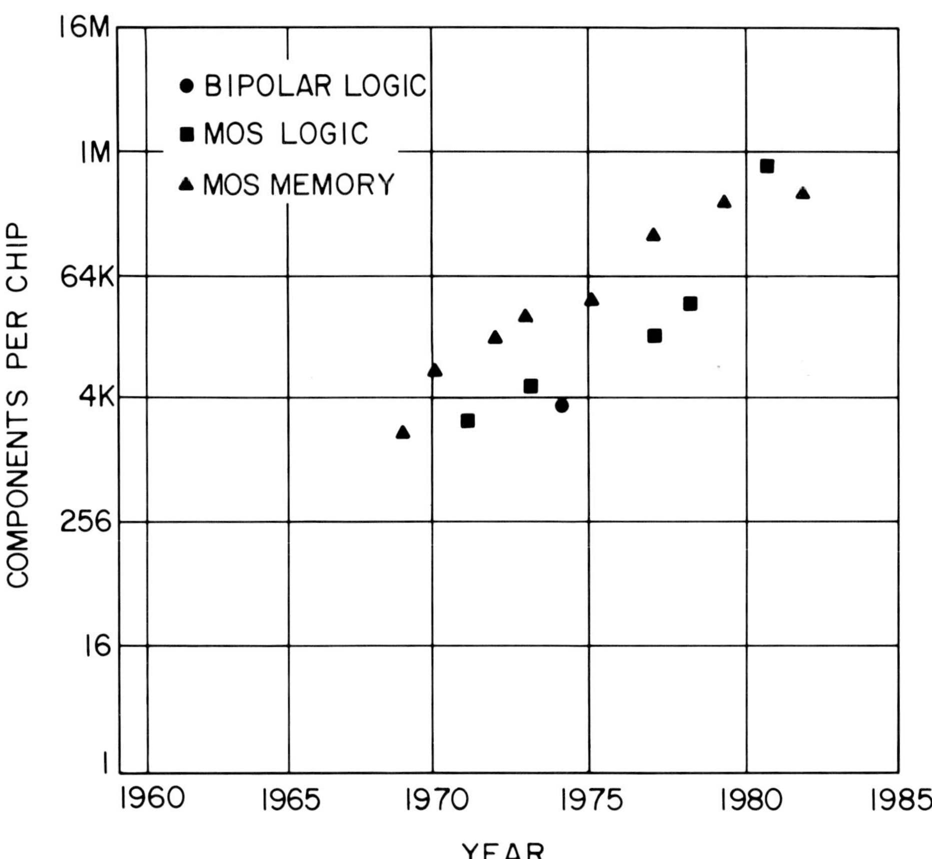

FIGURE 1. A plot of the trend in number of components per IC chip for three technologies (after Fichtner, Ref. 3).

the 1960s has gradually evolved to the 2–3-μm linewidth of today. Figure 2 shows the trend in minimum feature size on MOS memory chips produced by Bell System between 1974 and 1982, with extrapolation to 1987. Although submicron technology has already been achieved in a number of laboratories, the commercial production of submicron geometries has not yet been attained.

The future trend of minimum feature size is uncertain. Technological limitations to the production of submicron linewidths are gradually being overcome. The development of better means of producing small dimensions by optical, X-ray, and electron-beam lithographic techniques, better resists, and better means of etching patterns, all contribute to the emergence of submicron technology. However, the main uncertainty in the extrapolation of the curve in Fig. 2 is the question of industrial demand for submicron circuits. If the need is real and the demand for smaller circuits continues, there is little doubt that the necessary technology will follow.

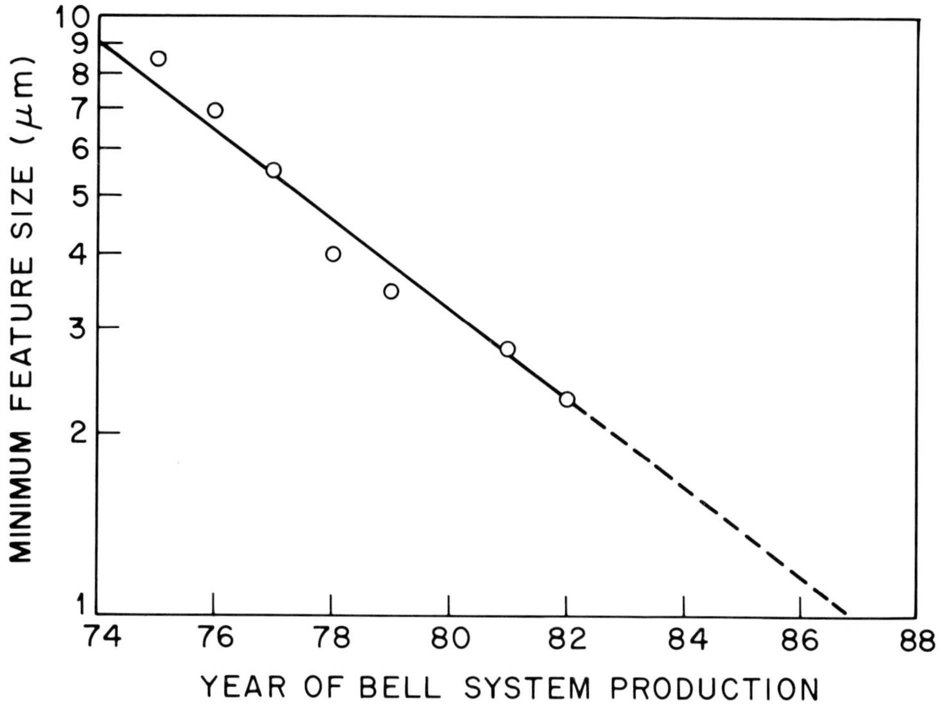

FIGURE 2. A plot of the trend in minimum feature size on MOS memory chips produced by the Bell System.

1.2 OPTICAL AND SCANNING ELECTRON MICROSCOPY

The optical microscope has played a key role in the evolution of silicon IC technology. Examination with a microscope usually provides the first measure of the quality of a lithographically defined or etched pattern and an initial indication of surface cleanliness. Also, examination with a microscope is often the first step in the search for the cause of an electrical failure. The key role of microscopy is illustrated by Fig. 3, which diagrams an ideal facility that develops and processes circuits with very large scale integration (VLSI). (VLSI circuits are those ICs containing over 10^5 components.) In Fig. 3 new device designs are fed into the processing line, the finished product is tested, and the test results are fed back to the designers. Information flow also occurs in a counterclockwise direction, and such feedback is just as important as the forward flow of information and product in obtaining a successful VLSI facility. A fourth functional group concerned with process development and failure analysis necessarily interacts with the other three groups, and again the interaction is in both directions. The starred areas in Fig. 3 indicate where microscopy is essential.

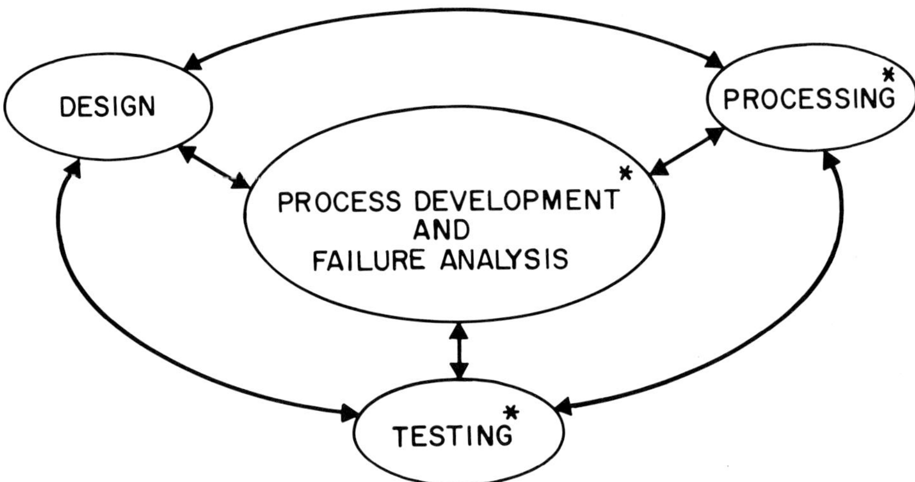

FIGURE 3. Functional organization of a VLSI process development laboratory. Microscopy is essential to the starred areas.

The optical microscope distinguishes topographic features on opaque substrates by variations in surface reflectivity, index of refraction, slope, and feature thickness. The vertical and horizontal spatial resolution of an optical microscope depends on the wavelength of the illuminating light λ and the numerical aperture NA of the objective lens, and varies as $0.61\lambda/NA$. Although 0.25-μm horizontal resolution is possible under optimum conditions, 1-μm resolution is more realistic for routine diagnostic work in a development environment. Since the resolution of the human eye is approximately 200 μm, the useful upper limit of magnification of an optical microscope is less than 1000×; at 1000× a 1.0-μm feature appears as 1.0 mm. Vertical spatial resolution (depth resolution) is enhanced by means of Nomarski interference contrast microscopy,[4] with a minimum resolvable height on the order of 30–50 Å. Thus, all of the discrete devices of the 1950s and IC features of the 1960s could be clearly imaged by optical microscopes.

The development of increasingly complex ICs meant that certain microcircuit features, which were highly relevant to successful chip operation, required more resolution than that offered by optical microscopy. As IC technologies expanded, more attention had to be focused on problems such as metallization continuity over steps and the shape of etched window features in an oxide layer. Concurrent with early IC development was a growing concern about the morphological aspects of materials used, such as the effect of processing on grain structure of metal films and on defect formation within the silicon single-crystal substrate.

The scanning electron microscope (SEM) made its first commercial appearance as a secondary feature to the electron probe. Manufacturers began designing and selling SEMs whose primary function was imaging, however, in response to the needs of biologists and physical scientists for an instrument that

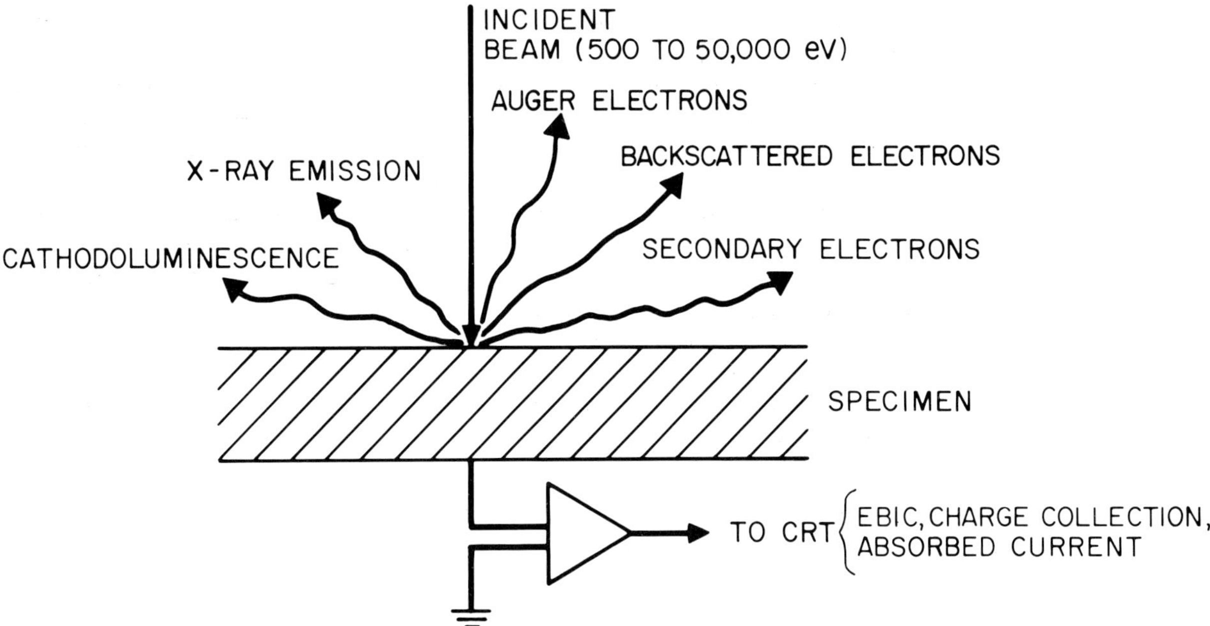

FIGURE 4. Various types of radiation and current are produced when an electron beam strikes a sample. The interaction most generally useful to the determination of surface morphology is the one producing secondary electron radiation.

could provide significantly better resolution. The semiconductor industry first began to use these instruments in the late 1960s. The essential feature of a SEM is the interaction of a rastering (scanning) electron beam of energy 1–40 keV with the surface of a sample, and the stimulated emission of various radiations and currents from this interaction.[5] Figure 4 shows the types of interactions that are of most interest to VLSI process development studies. For morphological studies the most generally useful type of interaction is one that produces secondary electrons (<50 V), which are detected and displayed on a screen. Because surface geometry and, to some extent, chemical composition modulate the secondary electron emission from a surface, the secondary electron image is similar to that obtained by Nomarski interference contrast optical microscopy. The high depth of field (-50 μm at 500× is typical) makes a SEM particularly useful for the study of VLSI device surfaces where film thicknesses rarely exceed 1.0 μm.

The broadening of the incident beam as it penentrates the sample contributes to resolution loss of the secondary electron image; only thin-film samples prepared for scanning transmission electron microscope study avoid this limitation. Features whose secondary electron work functions differ, such as those made from aluminum and SiO_2, are easily distinguished, and the ability of a SEM to resolve small inclusions (e.g., aluminum in an SiO_2 matrix) is limited by the condition of the instrument. A practical resolution limit of approximately 100 Å for a well optimized and moderately well maintained SEM is realistic.

The successful resolution of surface features that differ only in surface inclination depends upon the ability to detect differences in the flux of secondary electrons arriving at the detector from these regions. The effect on the secondary electron yield σ_θ of changes in the angle of incidence θ is given by $\sigma_\theta/\sigma_0 = e^{\alpha d(1-\cos\theta)}$ where α and d are the linear absorption coefficient and escape depth, respectively. A section of 1000-Å diameter silicon sphere with a vertical height of 100 Å relative to the surrounding silicon surface produces a σ_θ/σ_0 ratio that is too close to unity for the sphere to be seen in the secondary electron image. To generate a sufficient flux of electrons to produce contrast in the secondary electron image from regions that differ only slightly in surface inclination, a larger probe diameter is required and resolution degrades correspondingly.

SEMs have been and will continue to be indispensable in VLSI laboratories, although the resolution limits discussed above are at times a serious drawback to their use.

1.3 TRANSMISSION ELECTRON MICROSCOPY

Both the conventional transmission electron microscope (TEM) and the scanning transmission electron microscope (STEM) produce the primary image by differential loss of electrons from a beam transmitted through a thin-film sample. Figure 5*a* shows that in a conventional TEM, electrons from a source are focused by a condenser lens, passed through the sample, and imaged onto a fluorescent screen, photograpic film, or image converter plate. In a STEM (Fig. 5*b*), the electron beam is focused to a small diameter and rastered across the sample. The emerging beam current is used to modulate the brightness of an electron beam moving in synchronism in a cathode ray tube. For examination by both a TEM and STEM the sample must be thin enough to transmit the electron beam to preserve essential information caused by differences in sample thickness, phase composition, and crystal structure and orientation, as discussed by Hirsch et al.[6] and numerous other authors.

The maximum sample thickness that provides a useful, good quality microscopic image depends on a number of factors, including the accelerating voltage of the microscope. Figure 6 shows the limiting thickness for silicon as a function of accelerating voltage.[7] The criterion used for defining limiting thickness was the condition for the disappearance of fringe contrast from twin and stacking fault boundaries using a ⟨111⟩ reflection. The limiting thickness values in Fig. 6 are close to upper limits. In a VLSI analytical laboratory where TEM studies need to be performed relatively quickly and with a minimum of fuss, a more realistic value for silicon would be approximately 0.8 μm at 200-kV accelerating voltage. (This value was obtained from a wedge-shaped silicon sample using a ⟨220⟩ reflection at 200 keV.) The limiting thickness becomes much smaller when one considers the type of information usually sought in TEM studies of thin cross sections of VLSI device structures, as will be discussed in Section 2.2.

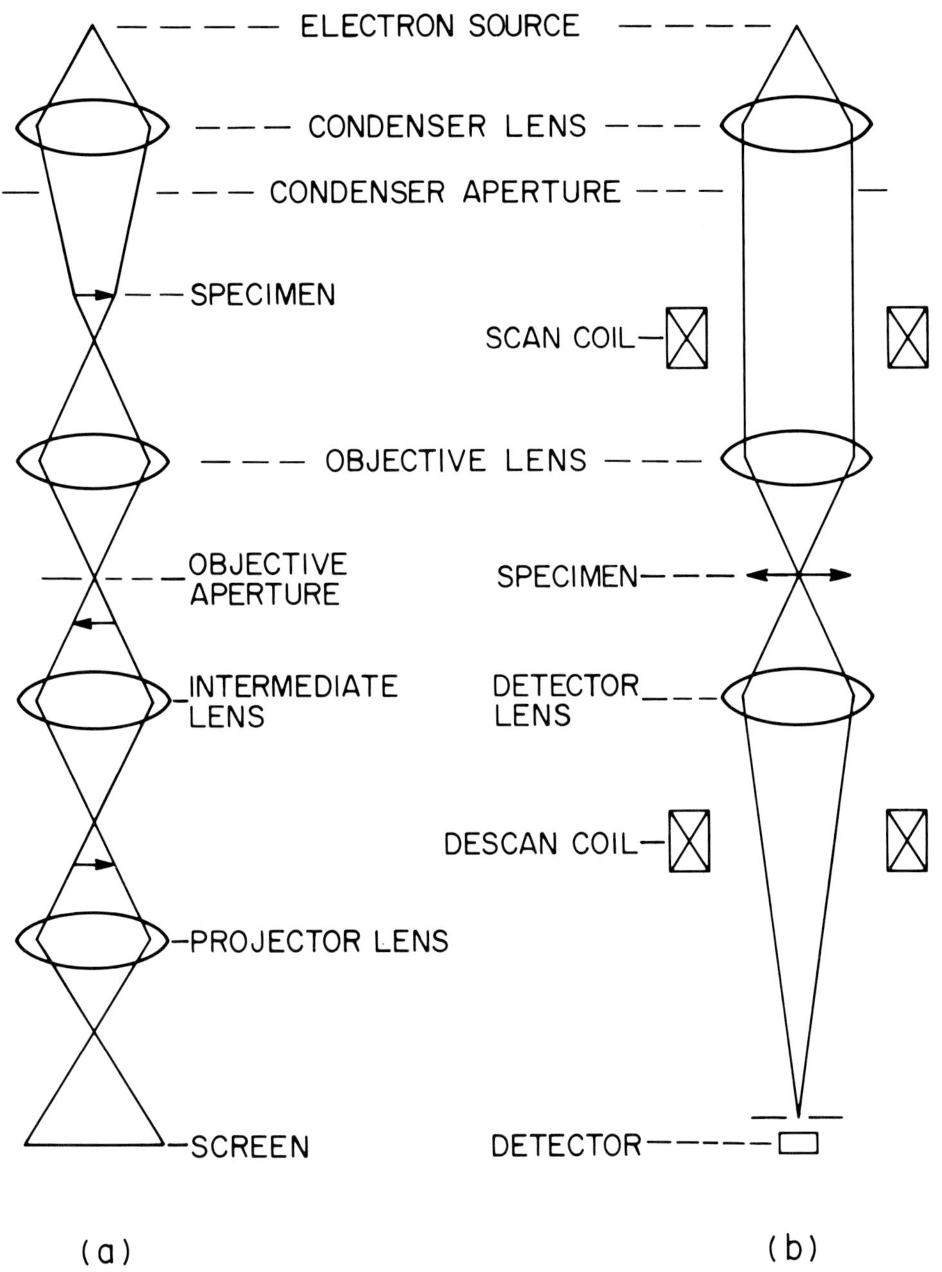

FIGURE 5. Electron-beam ray diagram. (*a*) Conventional TEM and (*b*) STEM operating in the transmission mode.

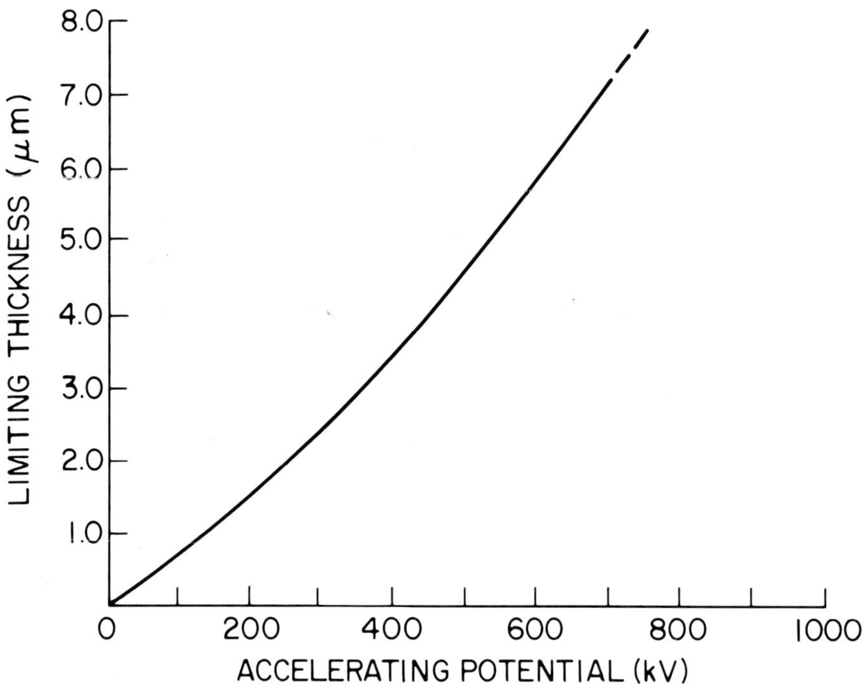

FIGURE 6. Limiting thickness of a silicon sample as a function of accelerating voltage (after Thomas, Ref. 7).

1.4 VLSI PROCESSING

Most of the micrographs in this book are taken from cross sections of *n*-channel MOS VLSI devices and from test structures relevant to *n*-channel devices. To help provide some perspective to the analyses that accompany these micrographs, we include here a brief review of the main features of the VLSI processing sequence used in this technology.

Both optical[8] and X-ray [9] lithography are used to fabricate *n*-channel VLSI device structures. Most of these *n*-channel device structures conform to 1.0-μm design rules, that is, the smallest feature dimensions are 1.0 μm. Figure 7 shows four views of an *n*-channel 4096-bit (4K) static random-access memory (SRAM),[10] which is only one of a number of *n*-channel MOS devices represented in this book. This chip measures 1.6 mm on a side. Figures 7*a* and 7*b* are optical micrographs that show the main functional units of the chip; Figs. 7*c* and 7*d* are SEM micrographs of the chip after polysilicon patterning and show in greater detail the features in a corner region of the memory cell array. Figure 7*d* shows a source that is common to two transistors; the gates are above and below the source.

The processing sequence used to fabricate the device in Fig. 7, as well as other *n*-channel MOS structures, is illustrated in Fig. 8.[11] The sequence shown is mildly generic, although this scheme is specific to *n*-channel MOS fabrication. A number of details that vary, depending on the device, are omitted, and

64 ROW DECODERS
AND WORDLINE DRIVERS
6 ROW ADDRESS
BUFFERS
MEMORY ARRAY
A726D
64-BIT DECODERS AND
BITLINE AMPLIFIERS
0.5 mm

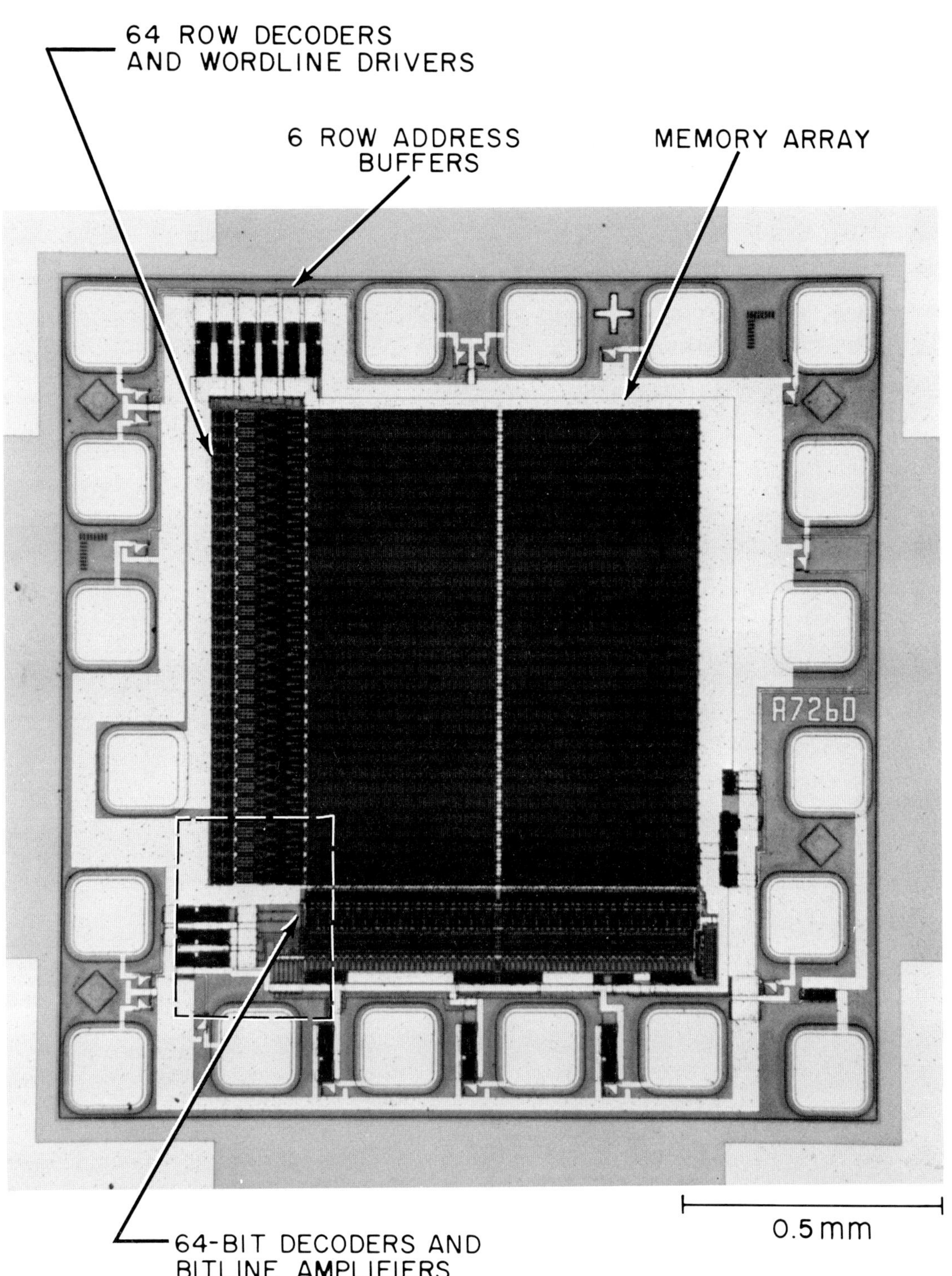

FIGURE 7a

ROW DECODERS
ROW DRIVERS
MEMORY ARRAY
3 BIT ADDRESS BUFFERS
50 μm

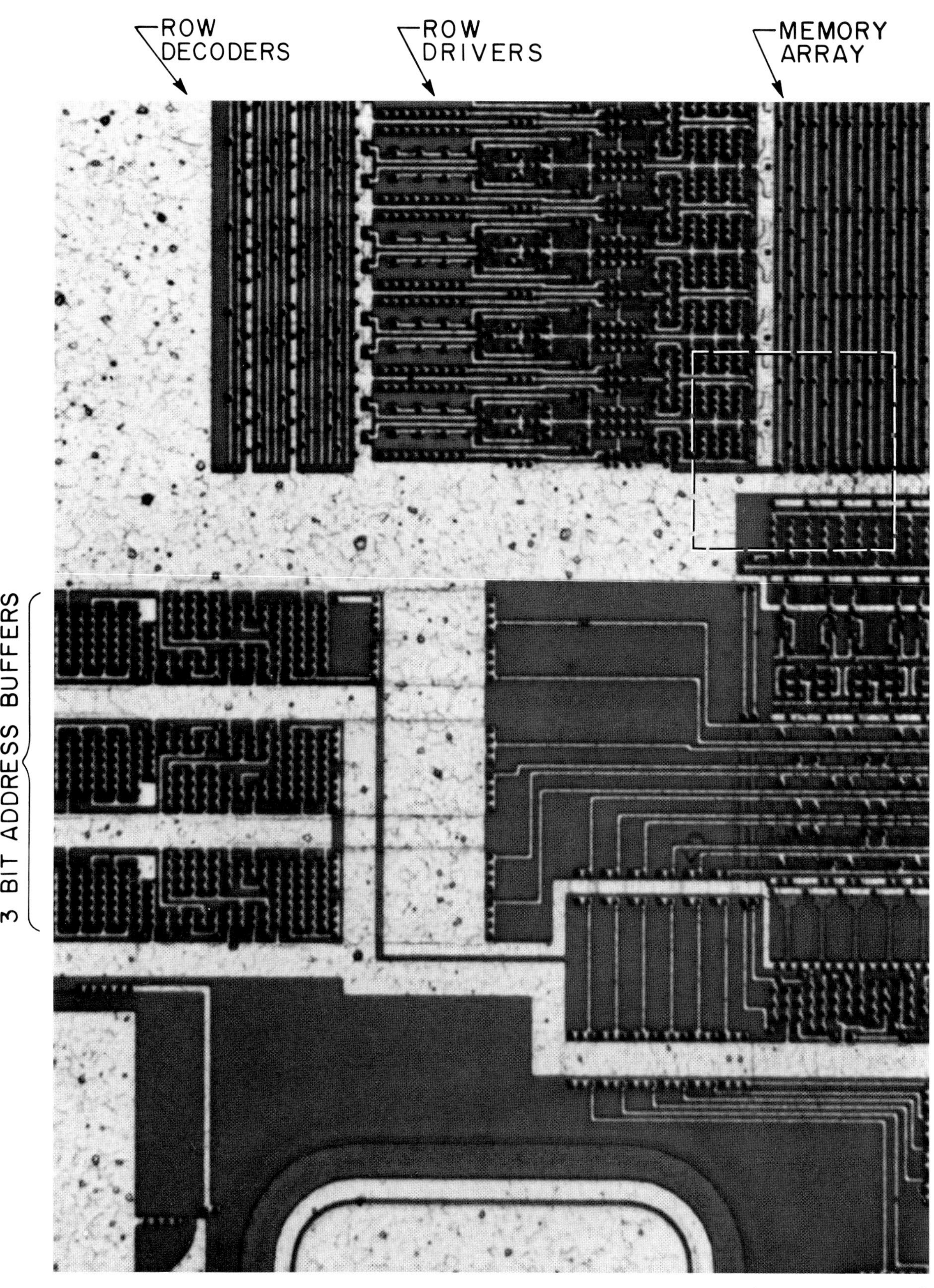

FIGURE 7b

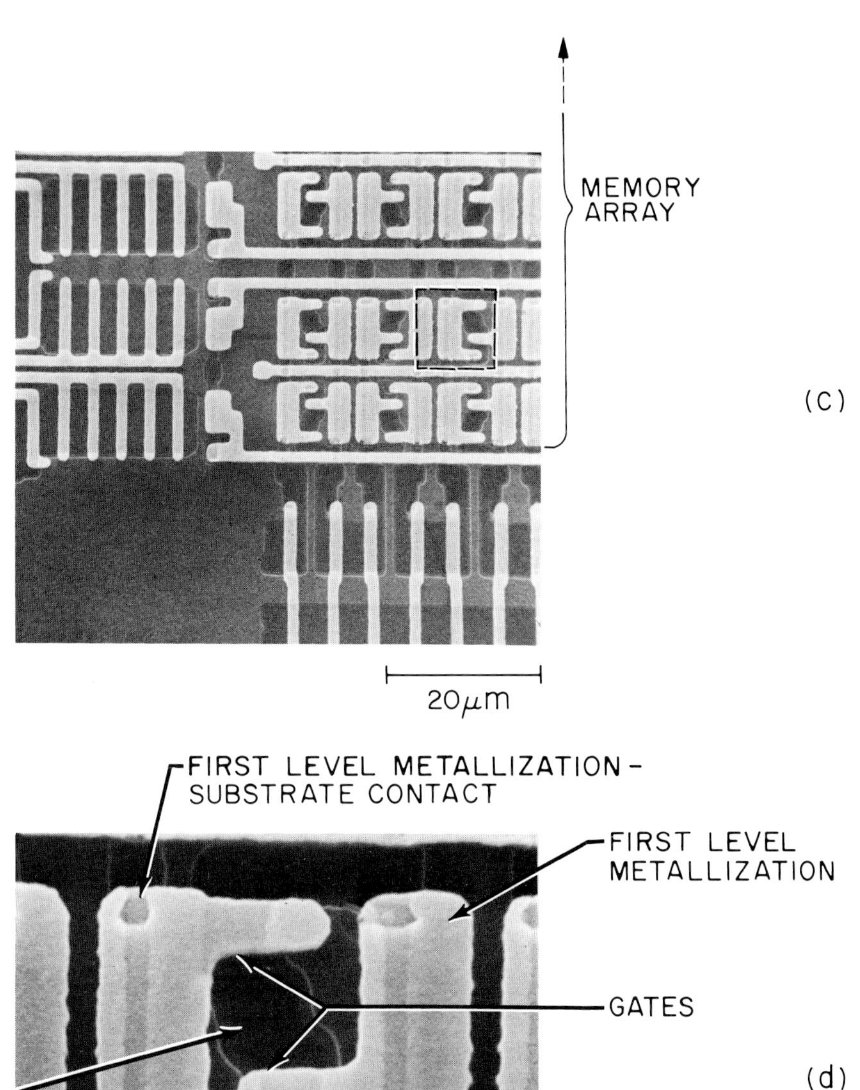

FIGURE 7. n-Channel 4096-bit SRAM made with 1.0-μm design rules; the chip size is 1.6 mm × 1.6 mm. (*a*) and (*b*) Optical micrographs of a fully processed chip and part of a chip as outlined in (*a*), respectively. (*c*) and (*d*) Scanning electron micrographs taken after the first level metallization patterning step showing a region at the lower-left corner of the memory array, as outlined in (*b*), and the single memory cell, as outlined in (*c*), respectively.

some of the processing steps shown can also be varied. In the labels of the processing steps, "patterning" implies both lithography and etching. An isolation oxide, sometimes called a *field oxide*, having a thickness of 3000–10,000 Å is grown on a *p*-type silicon wafer. Then the wafer undergoes boron implantation to bring the substrate doping level at the surface to a desired value (Fig. 8*a*). The gate and source and drain (GASAD) regions are patterned, and, after a short oxidation step, another implant is used to tailor the transistor threshold (Fig. 8*b*). Next, a gate oxide (200–500 Å) is grown and a conducting material, which serves the the first level of metallization, is deposited and patterned. Usually the first level metallization is either an electrically conducting polycrystalline silicon film (polysilicon) or a metal silicide film. In one processing variation using polysilicon, two layers (~1000–3000 Å each) are deposited and contacts to the substrate (polycon) are made as shown in Figs. 8*c* and 8*d*. A diffusion step assures good ohmic contact to the substrate. After patterning the polysilicon (or other material used for first level metallization) and implanting the source–drain regions (Fig. 8*e*), a layer (2000–8000 Å) of a chemical vapor-deposited (CVD) oxide or other insulator is deposited and planarized[12] to remove surface steps (Fig. 8*f*). Windows are opened to both the first level metallization and the substrates; a diffusion step is used to assure ohmic contact to the substrate (Fig. 8*g*), and a second level metallization is deposited and patterned (Fig. 8*h*). A final step (not shown in Fig. 8) is the deposition of a protective coating over the entire chip, often a silicon nitride layer.

Many of the micrographs in Chapters 3 through 6 show features obtained at processing stages that can be determined by referring to Fig. 8. In other cases, the micrographs are of features from different device processing schemes or from simplified processes for production of test structures, and Fig. 8 does not directly apply. In both cases the information from the micrographs is often of generic interest and can be applied equally to more than one technology. For example, problems of polysilicon/oxide texture (Figs. 28–32) or the ability to successfully delineate n^+p junctions (Figs. 60–64) clearly apply to unipolar as well as to bipolar VLSI technologies.

FIGURE 8. Processing steps in the fabrication of *n*-channel MOS device structures. (See Section 1.4 for an explanation of terms.)

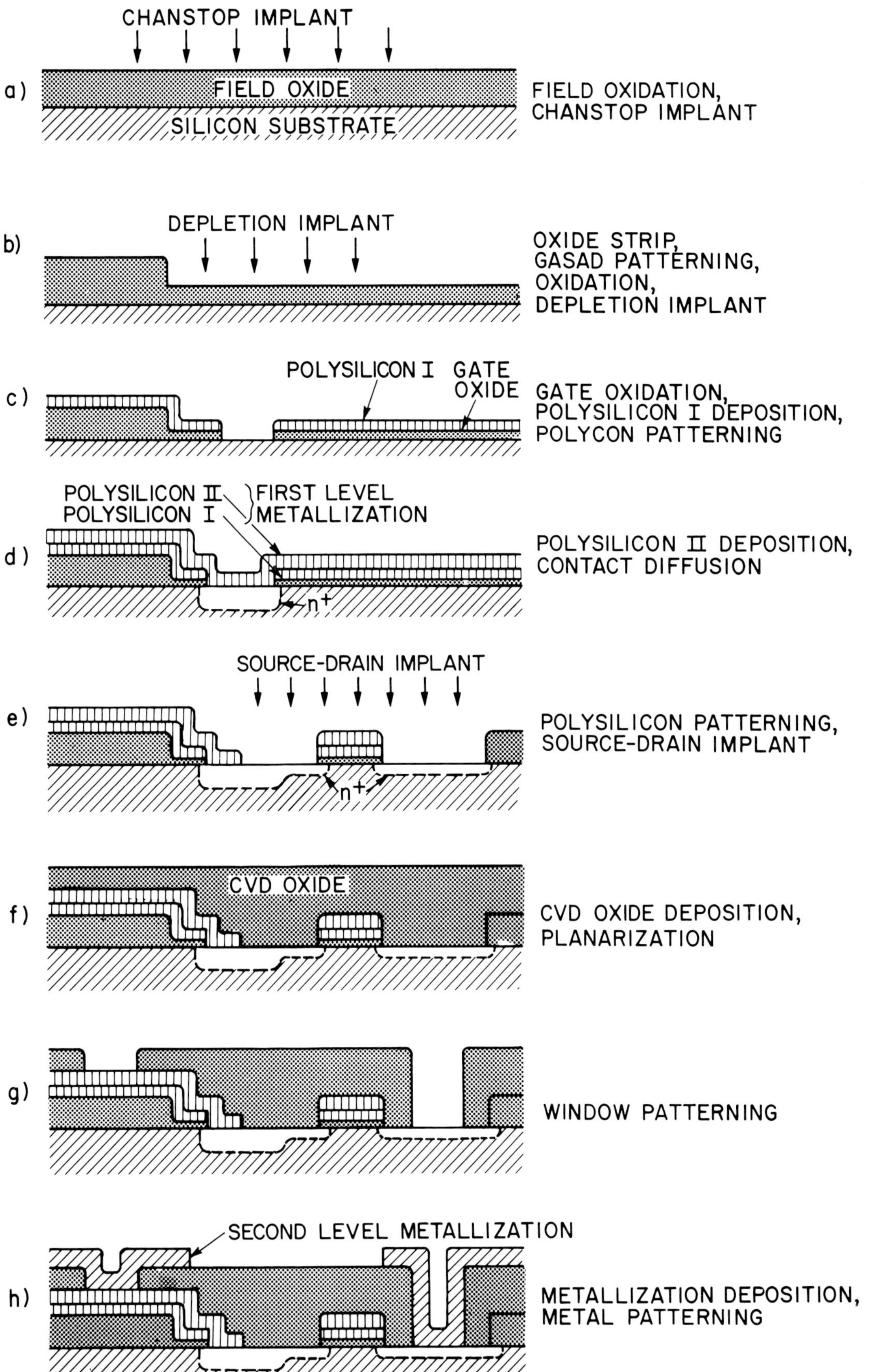
CHANSTOP IMPLANT
a)
FIELD OXIDE
SILICON SUBSTRATE
FIELD OXIDATION,
CHANSTOP IMPLANT
DEPLETION IMPLANT
b)
OXIDE STRIP,
GASAD PATTERNING,
OXIDATION,
DEPLETION IMPLANT
POLYSILICON I
GATE
OXIDE
c)
GATE OXIDATION,
POLYSILICON I DEPOSITION,
POLYCON PATTERNING
POLYSILICON II
POLYSILICON I
FIRST LEVEL
METALLIZATION
d)
n+
POLYSILICON II DEPOSITION,
CONTACT DIFFUSION
SOURCE-DRAIN IMPLANT
e)
n+
POLYSILICON PATTERNING,
SOURCE-DRAIN IMPLANT
CVD OXIDE
f)
CVD OXIDE DEPOSITION,
PLANARIZATION
g)
WINDOW PATTERNING
SECOND LEVEL METALLIZATION
h)
METALLIZATION DEPOSITION,
METAL PATTERNING

CHAPTER TWO

APPLICATION OF TRANSMISSION ELECTRON MICROSCOPY TO VLSI PROBLEMS

2.1 INTRODUCTION

Transmission electron microscope (TEM) studies provide high spatial resolution information on the morphology, crystallographic structure, and (with special attachments) chemical composition of materials used in VLSI circuits and chips. Although all three studies are essential in solving a variety of problems in VLSI technology, the most extensive use of TEM is in providing high spatial resolution for imaging, often better than 2 Å, for the determination of the morphology or shapes of patterned features and crystal grains. In practice, the most important factor that limits image resolution is the thickness of the sample. Therefore, sample preparation techniques are a critical concern in applying TEM methods to VLSI problems.

2.1.1 TEM Imaging

Contrast in a TEM image can be described for two situations. One is TEM study of crystalline materials (such as silicon, aluminum, polysilicon, and various silicides), and the other is the study of amorphous and nearly amorphous materials.

In crystalline materials, the incident electron beam is diffracted by the sample, and local variations in diffraction intensity produce contrast in the image from the undiffracted beam (bright field image) or from one or more diffracted beams (dark field image). The intensity of an emergent beam is periodic with sample thickness. The sample thickness corresponding to one period in silicon is 602 Å for a ⟨111⟩ reflection and 757 Å for a ⟨220⟩ reflection. Thus, a thicker region of the sample does not necessarily look lighter in the negative. A wedge-shaped crystalline sample produces a TEM image that has alternate light and dark bands called *thickness extinction contours*. Dark bands are also caused by a bent sample and are then called *bend extinction contours*. Abrupt changes in the thickness, phase structure, or crystallographic orientation cause corresponding abrupt changes in the contrast, and these regions can be easily imaged at high resolution.[6]

For nearly amorphous materials, contrast is determined by local changes in electron scattering, which result from differences in sample thickness or in chemical or phase composition. A sample region whose thickness varies continuously produces a corresponding continuous variation in image intensity, unlike the case of diffraction contrast. Therefore, TEM images from oxides, nitrides, and other amorphous materials are somewhat easier to interpret intuitively than images from crystalline samples. Chapter I, Section B, of Reference 13 discusses image formation from amorphous materials.

2.1.2 Crystallographic Structure and Chemical Composition

TEM studies produce electron diffraction patterns that provide information on both the crystal structure and the degree of preferred orientation of a polycrystalline overlay on a substrate. With the small beam of the STEM and some recent TEM instruments, areas as small as the beam diameter ($<$100 Å) can be analyzed. Chemical information can also be obtained through the use of X-ray emission spectrosopy (XES),[14] which uses energy dispersive X-ray detectors above the sample, or by the use of electron energy loss spectroscopy (EELS),[14] which requires energy analysis of the transmitted beam. In both XES and EELS methods, chemical information can be extracted from regions smaller than 100 Å.

2.1.3 Application of TEM to Device Problems

In a VLSI process development laboratory TEM studies are applied to two major classes of problems: those occurring during device processing and those arising during the development of new processes, such as new patterning procedures and new device materials systems.

Problems discovered during processing and testing programs (Fig. 3) are often caused by processing step errors or by errors in circuit design. In many cases TEM studies are essential in describing the morphological configuration at the site of the problem so that the problem can be understood and corrective measures can be applied. For example, improper channel lengths in a MOS structure might be caused either by changes in the slope of the edges of gate metallization (for self-aligned gate configurations) or by unnecessarily long or short diffusion times. High resolution microscopy by TEM can clearly reveal each of these mechanisms.

Process development programs exist to maintain and improve state-of-the-art technology and include studies of new materials systems, new configurations of existing materials systems, and new processing procedures. For example, the drive toward smaller feature dimensions is partly responsible for programs involving the generation of new metallization systems, and TEM studies are essential in describing the interactions of these metallizations with each other and with silicon and silicon dioxide. Another example is the development of new procedures for "dry" (plasma or reactive ion) etching of resists and VLSI materials. TEM studies of the configurations of walls at the edges of etched windows are essential for evaluating these schemes. This book contains many examples of the application of TEM to studies of new materials systems and new processing procedures.

2.2 SAMPLE THICKNESS

The problem of image doubling limits the useful thickness of TEM samples that are cross sections of VLSI structures. This problem can occur in TEM samples containing planar features that intersect both surfaces of a sample and is manifested as the appearance of images of both intersections. The linear images are separated by an amount related to the slope of the planar feature and to the sample thickness.

The factors responsible for image doubling can more easily be understood with the aid of Fig. 9. Figure 9*a*, a side view of a thin film sample, shows a region where the gate oxide joins the thicker field oxide. Figure 9*b* shows a TEM image of this sample; local thinning of the gate oxide (see Figs. 22 and 23) is exaggerated. If the sample is misoriented at an early stage in TEM sample preparation (Fig. 9*c*) or is unknowingly tilted during viewing in the TEM (Fig. 9*d*), a double image forms that corresponds to the planar projection of the intersection of phase boundaries with both surfaces of the thin sample (Fig. 9*e*). Figure 9*f* shows an image caused by additional tilting; here, the error is generally caused in part by improper orientation in the TEM rather than misorientation during an early stage of processing. Figure 10 shows a plot of the double

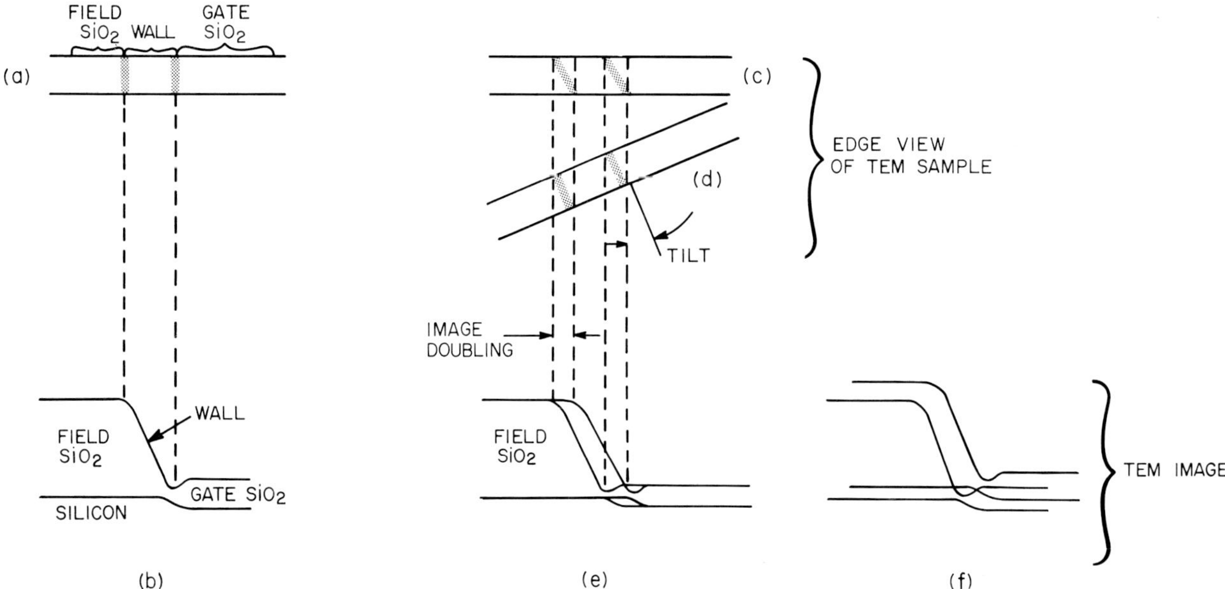

FIGURE 9. Illustration of image doubling that occurs when a VLSI sample is improperly oriented. (*a*) Side view of a well-oriented, thin cross-section sample. (*b*) The corresponding TEM image. (*c*) Misorientation occurring during the early stage of sample preparation and (*d*) misorientation occurring because of tilting of the sample during viewing. (*e*) Image doubling in one dimension as a result of (*c*) or (*d*). (*f*) Image caused by additional tilting.

image width versus specimen thickness when the specimen is tilted at three different angles. (Tilt is defined in Fig. 9*d*.) By using a special stage, called a *double-tilt* stage, a sample can be tilted about two mutually orthogonal horizontal axes, thus correcting many tilt errors. Without a double-tilt stage, a sharper image is more likely with a thinner specimen.

A double image is hard to avoid in some TEM studies of cross sections of VLSI chips, since many VLSI features are not planar on a scale of 100 Å or less. For example, the silicon/SiO_2 interface is planar, but the field oxide wall (normal to the plane in Fig. 9*b*) undulates due to the texture at the edge of the resist used to define the wall (Figs. 80 and 90). Other surfaces may not be planar because crystallites in a polysilicon or aluminum film may have different orientations and, consequently, different etch rates. These "built-in" deviations from planarity cause features such as a field oxide wall boundary to assume (in part) the orientation shown in Fig. 9*c*, and image doubling results. Since deviations from planarity are usually less than 2 degrees in any given TEM cross-section sample, sample thickness must not exceed approximately 500 Å if image doubling is to be less than approximately 20 Å (Fig. 10).

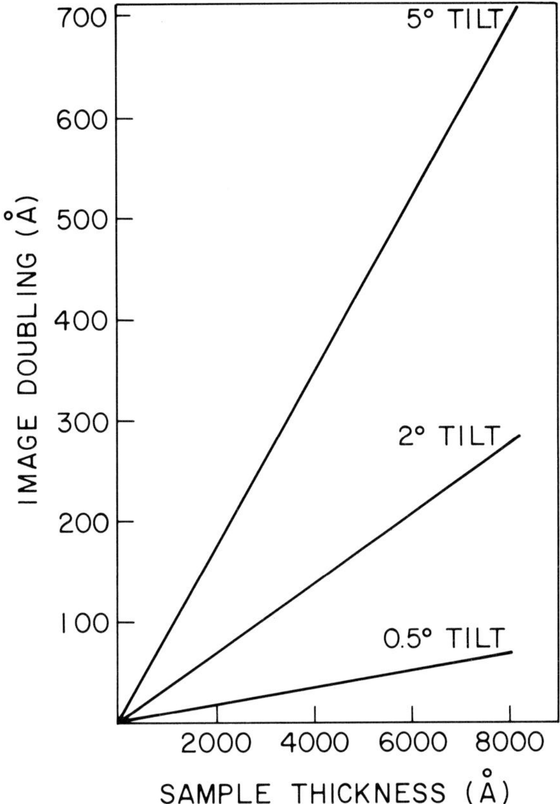

FIGURE 10. Image doubling as a function of sample thickness and tilt angle. The tilt angle is defined in Fig. 9*d*.

2.3 SAMPLE PREPARATION

TEM study of VLSI circuits uses three types of samples: a *replica* of the surface of the device, a thin section of the device parallel to the surface and containing the surface region of interest (*horizontal section*), and a thin section of the device perpendicular to the surface, capturing a cross section through layers in the region of interest (*vertical section*). Special methods can be used with both horizontal and vertical sections to enhance the appearance of certain features (*feature enhancement*). Finally, a special test pattern chip (TEM test pattern), which is especially useful for the preparation of vertical sections, ensures that the desired region is included in the sample. These procedures are discussed below in detail.[15]

2.3.1 Replica Methods

This method of TEM study involves preparation of a replica of a surface and TEM study of the replica. A replica can be made by first depositing a small amount of heavy metal at an oblique angle of incidence to the sample surface; the metal deposit contains thickness variations that depend on changes in surface geometry. A carbon film is deposited over the surface and then withdrawn, extracting the heavy metal deposit. The film with regions of varying metal deposit is a replica of the sample surface and can be examined in the electron microscope. Local variations in metal thickness produce variations in electron scattering. These variations result in an image that is closely related to the texture of the original surface.

In another replication method, advantage is taken of the fact that multilayer VLSI structures generally do not fracture cleanly through all layers, and one of the layer edges at an interface is usually set back slightly. Therefore, a shadowed replica can be made of the exposed interfacial surfaces and fracture surface simultaneously.

Figure 11 illustrates a method for preparing this latter type of replica from a 508-μm (0.020-in.) thick silicon wafer fractured through a region containing field oxide and polysilicon. Two collinear marks on opposite sides of the region of interest are scribed, and the sample is fractured by sandwiching the wafer between glass slides with the scribe marks lining up along the slide edges. When a downward force is applied to the wafer, the working surface (containing the scribe marks) is the first to break (Fig. 11*a*). Next, a very thin film (-10–30 Å) of a heavy metal is deposited at an inclination of approximately 15 degrees to both the sample and fracture surfaces (Fig. 11*b*). Palladium, platinum, gold, and various alloys work quite well; the main requirement is that the grain size of the shadow material be smaller than the required resolution and that the metal be a good electron scatterer. Dissolving the underlying material strips the shadowed replica from the wafer (Fig. 11*c*). The chemical etchant for dissolving the material must act slowly since violent dissolution with bubble formation can destroy the replica. For samples of oxide and polysilicon layers on silicon, a convenient etch is HOAc:HF:HNO_3 = 6:6:10, diluted 1 part to 10 parts HOAc (acetic acid). Figure 51 is an electron micrograph of a replica made in this manner.

2.3.2 Horizontal Sections

The preferred method for preparing horizontal-section samples is chemical thinning, since chemical thinning of silicon and SiO_2 is considerably faster than ion milling. Chemically etching from the back (silicon substrate) side with the HOAc:HF:HNO_3 etch described above thins large areas sufficiently for TEM study. When one layer of a sample has a significantly different etch rate than silicon and needs to be removed or thinned, such as a layer of SiO_2, aluminum, or a silicide, then ion milling is used to continue the thinning process. In some cases where a horizontal section is to be prepared from a region below the wafer surface, the front surface is ion milled prior to chemical etching. Satisfactory samples are produced using argon for ion milling at an ion current of 100 μA

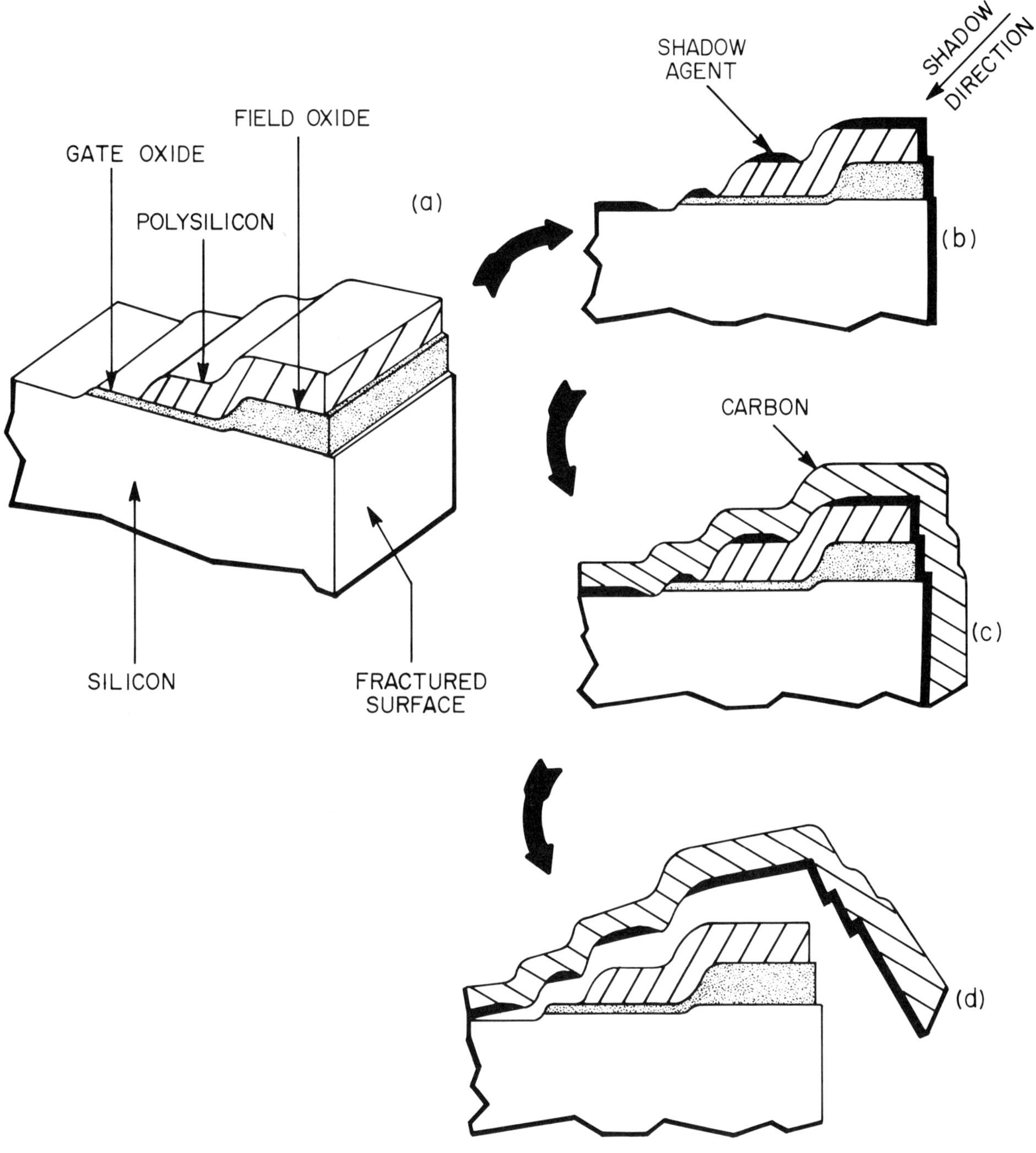

FIGURE 11. Illustration of a method for preparing a shadowed replica of a fractured VLSI chip for TEM study. (*a*) The fractured sample. (*b*) Deposition of a heavy metal shadowing agent. (*c*) Deposition of carbon film. (*d*) Stripping the replica that contains the shadowing agent.

and voltage of 6 kV per gun. The ion beam strikes the sample at 15 degrees to the surface, and the sample slowly rotates about an axis normal to the sample surface.

Figure 12 illustrates part of the procedure for preparing a horizontal section. The sample is lapped to less than 100-μm thickness using 400-grit paper. A disk with the area of interest approximately at the center is created by trimming the sample with a sharp knife. The disk should be cut to fit the sample holder of the electron microscope, which is usually 3.0 mm in diameter. Next, the disk is mounted face down on a clear sapphire plate covered with wax, except for a small area that is left exposed (Fig. 12*a*). The wax protects the front surface, the edge, and part of the back surface on the disk. The sample is then chemically etched for 2–3 mins, which leaves a shallow depression (Fig. 12*b*). The wax is dissolved and then reapplied, which leaves a larger area exposed; etching is allowed to continue until a small hole forms (Fig. 12*c*). The transparency of the sapphire aids in determining the end point of the thinning process by visual inspection. The wax is dissolved in a solvent, and the sample removed and cleaned. Usually the taper at the edge of the hole is sufficiently small so that a sample transparent to 100-keV electrons and extending 50–100 μm back from the hole can be produced. The two-step etching procedure is necessary because of "trenching," or anomalously fast etching that occurs at the edge of the wax mask. To avoid formation of an annular hole, etching is stopped after a short interval and continued only after changing the position of the wax mask.

2.3.3 Vertical Sections

Most of the useful TEM studies of VLSI device structures have been performed on vertical-section samples. Although vertical sections are somewhat more difficult to prepare than horizontal sections, a method has been devised for preparing them routinely and easily. This method, described here and illustrated by Fig. 13, simultaneously forms many vertical-section samples in one TEM preparation. The end of this section describes a variation of this method that is simpler to use and suitable for hot stage TEM studies, but limited to one sample per preparation.

Figure 13*a* shows a 3-in. diameter silicon wafer that contains two special TEM test pattern chips that will be described later; an enlargement of one of these chips is shown in Fig. 13*b*. Samples measuring about 1.5 mm by 3 mm are cut or cleaved from the wafer(s) and lapped to less than 100-μm thickness using 400-grit abrasive paper. These pieces are stacked face-to-back (Fig. 13*c*) and bonded together with epoxy. The epoxy should be squeezed out between the pieces so that only thin (approximately 0.1 μm) epoxy films of uniform thickness separate the pieces. This step is important, since too large an epoxy thickness will give a poor quality, vertical-section sample after ion milling. Figure 14 shows a jig that is useful for clamping the sample pieces evenly. Clamping the stack of sample pieces in the Teflon*-coated chamber and heating for 3 hr at

*Trademark of Du Pont Inc.

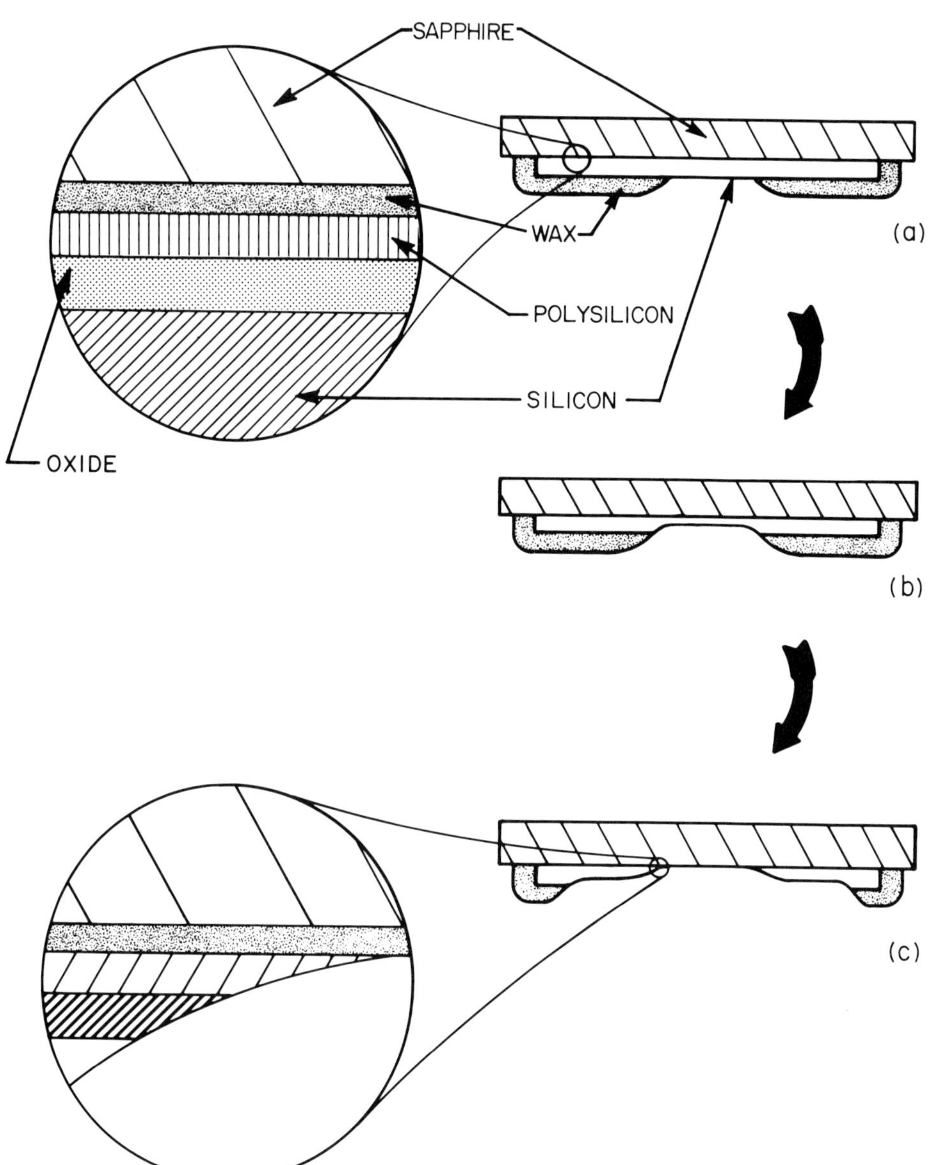

FIGURE 12. Illustration of method for preparing a horizontal section of a VLSI device structure for TEM study. (*a*) Sample disk is mounted with wax on a sapphire plate; inset shows bonding details for an oxidized silicon sample. (*b*) Shallow depression after brief chemical etch. (*c*) The mask is enlarged, and a hole forms after additional etch; inset shows specimen taper at edge of hole.

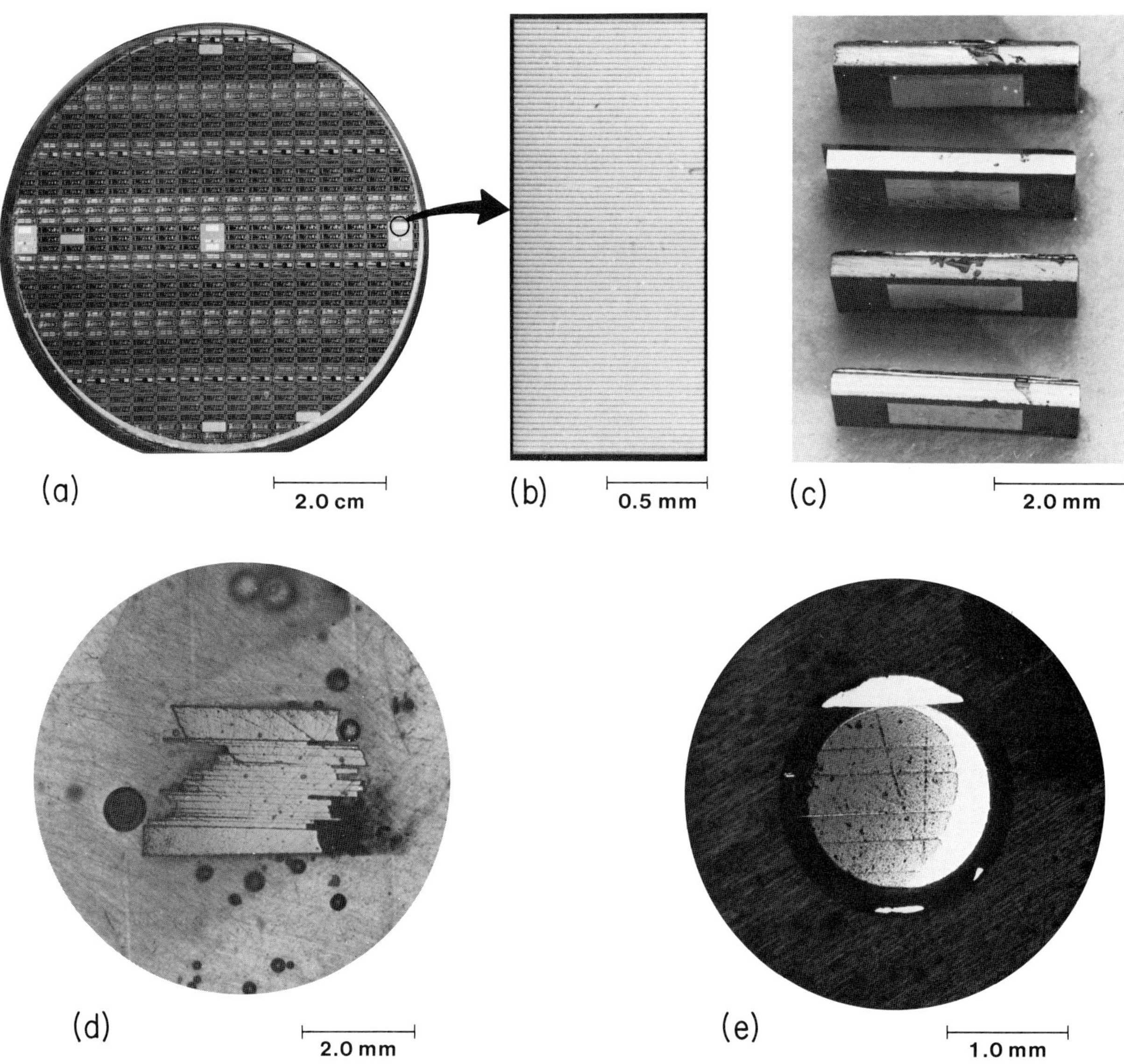

FIGURE 13. Illustration of method for preparing vertical sections of VLSI device structures for TEM study. (*a*) A device wafer with four TEM test patterns. (*b*) Enlargement of one of the test patterns. (*c*) The separately cleaved pieces containing the test pattern or other device regions arranged together in the orientation shown for bonding. (*d*) The separately cleaved pieces bonded together with epoxy using a jig shown in Fig. 14 and embedded in an epoxy button. (*e*) After thinning, the preparation is bonded onto a molybdenum ring and ion milled until a hole appears.

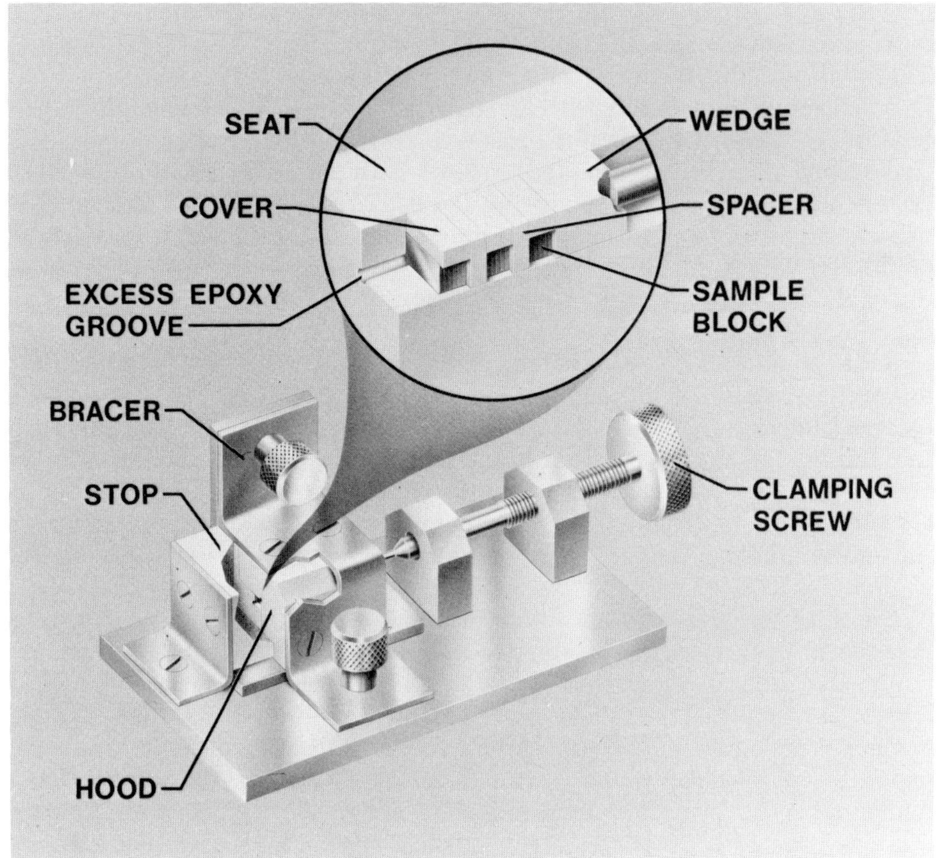

FIGURE 14. Jig for clamping samples for preparation of vertical cross sections. The inset is an enlargement of the Teflon® (Du Pont trademark) coated chamber.

80°C cures the epoxy. Next, the stack is embedded in an epoxy "button" using a round, silicone rubber mold. Figure 13*d* shows 18 sample pieces in such a button. Both sides of the button are lapped and then polished to a final thickness of less than 100 μm using, in succession, 400-grit abrasive paper, 600-grit abrasive paper, and 3-μm polishing compound. A disk is cut with the area of interest approximately at the center. To protect this fragile assembly, the disk can be bonded with epoxy to a small molybdenum ring. The outside diameter of the ring corresponds to the size of the sample holder for the electron microscope. Both surfaces are then ion milled simultaneously until a small hole develops in the sample. Figure 13e shows a TEM preparation after ion milling that contains six cross-section samples. Regions extending back 50–100 μm from the edge of the hole are usually thin enough for TEM study at 100–200-keV beam energy.

Three important points should be noted in using this technique. First, epoxy mills at a faster rate than most materials used in VLSI technology. If the epoxy between adjacent pieces in a preparation is preferentially milled away, the outer surface of the specimen suffers some erosion, and this material is lost to TEM study. Coating the wafer with a 0.5-μm layer of polysilicon, SiO_2, or Si_3N_4 prior to sample preparation prevents this loss. Many of the micrographs in this book are of samples having this *protective polysilicon* layer.

A second point is that samples prepared with the method just outlined are unsuitable for experiments involving the use of a hot stage for *in situ* TEM experiments, such as studies of polysilicon recrystallization, the reactivity of aluminum with a silicon substrate, or the annealing behavior of implant-induced defects. Using epoxy in sample preparation precludes *in situ* thermal experiments. A method has been devised for vertical-section sample preparation that eliminates the use of epoxy but is limited to the preparation of only one vertical-section TEM sample at a time.[16] With this method, a glass plate is electrostatically bonded[17] to the device surface; sectioning, mechanical polishing, and ion milling follow as described above. The top electrode (Fig. 15) is a gold wire touching the glass. The progression of electrostatic bonding is seen as a "wet" region that begins under the top electrode and slowly spreads across the entire interface. Details of the bonding procedure are given in the Fig. 15 caption.

The third important point to note in vertical-section preparation concerns the probability of locating the region of interest in the TEM image. If the sample is prepared from a wafer region with a high repetition of the morphological feature to be studied (such as a gate oxide-field oxide step in a high density memory array), then there is some probability of finding the desired feature in a region that is electron transparent. All too often, though, the feature is not found in the TEM sample region that is thin enough for viewing, and additional samples have to be prepared. Using a TEM test pattern avoids this costly problem.

2.4 TEM TEST PATTERN

Horizontal- and vertical-section sample structures usually contain regions thin enough for TEM study over a 50–100-μm distance from the edge of a hole. The probability that a final TEM preparation contains a specific feature in the electron transparent region increases to nearly 100% when a special test pattern is incorporated in the wafer and the TEM sample is prepared from this test pattern. The test pattern contains all the essential morphological features (contacts, steps, oxides, etc.) compressed in a region 20–40 μm along the surface and extending 1 mm or farther in the orthogonal surface direction. The 40-μm (or smaller) unit is repeated a sufficient number of times to cover a total distance of 2 mm. Figure 16 shows a TEM test pattern appropriate to NMOS technology.

TEM test pattern chips are small and can be located next to an alignment feature so that little active device area is used (Fig. 13*a*). The TEM test pattern in Fig. 13*b* and diagramed in Fig. 16 has a 29.0-μm repeat unit, which appears 70 times in the chip. The main requirements in designing such a test chip (other than incorporating all the essential processing features) are: (1) the total area is similar to that described in Fig. 16, that is, 1 mm by 2 mm with the

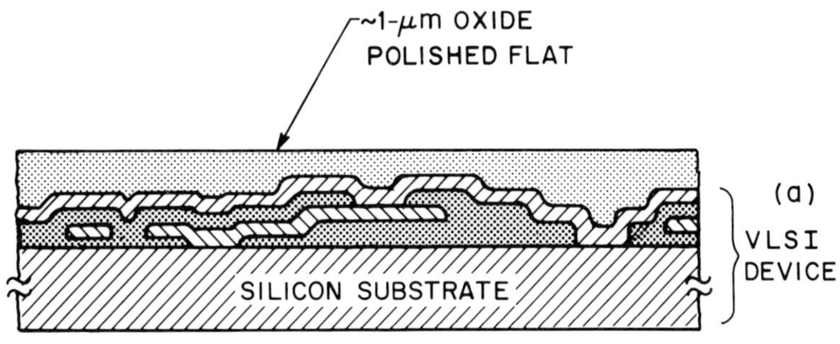

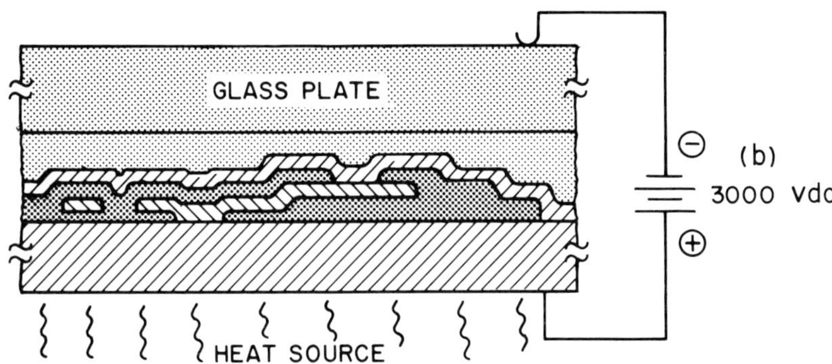

FIGURE 15. Method for preparing vertical section of a VLSI device structure for TEM study without use of epoxy (after T. T. Sheng, Ref. 16). (*a*) Plasma deposition of 1-μm SiO_2 over the top surface which is then polished smooth. (*b*) A glass plate is laid over the surface, the sample is heated to 250°C, and a potential of 3 kV is applied with the top electrode negative; this procedure bonds the glass plate to the oxide. A sample is diced from this preparation, lapped, polished, and ion milled.

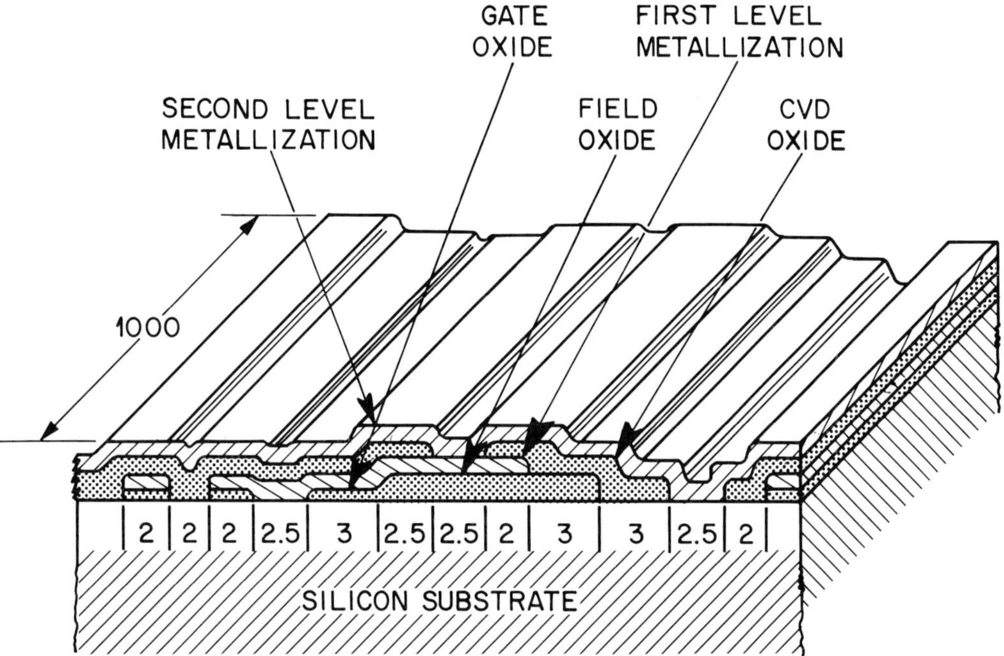

FIGURE 16. A schematic of a TEM test pattern for NMOS device technology showing a 29-μm repeat unit. All dimensions are in microns.

repeat direction running parallel to the 2-mm axis; (2) the repeat unit is less than 40 μm; (3) at least three test chips appear on a wafer; and (4) the test chips are separated rather than bunched together since each chip needs to be separately cleaved or cut from the wafer.

The importance of the TEM test pattern cannot be overemphasized. Vertical sections of active device areas can be prepared, only to discover that during TEM examination the region of interest was missed, and sample preparation has to begin again. The full potential of TEM applied to VLSI analysis programs is best realized when mask sets of devices and test structures contain intelligently designed TEM test patterns.

2.5 FEATURE ENHANCEMENT

Special techniques are sometimes needed to enhance the image contrast of features that would otherwise be difficult to see with the electron microscope. Image contrast from amorphous material is caused by variations in thickness or chemical composition or density that produce corresponding changes in electron scattering. Adjacent amorphous layers, such as CVD (chemical vapor deposited) SiO_2, plasma SiO_2, or thermal SiO_2, are sometimes difficult to distinguish without a special etch treatment of the sample. Since unhardened CVD SiO_2 etches 3–6 times faster in buffered HF (BHF) than thermal oxide, a 5–10-sec dip of the TEM preparation in BHF can produce a significant change in the thickness of the two layers that is easily detected in the TEM photograph. A number of micrographs in this book contain contiguous oxide regions that are distinguishable because of BHF treatment. Also, a brief treatment in BHF solution has been found to slightly enlarge small oxide voids, thus making them easier to see in the TEM image. Figure 38 shows an example of the application of this image enhancement procedure.

With single-crystal material as well as with various metallizations including metal silicides, image contrast is controlled by diffraction effects, which are determined by crystal structure and orientation, thickness, and defect structure. The ability to resolve small features in these materials, including the definition of interfaces, although usually not a problem, is limited mainly by the condition of the electron microscope and by the sample thickness. Variations or abrupt changes in doping level in silicon are generally not discernible, however, and special methods must be derived to reveal *pn* junctions.

Junctions formed in silicon by arsenic or phosphorus doping into a *p*-type substrate can be delineated by chemically treating the final TEM preparation in a manner that preferentially removes n^+ material.[18] The delineation of n^+p junctions made with phosphorus occurs at a region where the doping level falls below about 1×10^{19} cm^{-3}, so a correction term must be added to the measured delineation depth to arrive at a junction depth. The chemical treatment consists of (1) 5–50 sec in BHF to clean the surface and bring the various oxides down to a thickness than is comparable to the silicon thickness after the silicon etch, (2) rinse and dry, (3) 5–10-sec etch in a solution of 0.5% HF in HNO_3, and (4) rinse and dry. Figures 60 through 64 show micrographs of delineated junctions. These micrographs provide information on junction depth as well as on the lateral extent of junction penetration under gate regions.

CHAPTER THREE

OXIDES

3.1 INTRODUCTION

Silicon dioxide is used to provide physical and electrical isolation between conducting elements in VLSI circuits. "Thermal" oxide is grown at elevated temperatures (900–1100°C) on either single-crystal silicon substrates or polycrystalline silicon (polysilicon) in dry oxygen, in oxygen saturated with water, or in steam; HCl is often added to dry oxygen to improve gate oxide quality. Deposited oxides are used when free silicon is unavailable for thermal oxidation, or when temperatures must be kept low to avoid unwanted diffusion within the device. CVD oxides often use a mixture of SiH_4 and O_2 at 0.5–1.0 torr pressure and a temperature range of 350–450°C. These oxides can be doped with phosphorus through the addition of controlled amounts of PH_3 or other phosphorus-containing compounds. "Plasma" oxides are deposited at 100–340°C from a SiH_4 + N_2O plasma.[19]

3.2 THERMAL OXIDES

Flat silicon surfaces can be oxidized uniformly and then patterned, or oxidized through a patterned mask that has already been defined on the surface, resulting in a semirecessed oxide. This section discusses problems arising in semirecessed oxidation, oxidation at the edges of windows, oxidation at steps, and reliability problems caused by the oxidation behavior of polysilicon.

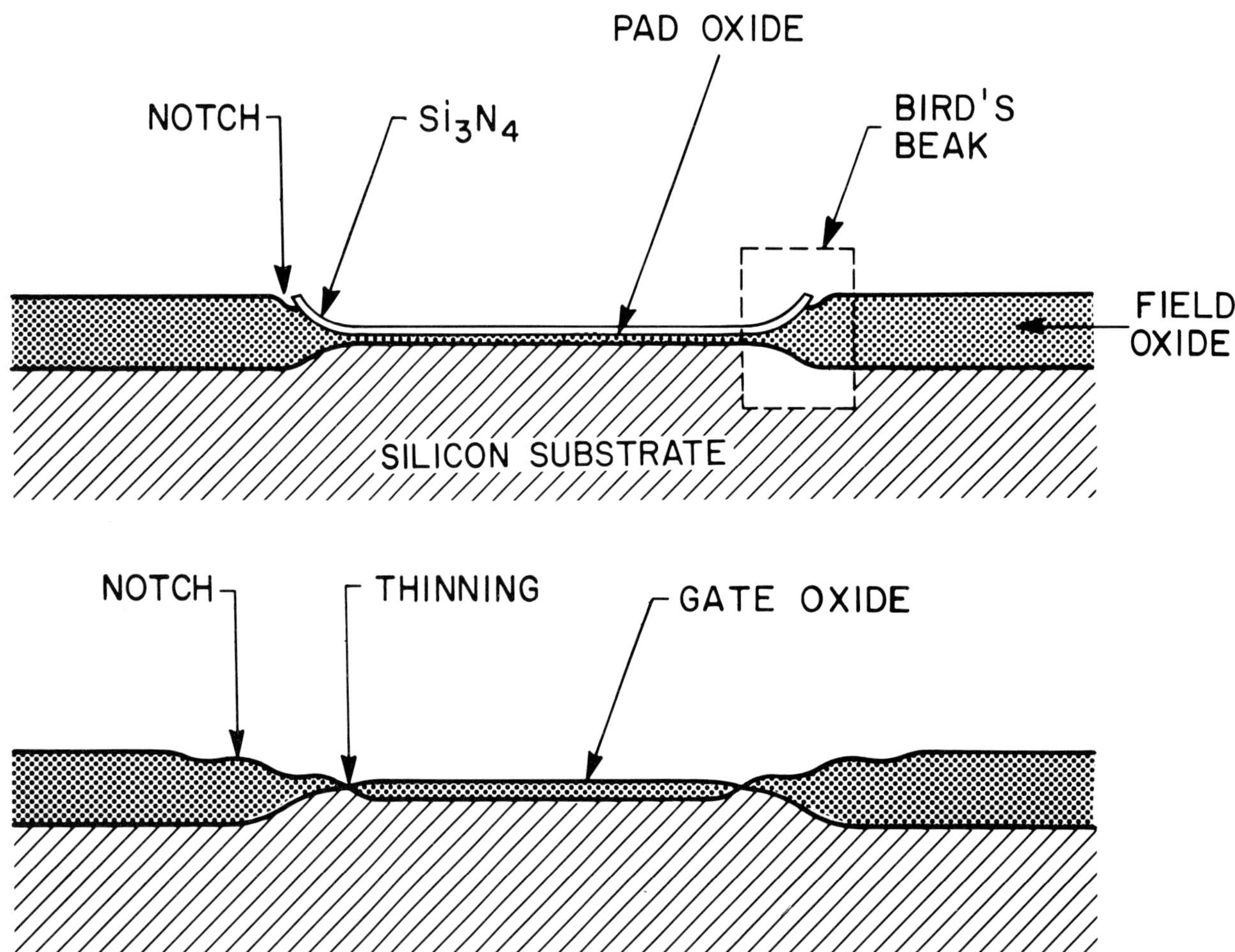

FIGURE 17. Two stages in semirecessed oxidation processing: The upper figure shows the stage following field oxidation; the lower figure shows a subsequent step after gate oxidation. Thinning of gate oxide can be seen.

3.2.1 Semirecessed Oxide

An advantage of this approach is that the resulting oxide is half-buried within the silicon substrate, and the step is only half as high as the step obtained by oxidization followed by patterning. A smaller step is a decided advantage in achieving good metallization step coverage at a later processing step. This topic is covered more thoroughly in Chapter 4. An example of semirecessed oxide formation is the use of a patterned Si_3N_4 layer as an oxidation mask. After field oxidation, a characteristic configuration at the edge of the mask called a "bird's beak" appears,[20,21] and subsequent gate oxidation often results in local thinning of the gate oxide next to the field oxide just beyond the tip of the bird's beak.[21] TEM studies of samples extracted at various steps during recessed oxidation processing more clearly show the nature of this important reliability problem.[22] Figure 17 shows cross sections of samples after field oxidation (upper), and

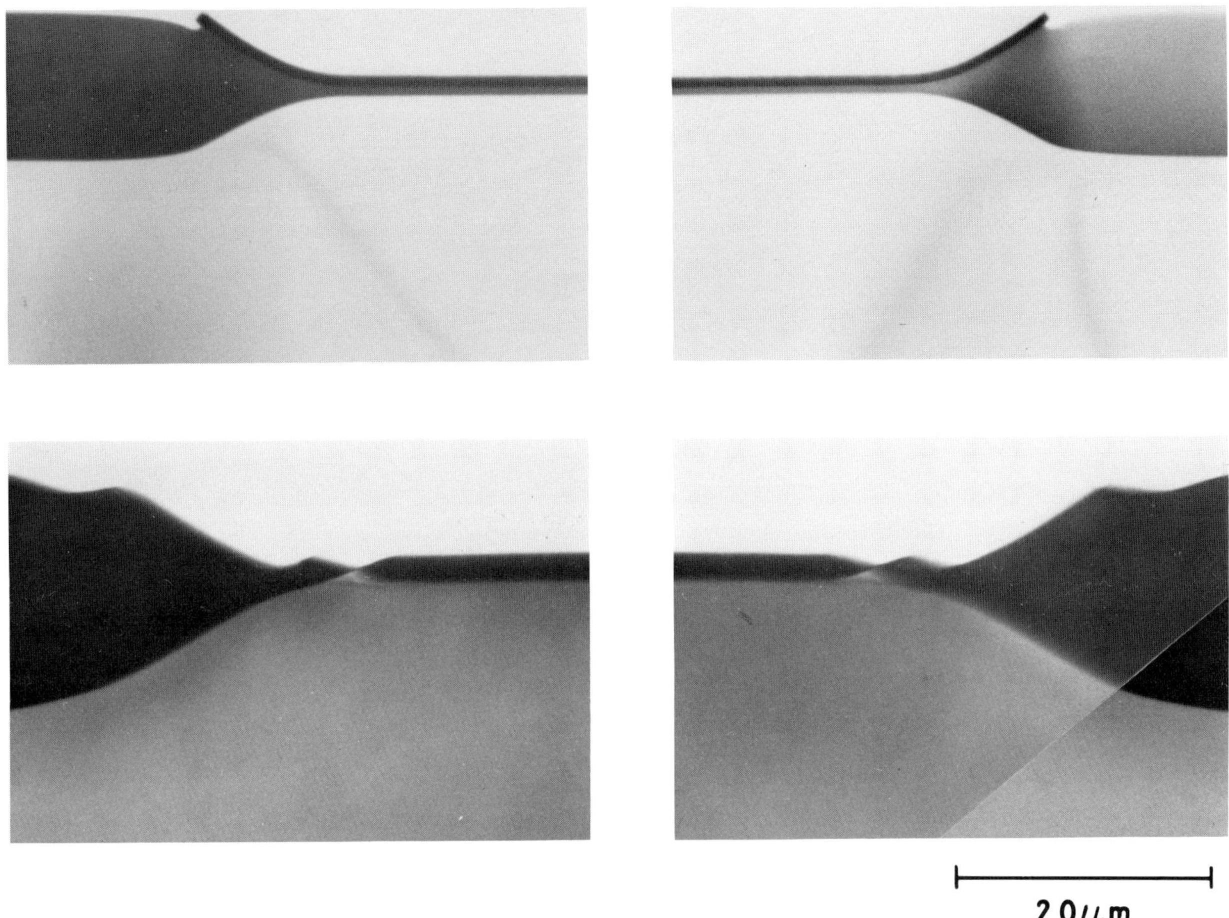
2.0μm

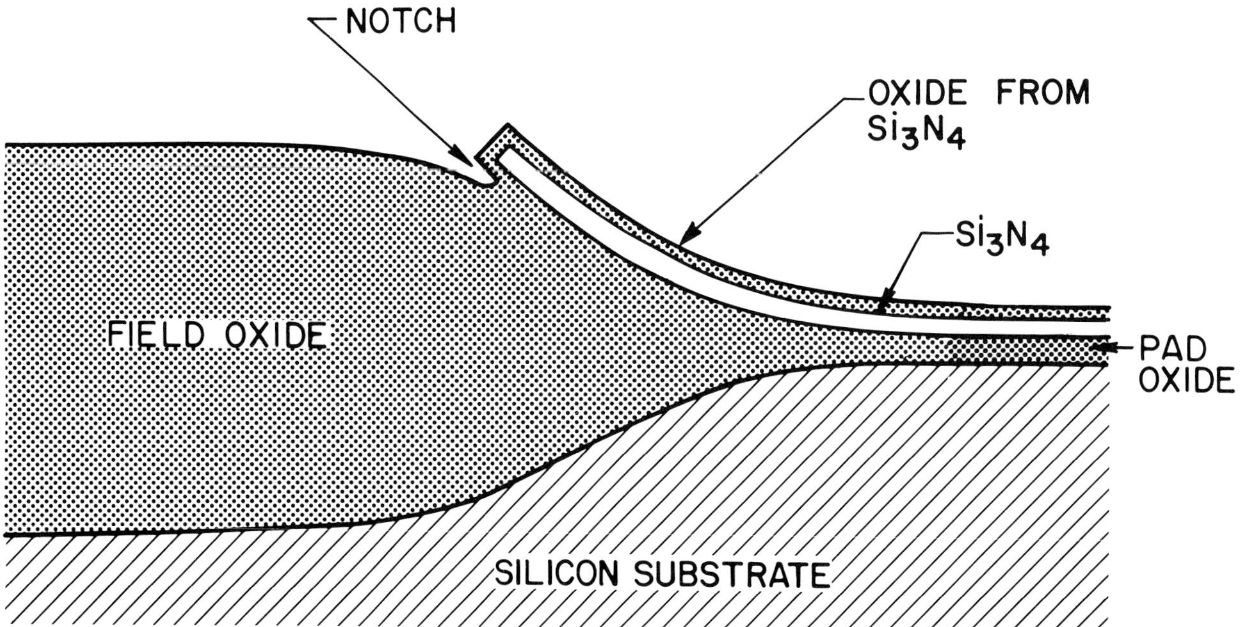

FIGURE 18. The bird's beak region after growth of field oxide. Increasing field oxidation (top to bottom) results in increasing thickness of oxide grown by conversion of the Si_3N_4 masking layer and decreasing severity of the notch at the Si_3N_4 edge.

after Si_3N_4 strip and gate oxidation (lower). A notch on the field oxide surface forms at the edge of the Si_3N_4. The bird's beak region contains this notch (Fig. 17, upper). When gate oxidation immediately follows stripping of the oxide pad, a region of thin gate oxide near the tip of the bird's beak forms due to local oxidation inhibition (Fig. 17, lower).

Figure 18 shows that the severity of the field oxide notch is directly related to the extent of field oxidation. This figure also shows the increasing oxide thickness converted from Si_3N_4 with increasing field oxide thickness. Oxidation time, temperature, and pressure control the overall shape of the bird's beak. The photographs in Fig. 18 are of samples oxidized at 1-atm pressure in wet oxygen; when oxidation is performed in a steam bomb at 725–760°C and at 20-atm pressure, the beak becomes severely elongated[23] (Fig. 19).

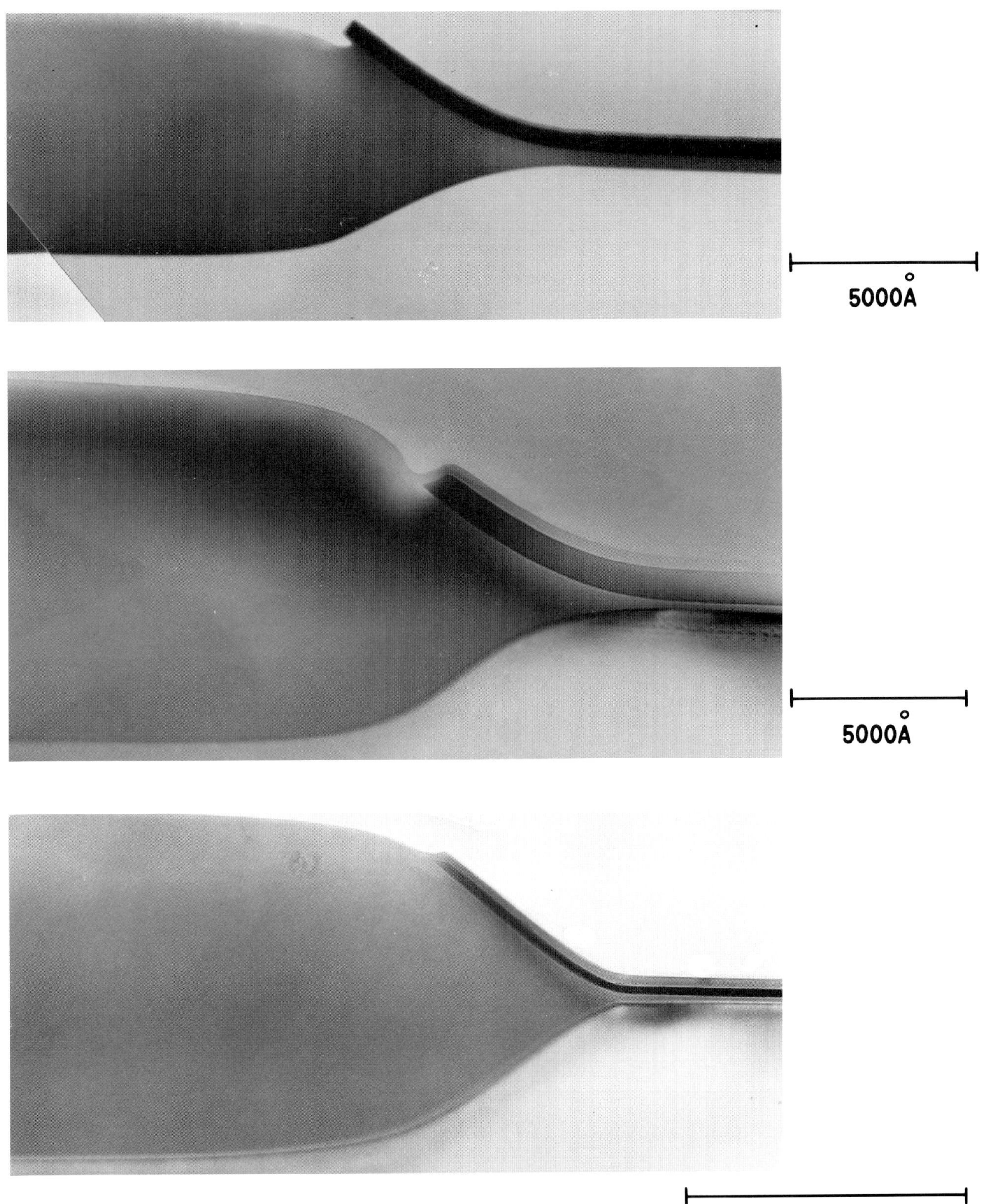
5000Å
5000Å
1.0μm

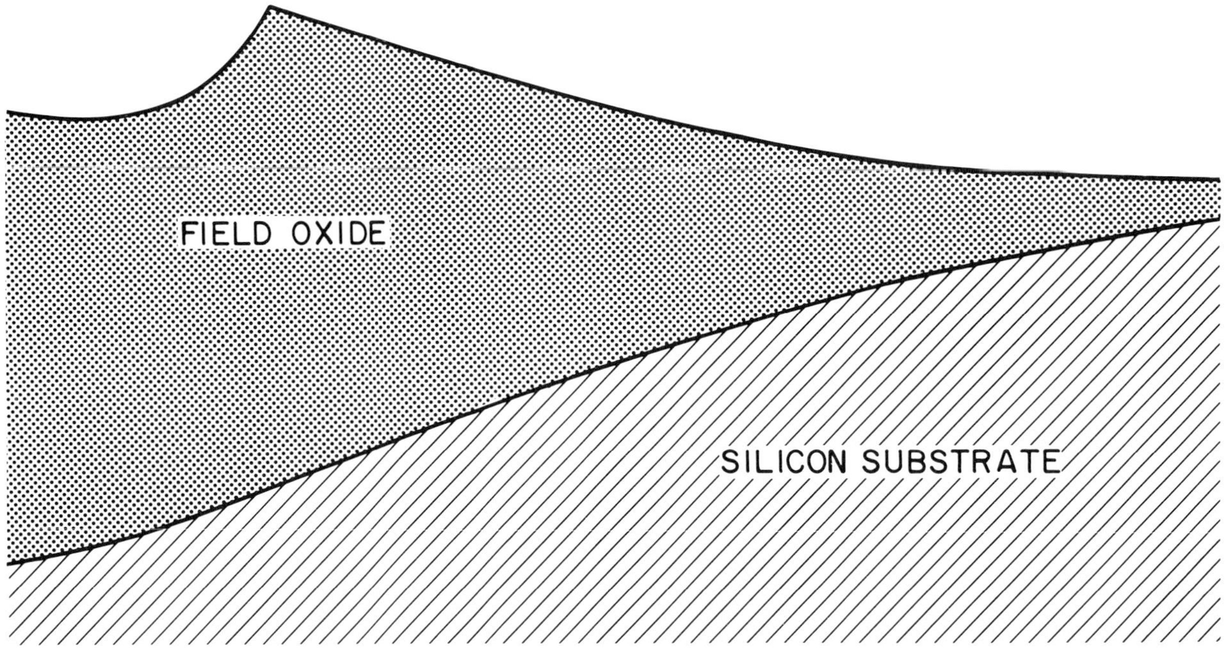

FIGURE 19. Elongated bird's beak due to oxidation at 20-atm pressure in a "steam bomb."

5000Å

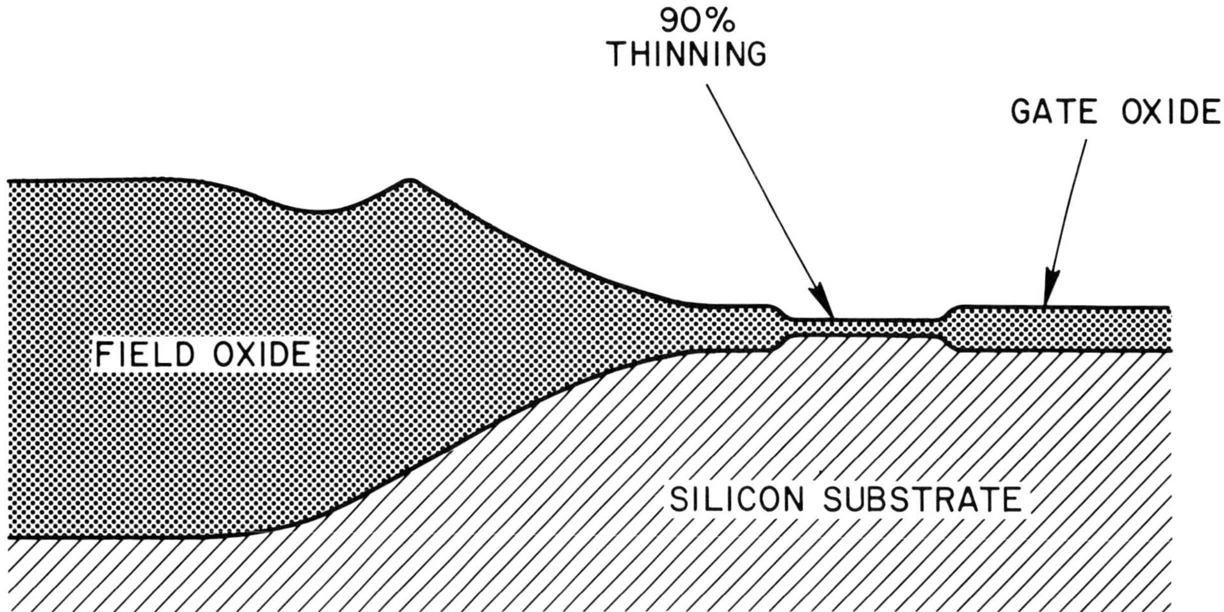

FIGURE 20. The bird's beak after Si_3N_4 stripping (upper) and gate oxidation (lower).

The gate oxide thinning problem is clearly illustrated in Fig. 20, where portions of the same device wafer are shown after field oxidation at 1000°C in wet oxygen and Si_3N_4 strip in the upper photograph, and after gate oxidation at 1000°C in dry oxygen in the lower photograph. The second H_3PO_4 treatment had been used to ensure the removal of any residual Si_3N_4 which might act as a local barrier to oxidation during gate oxide formation. In fact, the amount of gate oxide thinning was found to be independent of the nitride strip treatment. The oxidation barrier was assumed to be a slowly oxidizable silicon oxynitride that was formed during field oxidation. The oxidation barrier was successfully removed by interposing a wet reoxidation after Si_3N_4 removal, followed by a 90-sec BHF treatment and gate oxidation.[22] Both micrographs in Fig. 21 are of samples receiving the second oxidation and show the absence of gate oxide thinning. The lower photograph shows an additional notch at the tip of the bird's beak; the origin of this notch, seen on a number of samples, has never been successfully explained.

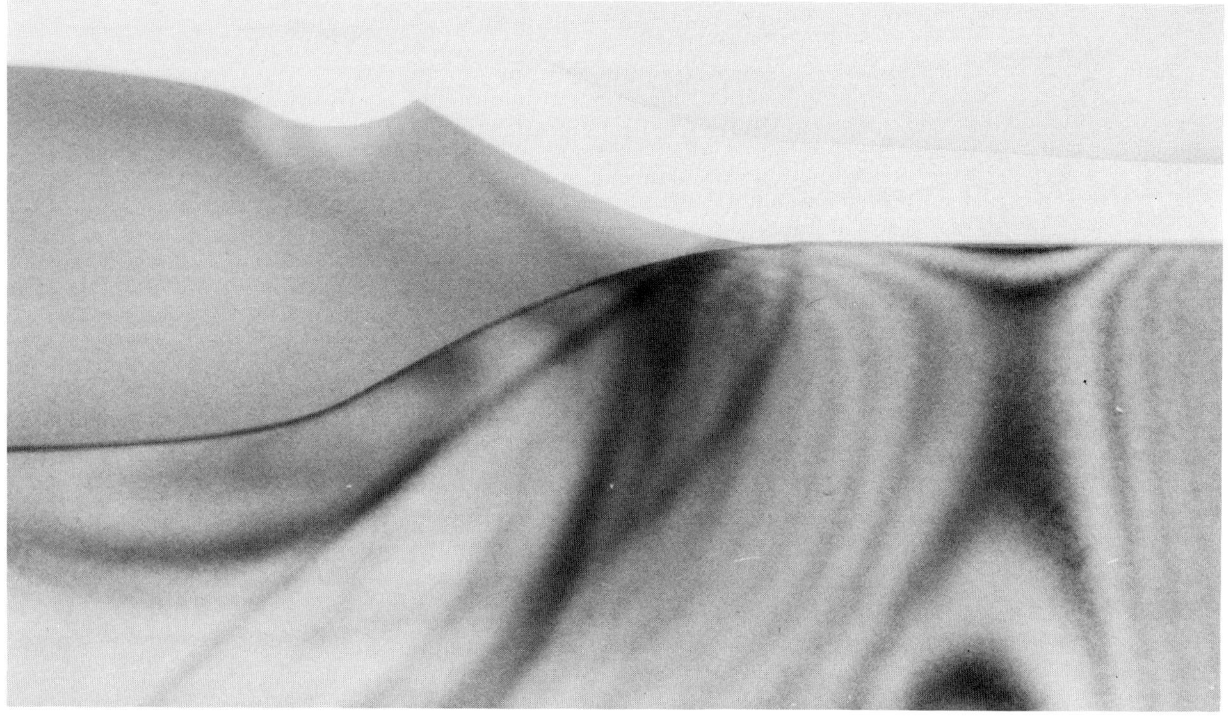

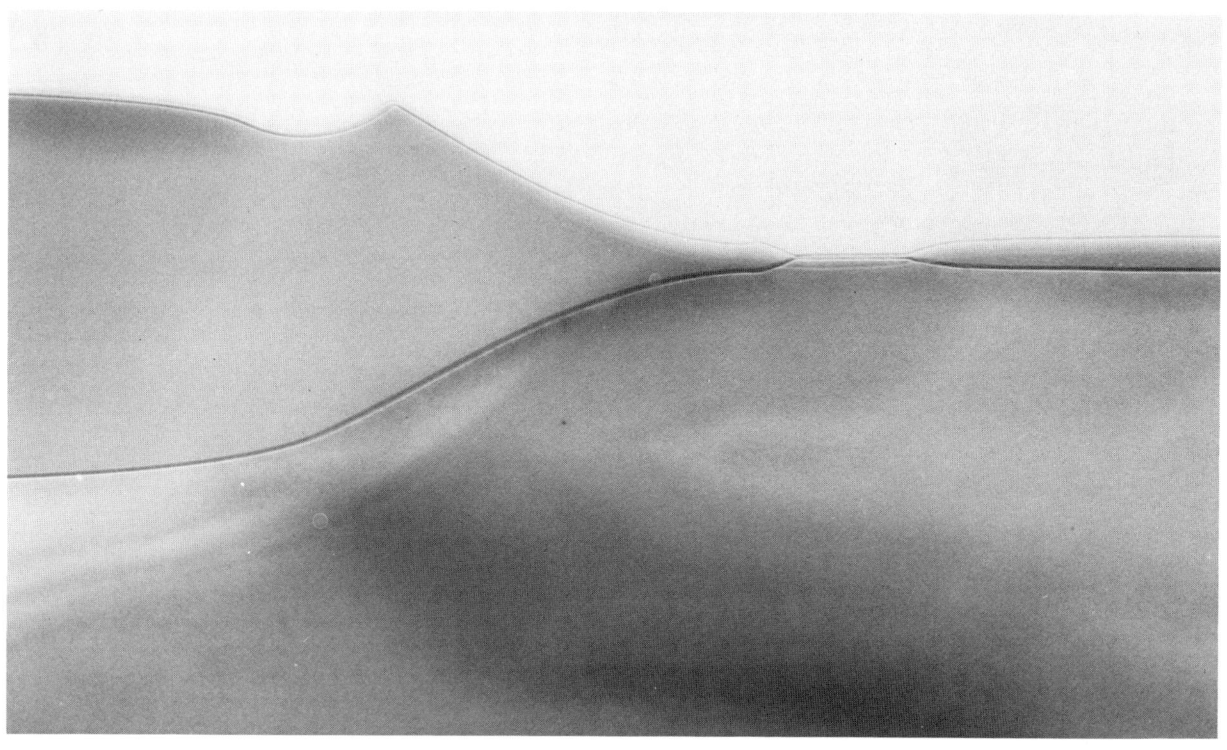

1.0μm

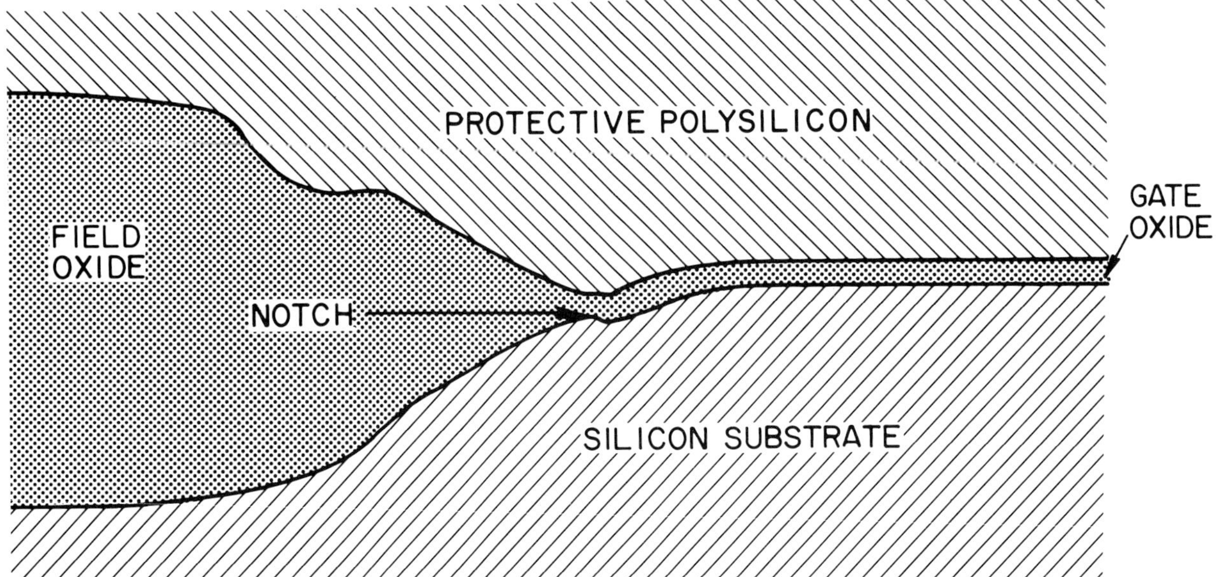

FIGURE 21. The bird's beak after Si_3N_4 stripping, additional oxidation and oxide stripping to remove the oxidation barrier, and gate oxidation. The origin of the notch feature is not understood.

3.2.2 Oxidation at Window Edges

BHF has been used for many years as a satisfactory oxide etchant for window dimensions larger than approximately 2 μm. For smaller window diameters the increase in width at the upper surface of the window due to isotropic oxide etching can be a serious problem, particularly for closely packed device configurations. Plasma and ion etch methods, which etch anisotropically, circumvent this problem. BHF etching may still be used for patterning large window areas, which are subsequently oxidized to form the gate and source and drain regions in MOS structures.

The use of BHF for oxide patterning produces another type of gate oxide thinning phenomenon, which, in its severe form, also becomes a device reliability problem. Close examination of the region where gate oxide joins field oxide sometimes shows a local thinning of the gate oxide (Fig. 22). The extent of gate oxide thinning (8% for the 22-degree field oxide slope in Fig. 22) was found to relate to the slope of the field oxide wall, which in turn is controlled

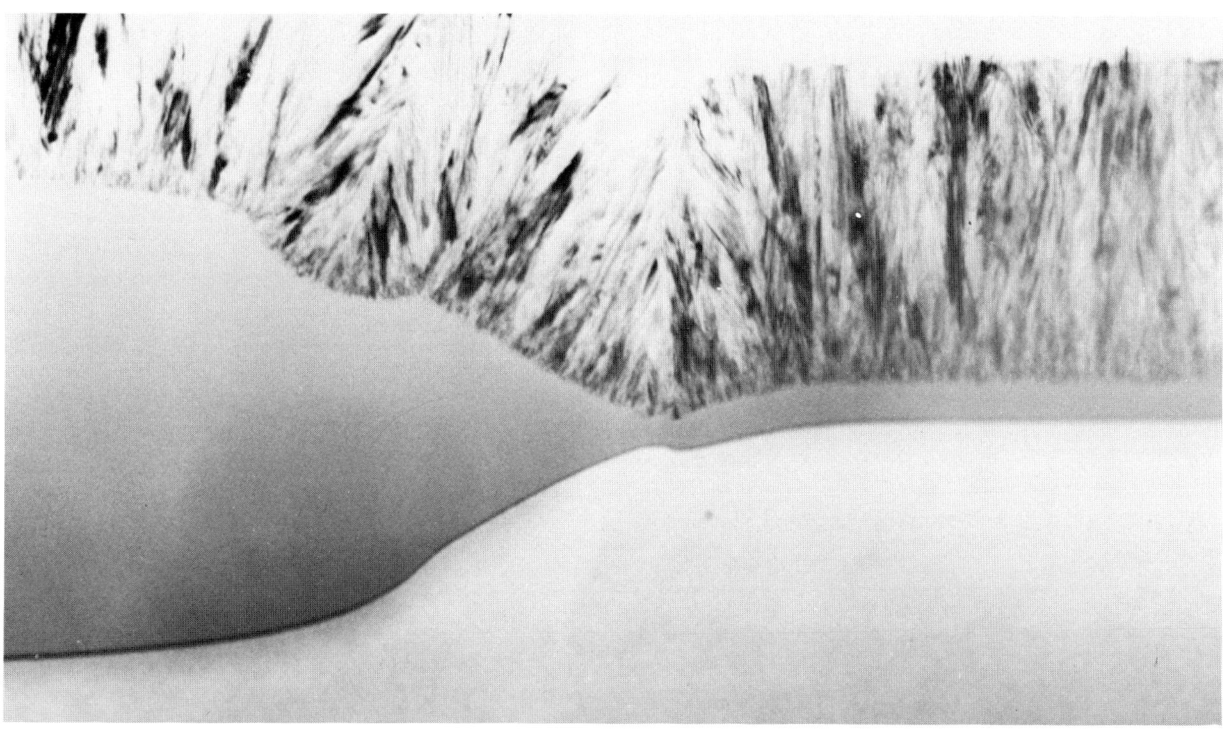

5000Å

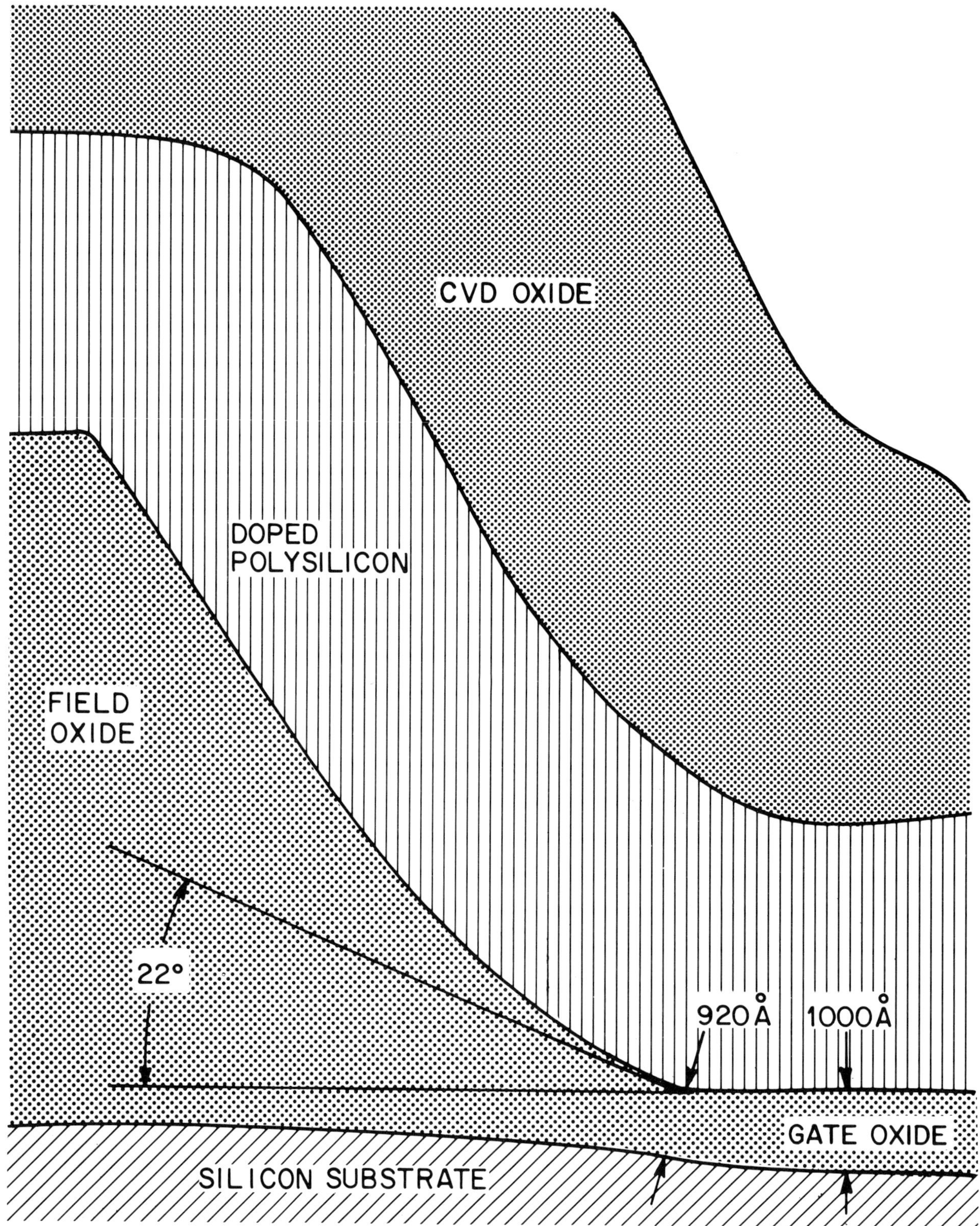

FIGURE 22. An edge of field oxide after gate oxide formation, polysilicon deposition, and doping with phosphorus. An 8% thinning of gate oxide occurs at the step on the silicon surface.

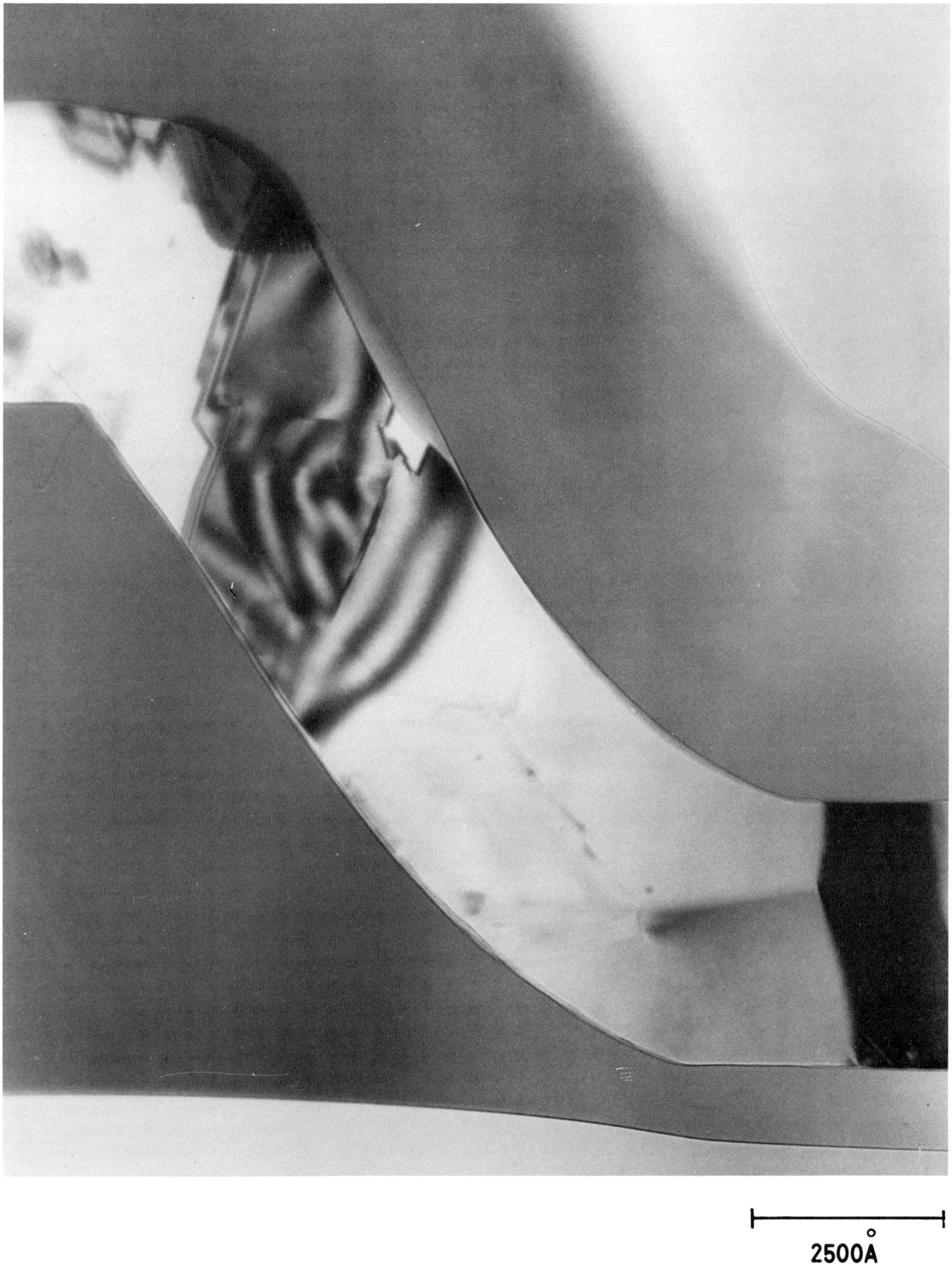

2500Å

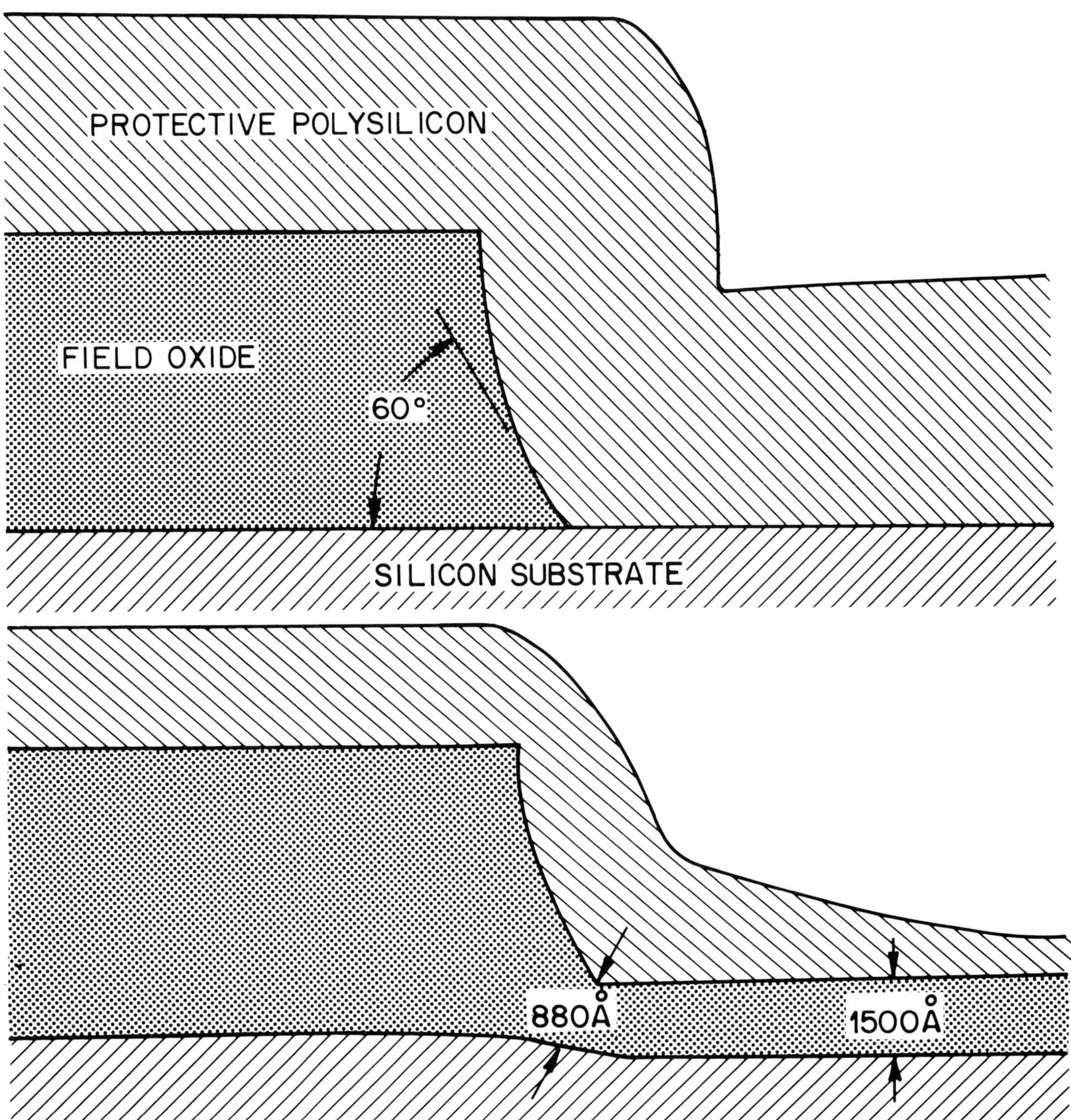

FIGURE 23. A steep wall (60°) at an edge of field oxide before (upper) and after (lower) gate oxidation. A 41% decrease in gate oxide thickness is observed at the step.

by the degree of overetch in BHF.[24] Small diameter windows tend to take about 10% longer to open than larger windows, and overetching is commonly performed to ensure that all windows on a wafer are open. Variations in overetching times create variations in oxide wall slopes, as shown in the upper micrographs of Figs. 23 and 24. Zero overetch results in a gradually tapering oxide wall (Fig. 24, upper), and 100% overetch results in a much steeper wall (Fig. 23, upper). The 100% overetched window, which produced a 60-degree wall slope,

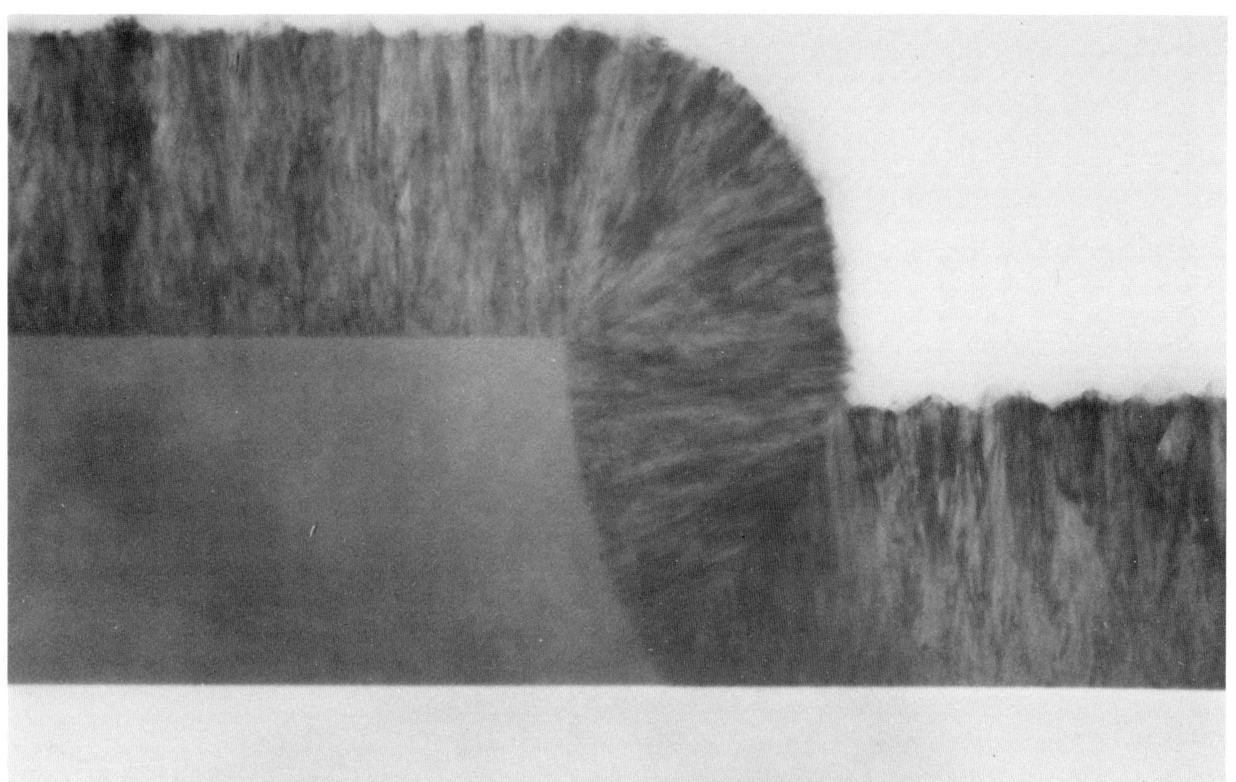

5000Å

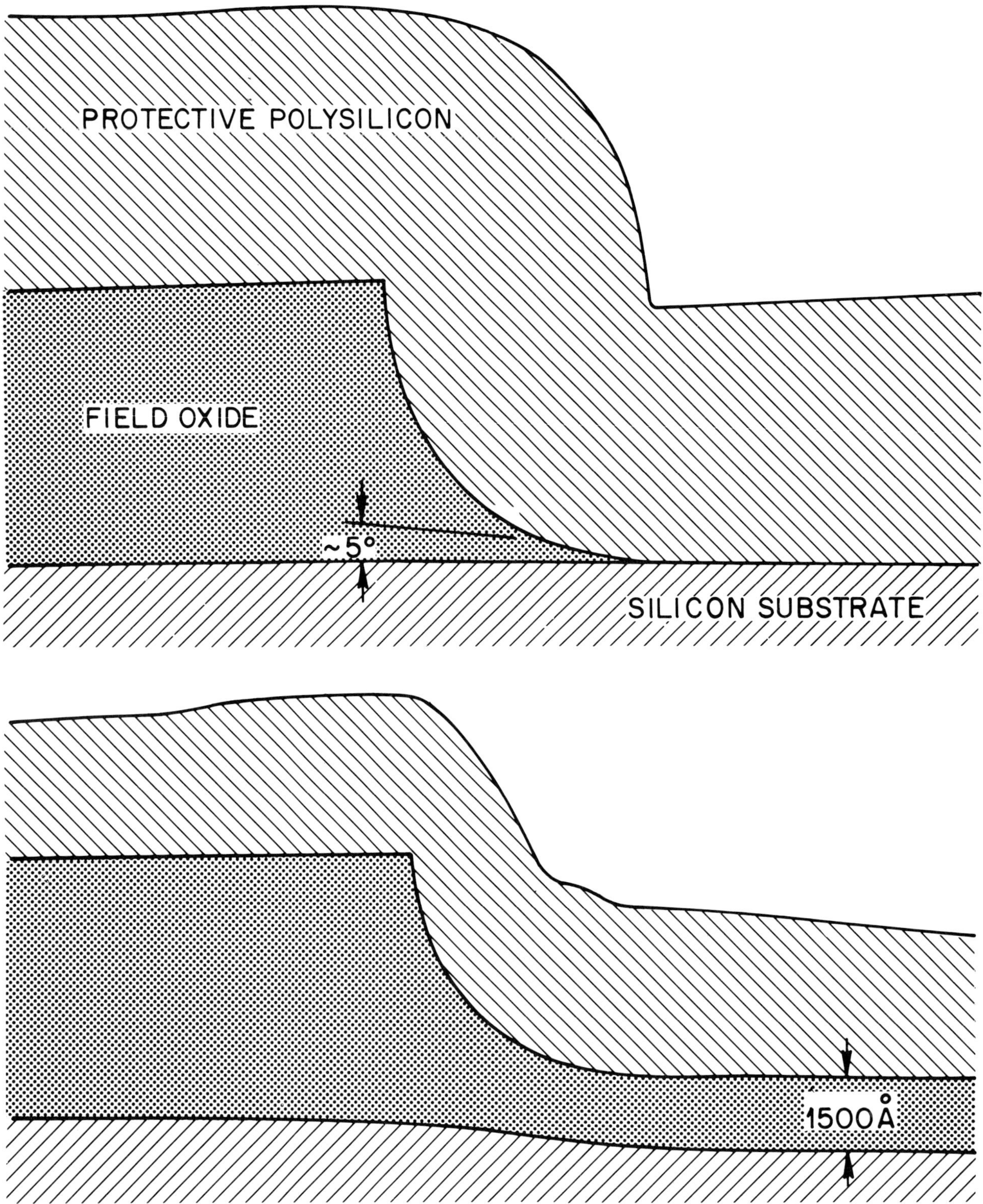

FIGURE 24. A gently sloping wall (5°) at an edge of field oxide before (upper) and after (lower) gate oxidation. No decrease in gate oxide thickness occurs at the step.

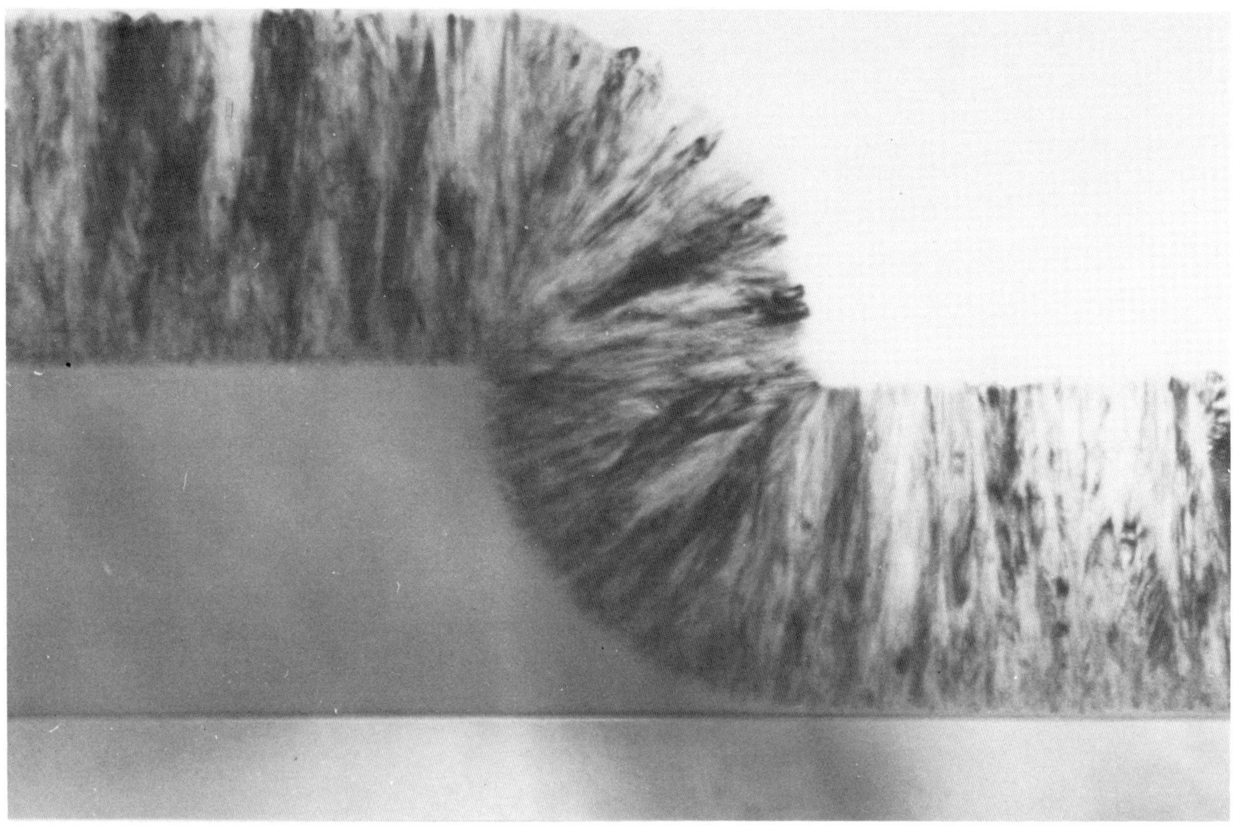

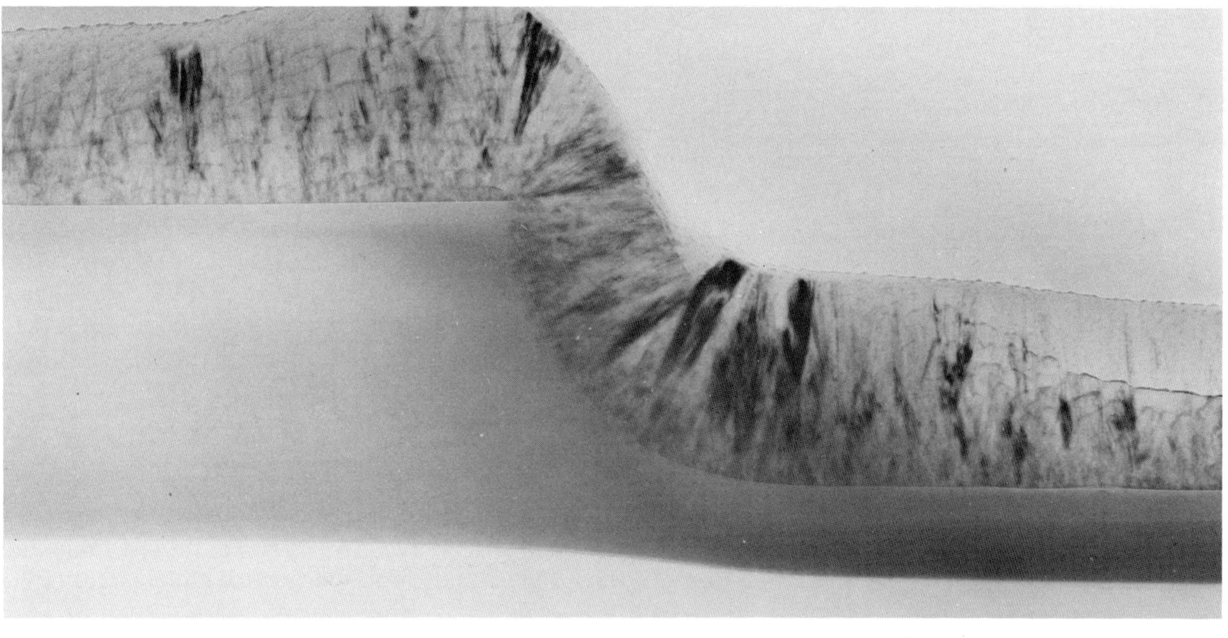
5000Å

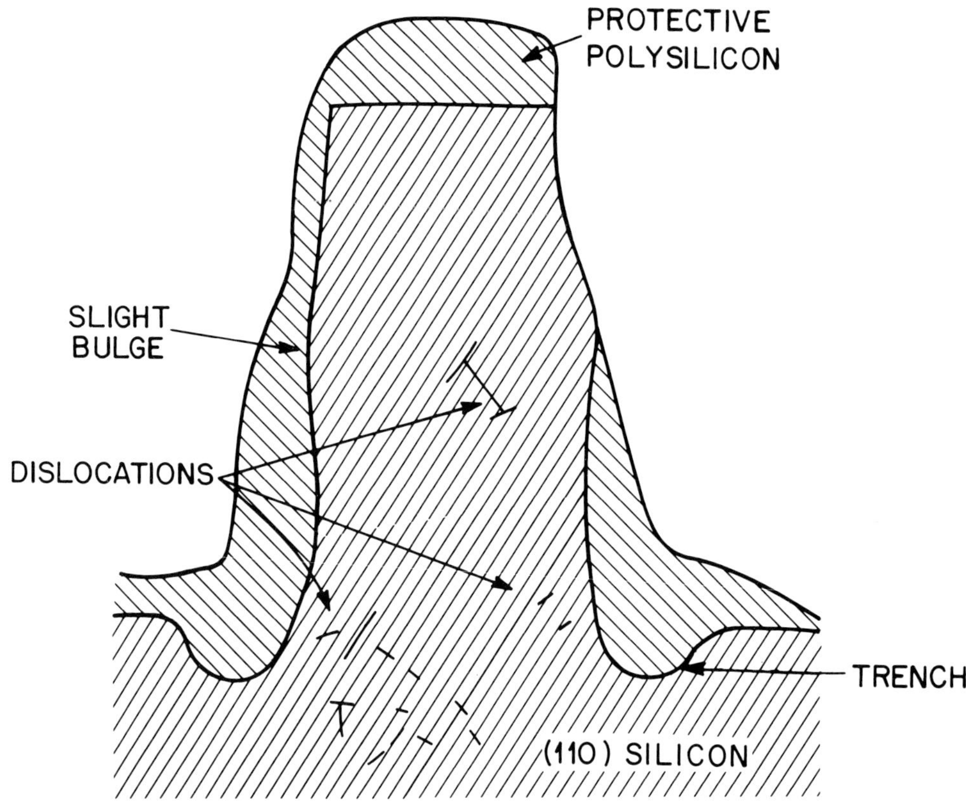

FIGURE 25. Silicon step structure prepared for oxidation studies. The dislocation network, slight bulge, and trenches are formed by reactive ion etching.

resulted in 41% gate oxide thinning (Fig. 23, lower) while the window that was not overetched, which had a 50-degree wall slope, produced no discernible oxide thinning (Fig. 24, lower).

This gate oxide thinning phenomenon has been attributed to the local inhibition of oxygen due to the diffusion barrier resulting from the presence of the field oxide wall. A two-dimensional model shows excellent qualitative agreement with TEM studies, but predicts a magnitude of gate oxide thinning only one-third of the amount observed.[25] This lack of quantitative agreement may be due to insufficient modeling of the oxidation process, particularly in the earliest stages and for shaped surfaces.

3.2.3 Oxidation at Steps

Early modeling of the oxidation of silicon by Grove[26] has essentially remained unaltered to the present time. Refinements have focused on understanding the stoichiometry and structure at the Si/SiO_2 interface and the events during the initial oxidation period τ in the oxidation-rate expression $x^2 + Ax = B(t + \tau)$,

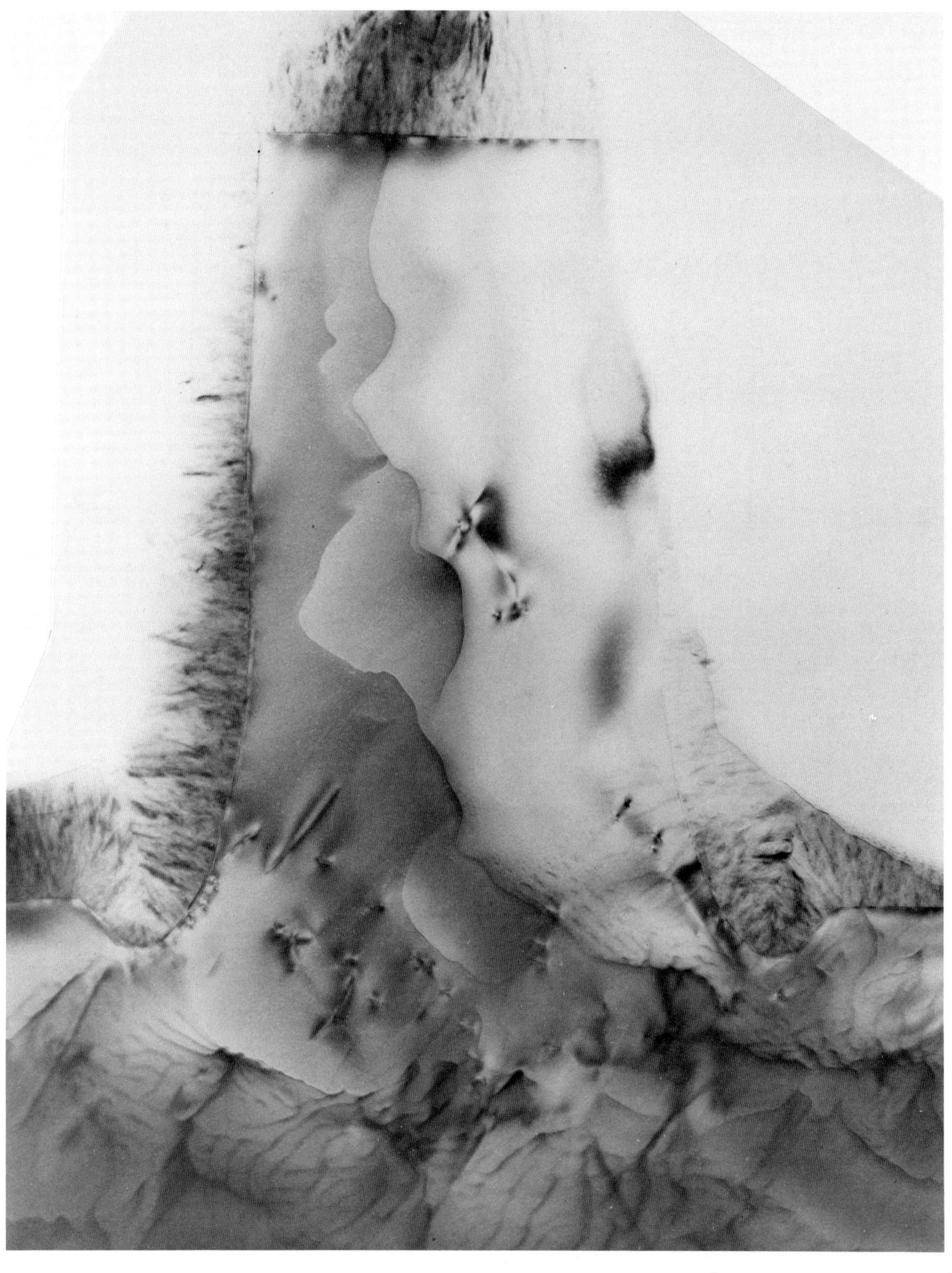
5000Å

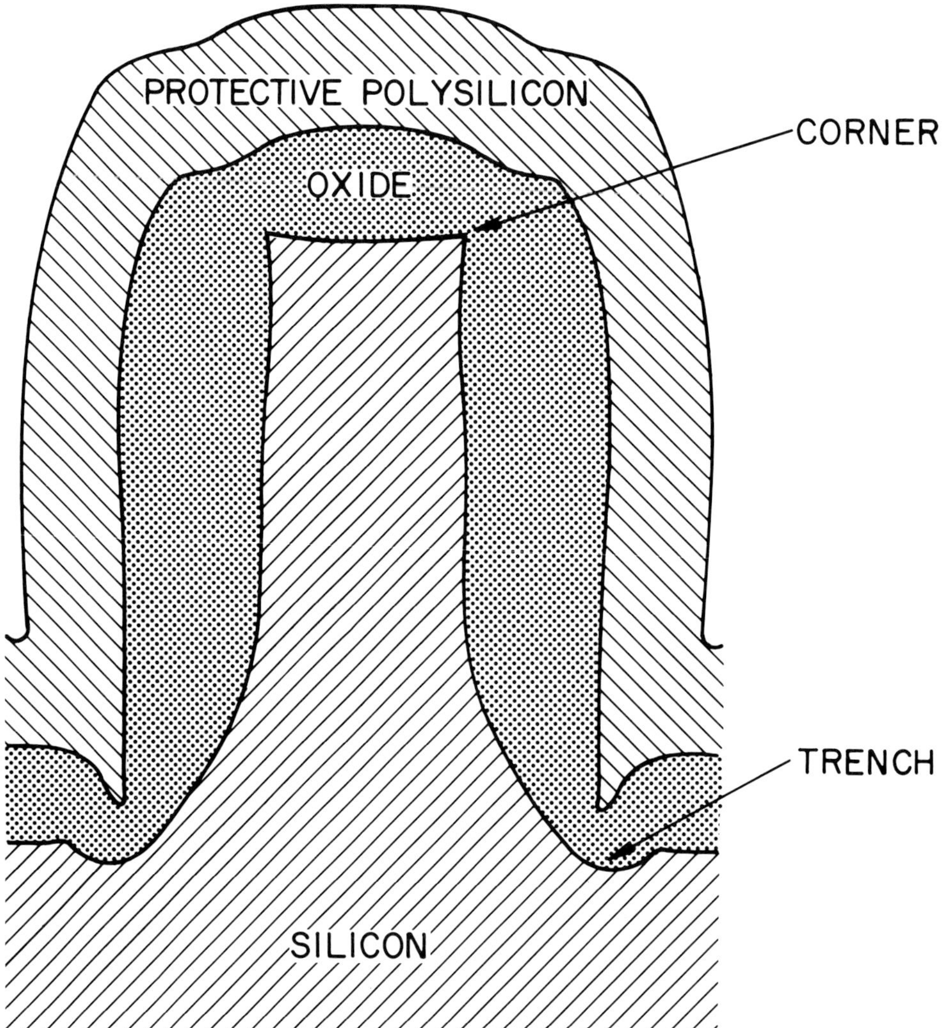

FIGURE 26. Oxidation of stepped silicon surfaces in wet oxygen at 900°C (left) and 950°C (right). Oxidation inhibition occurs at the corners and trenches.

in which x is the oxide thickness, t is the oxidation time, and A and B are constants.[27] Almost all oxidation studies to date have been concerned with flat silicon surfaces. Newer VLSI technologies using etched silicon grooves,[28] which require oxidation of silicon corners, and problems associated with anomalous oxidation characteristics of the edges of polysilicon films (see Chapter 4) both point to the need for the study of the oxidation behavior of nonplanar surfaces. The modeling of oxidation behavior at a shaped (nonplanar) surface has been attempted for the region at the edge of an oxide step,[25] but not for the region at a step on a silicon surface.

TEM studies of stepped silicon surfaces show a surprising oxidation behavior that indicates the importance of local oxide stresses on the oxidation process.[29] Arrays of steps 2 μm high and 1 μm wide were made on (001) silicon wafers by reactive ion etching, as shown in Fig. 25; these steps extend across the wafer surfaces. Figures 26 and 27 show the results of oxidation of these

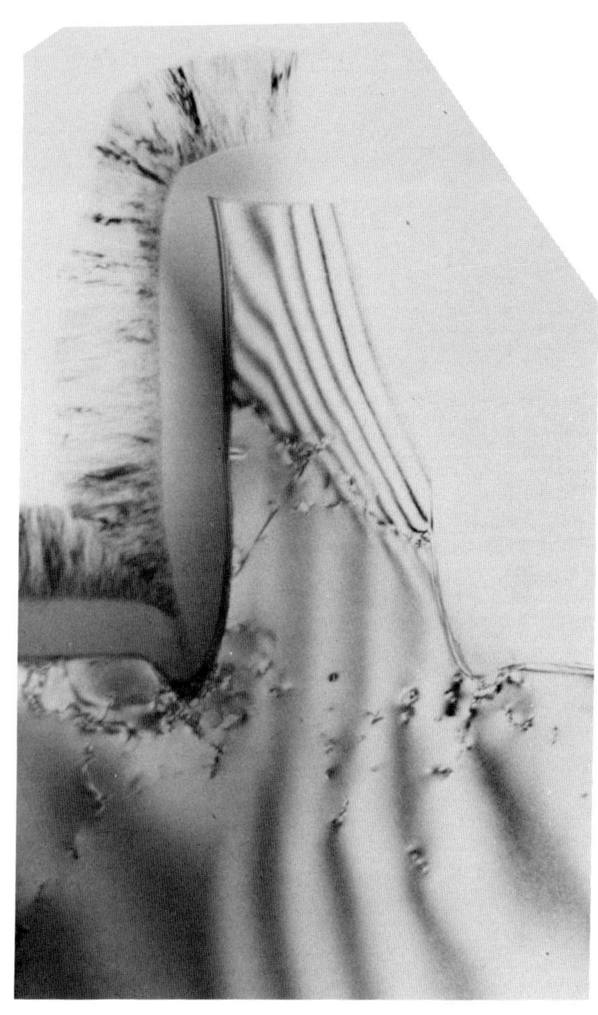
1.0μm

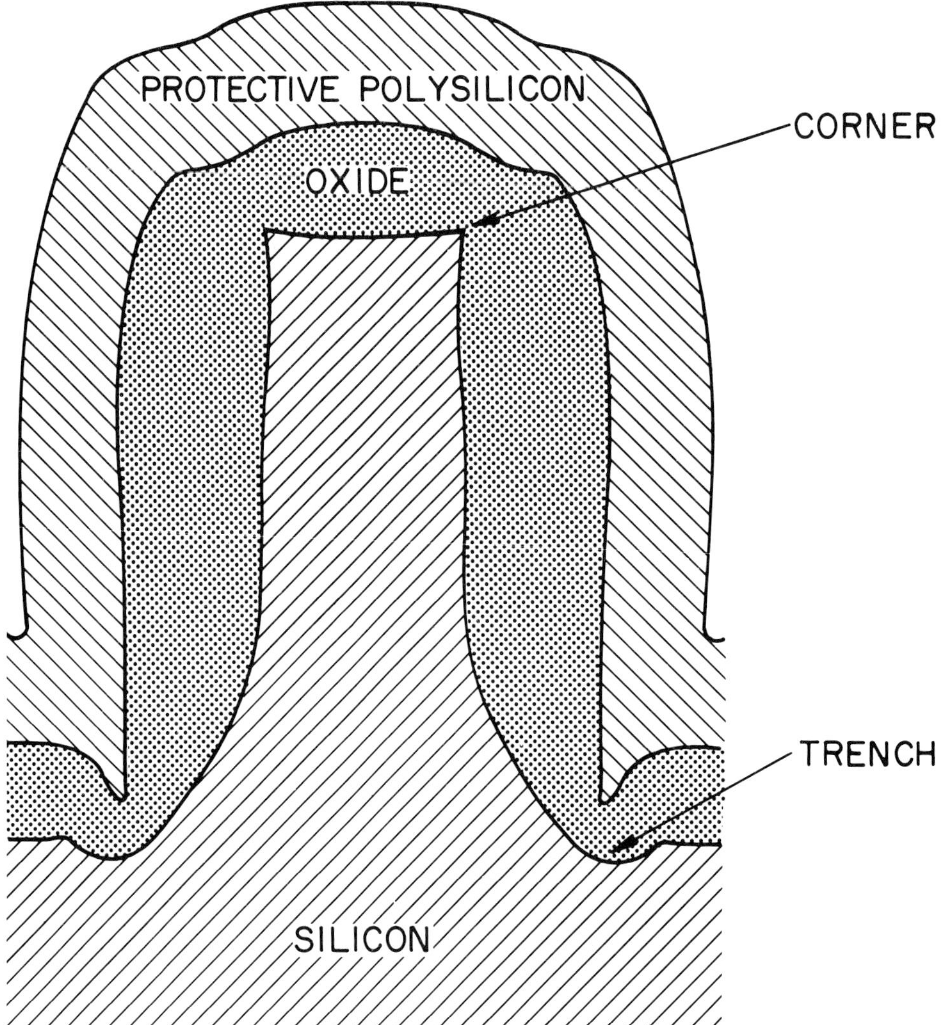

FIGURE 27. Oxidation of stepped silicon surfaces in wet oxygen at 1050°C (left) and 1100°C (right). Less oxidation inhibition occurs at the corners at higher oxidation temperatures.

wafers in wet oxygen at temperatures from 900–1100°C. The ion-etching step introduces dislocations and trenches at the base of the side walls and a slight bulge at the step center (Fig. 25). Oxidation is inhibited at the trench site by the limitation of available oxygen caused by the sample shape in the trench region. Oxidation behavior at the corner shows increasing local inhibition of oxidation with decreasing temperature (Table 1), leading to the formation of "horns" of silicon at these sites.

Oxidation inhibition at a corner is attributed to the effect of locally high, intrinsic oxide stress on the oxidation process.[29] The stress, and consequently oxide inhibition, is reduced at higher temperatures by the appearance of viscous flow of the oxide above 950–1000°C.[30]

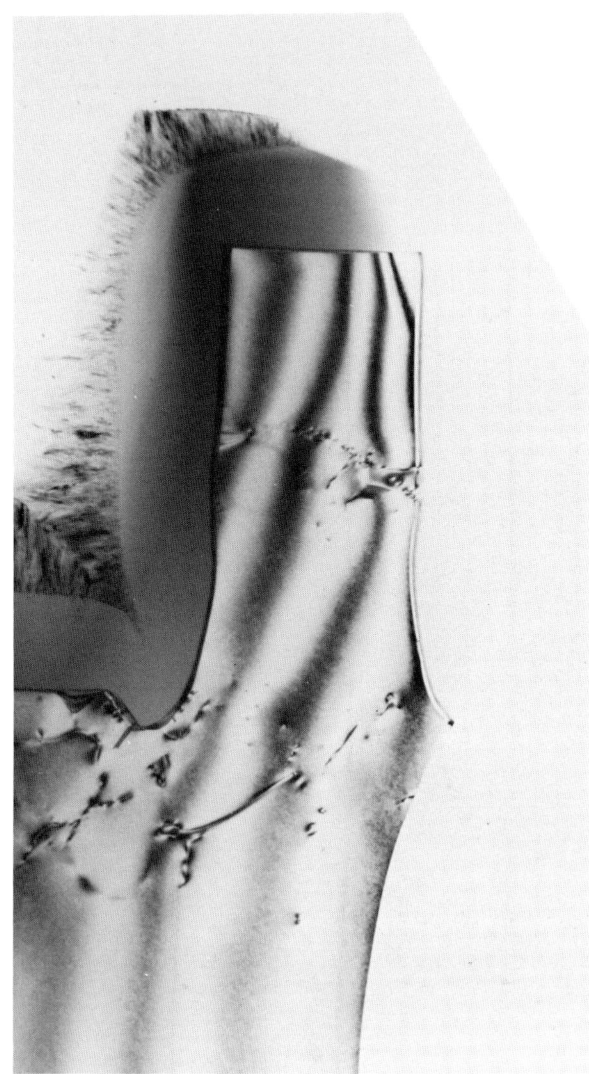

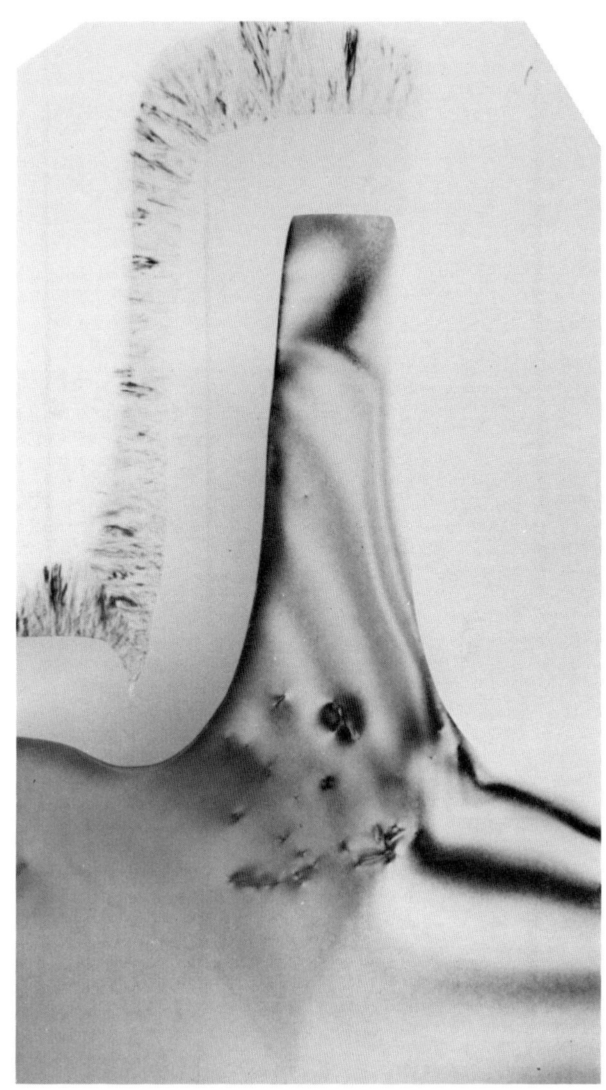
1.0 μm

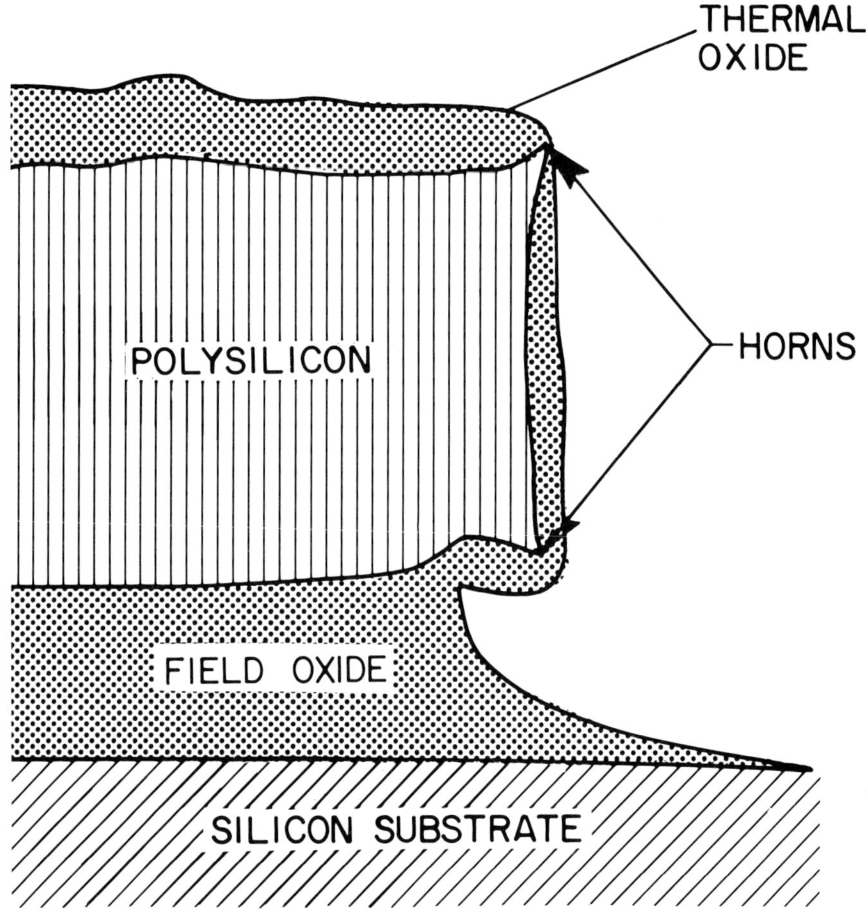

FIGURE 28. Patterned polysilicon that has been oxidized to grow 500-Å SiO_2. Oxidation often forms horns of less oxidized silicon at the corners of polysilicon.

TABLE 1 Dependence of Oxidation Inhibition at a Silicon Corner on Oxidation Temperature

Oxidation Temperature (°C)	Decrease in Oxide Thickness at Corner (%)
900	28
950	31
1050	14
1100	~0

3.2.4 Polysilicon Oxidation

When the corner region of a patterned polysilicon circuit element consists of a single crystal of silicon, then local oxidation inhibition occurs according to the mechanism described above, forming horns that project into the oxide (Fig. 28).

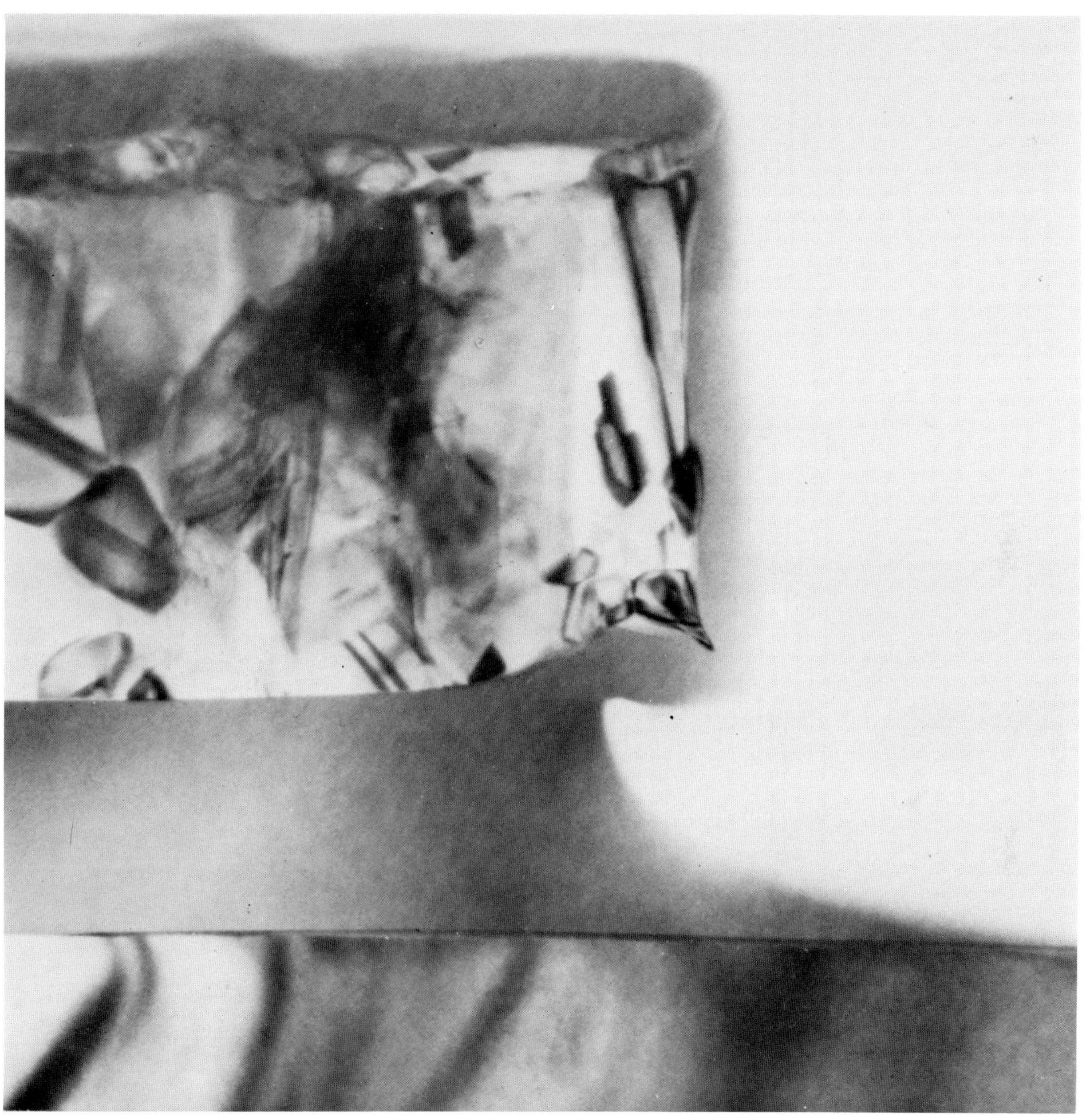

2000Å

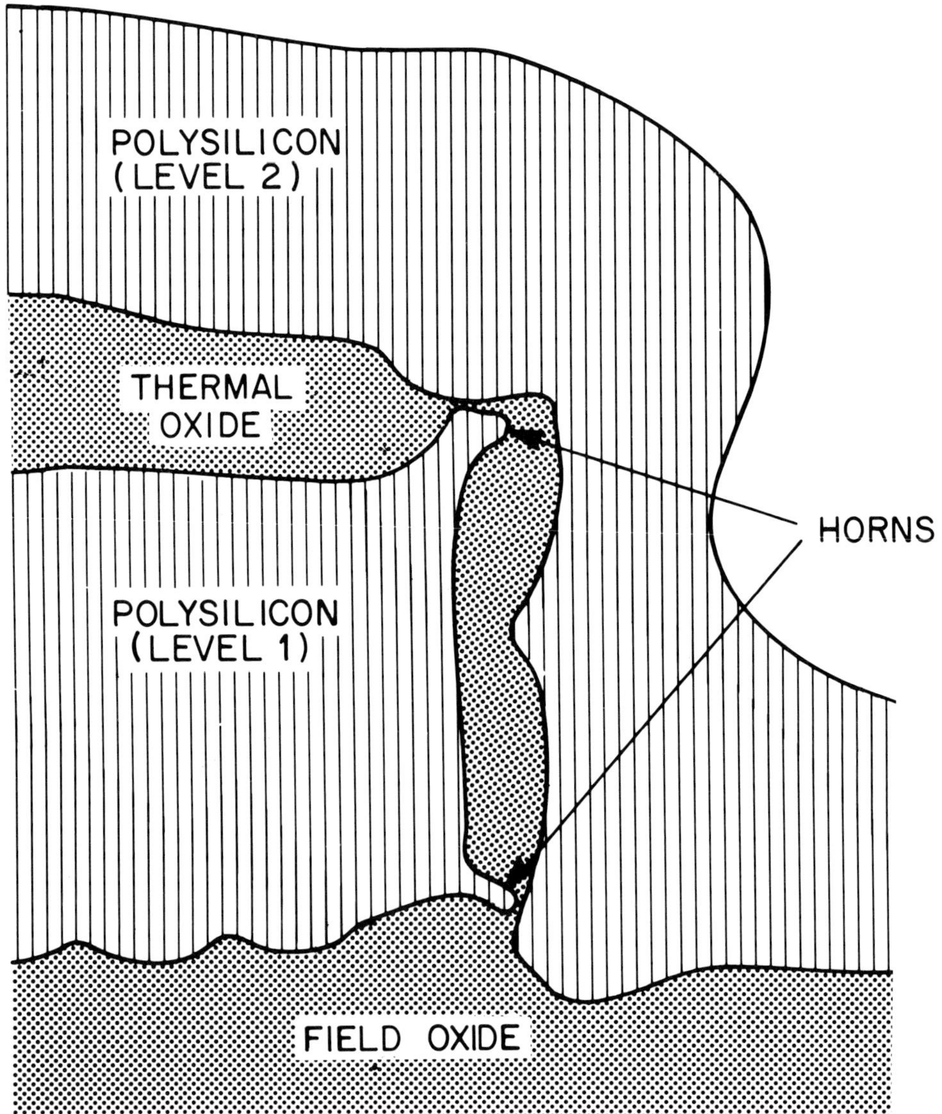

FIGURE 29. Two-level polysilicon device structure showing horns at the corners of polysilicon (level I). The thin oxide separating the two polysilicon layers at the horns is a potential device failure site.

These horns cause high leakage and electrical shorts when a second metallization layer covers the oxide step (Fig. 29), since the oxide thickness is severely reduced at these regions.

Another manifestation of oxidation inhibition, caused in part by oxide stress, is the formation of protuberances and inclusions in polysilicon. Thermal oxides grown from polysilicon films are weaker dielectrics than oxides grown from single-crystal silicon. This weakness is attributed to the presence of a highly textured, polysilicon/oxide interface that causes local increases in the electric field at the sites of specific disturbances.[31,32] Close study of the polysilicon/oxide interface reveals detailed information on the morphology of these

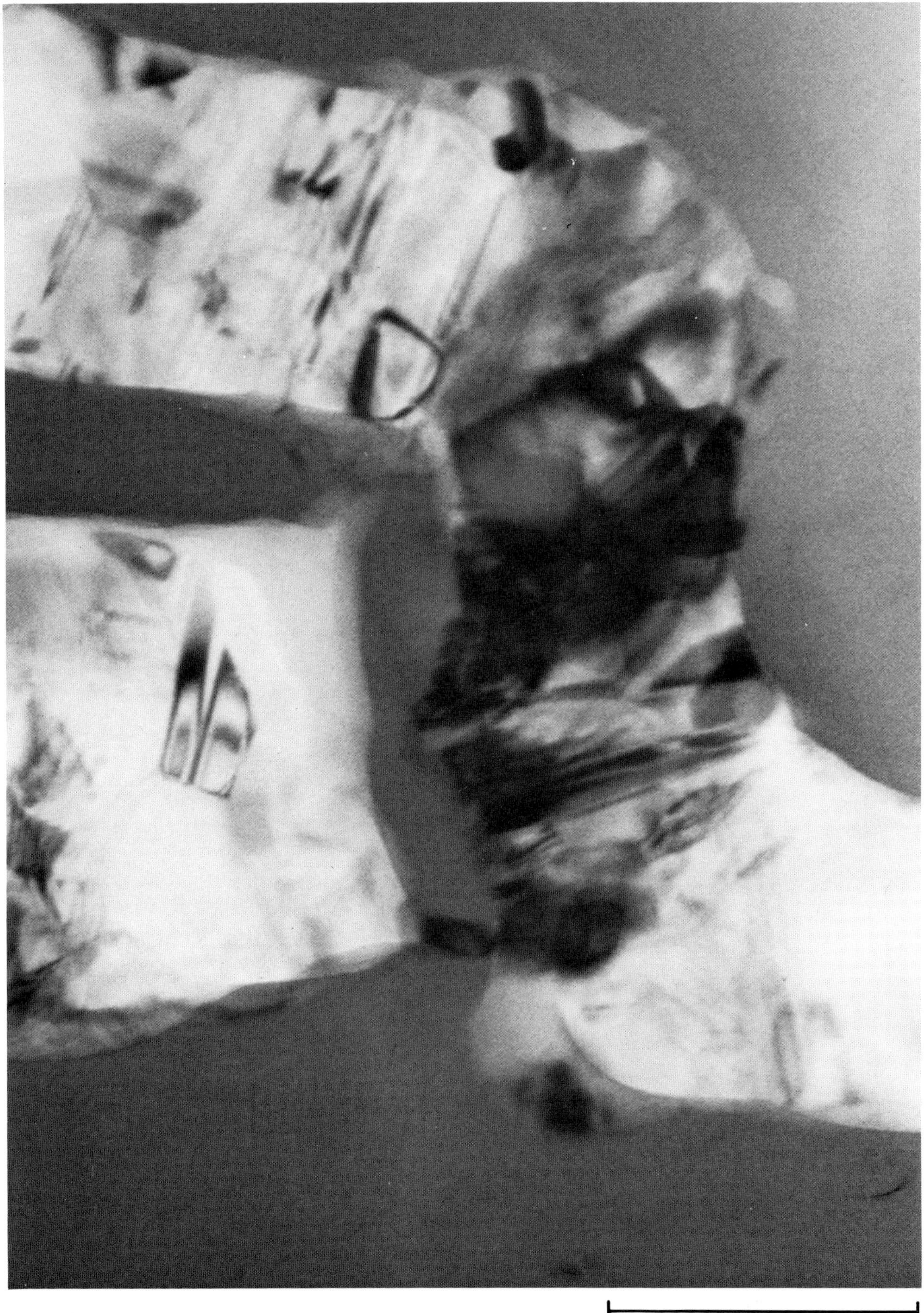
2000Å

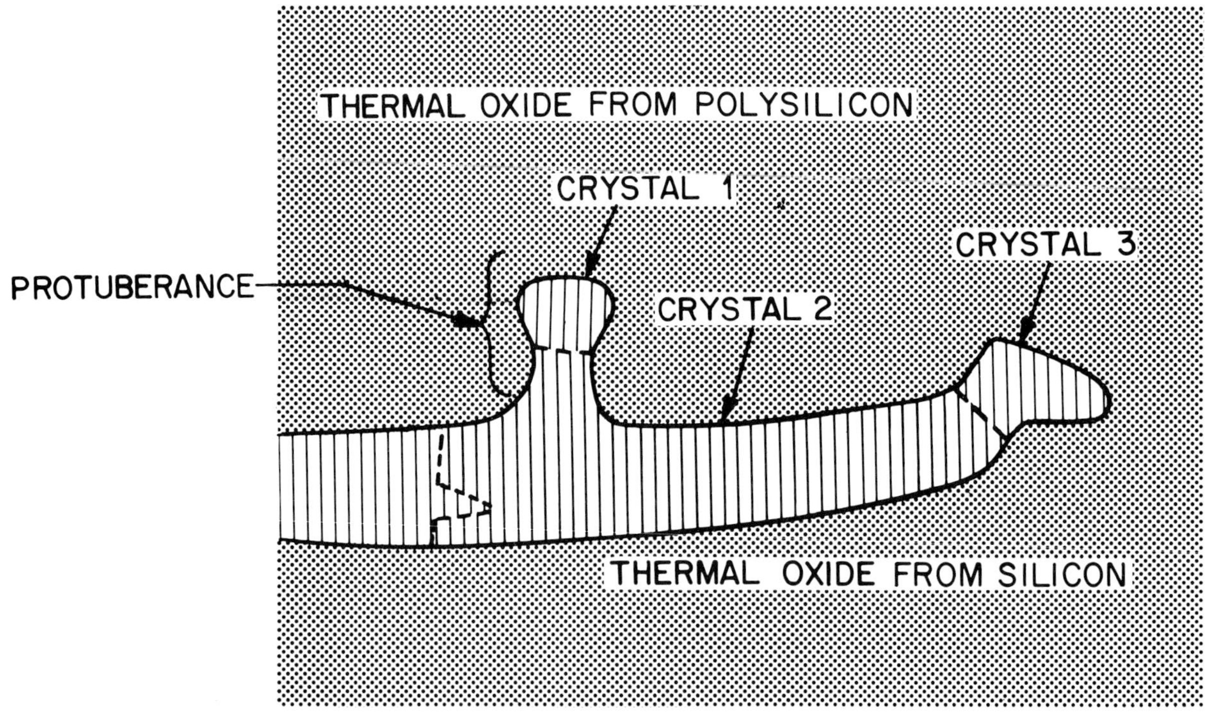

FIGURE 30. Micrographs taken under two different diffracting conditions of doped polysilicon oxidized at 950°C showing protuberance and horn formation. The crystallite at the top of the protuberance (crystal 1) becomes an inclusion with continued oxidation; crystal 2 forms part of the body of the polycrystalline film; crystal 3 has developed a horn.

disturbances and some insight into how they form. Figure 30 shows a patterned polysilicon film photographed under two different diffracting conditions; the film was 4000 Å thick before phosphorus doping and wet oxidation at 950°C. The protuberance is a region of local oxidation inhibition that "grows" in response to this inhibition. This protuberance is one of four types of commonly observed textural features found at the polysilicon/oxide interface.[33] The model, which was developed to explain the growth of this feature, proposes that the crystallite at the tip (crystal 1) initially oxidizes at a slower rate because its orientation presents a slow oxidizing face to the ambient. It then continues to oxidize at an even slower rate due to high intrinsic stresses created in the growing oxide. Eventually crystal 1 separates because of loss of oxidation inhibition at the grain boundary and forms an inclusion of silicon within the oxide. Figure 30 also shows a horn, which consists of a single grain of silicon (crystal 3), forming at the polysilicon edge. This horn has a somewhat modified appearance partly because the edges of the polysilicon have lifted in response to oxidation of both its own undersurface and the surface of the silicon substrate.

A second type of asperity at the polysilicon/oxide interface arises from a completely different mechanism.[33] Bumps of locally thick polysilicon are often produced during polysilicon deposition because of local sites of anomalous nucleation and growth; after doping and oxidation these polysilicon bumps remain, and oxide thickness is essentially constant across these regions (Figs. 31 and 32).

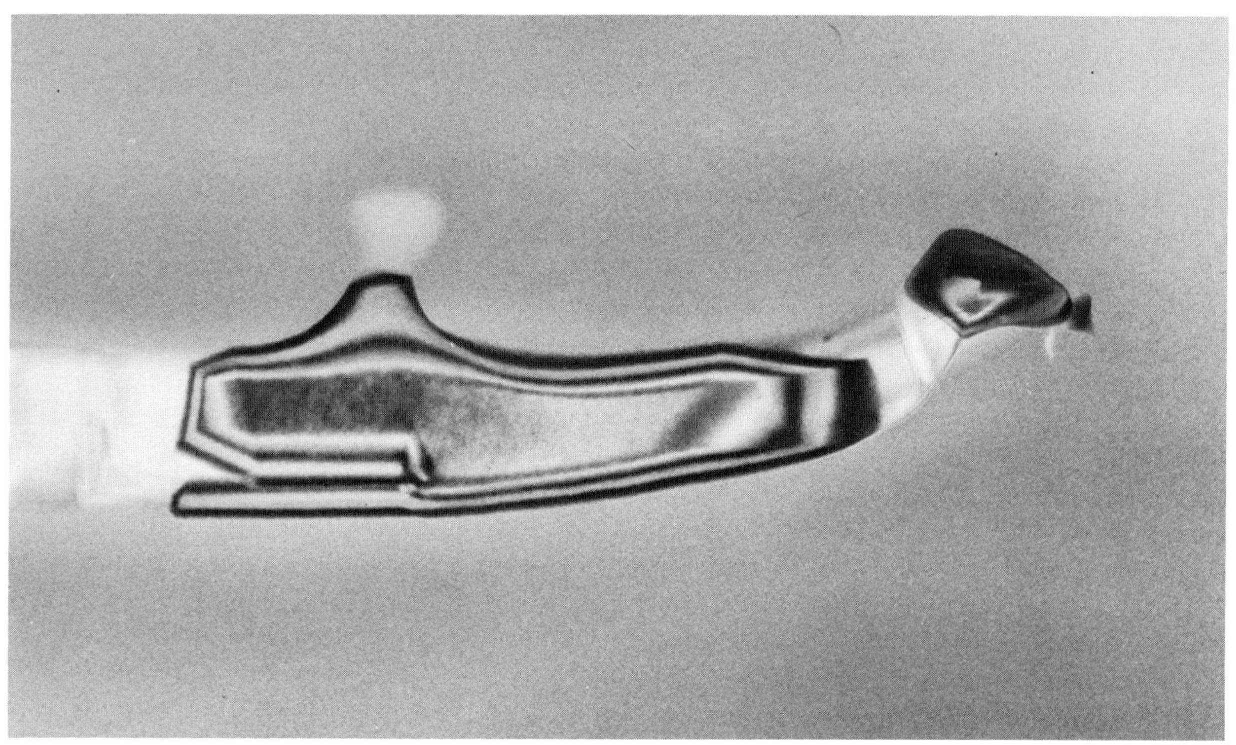

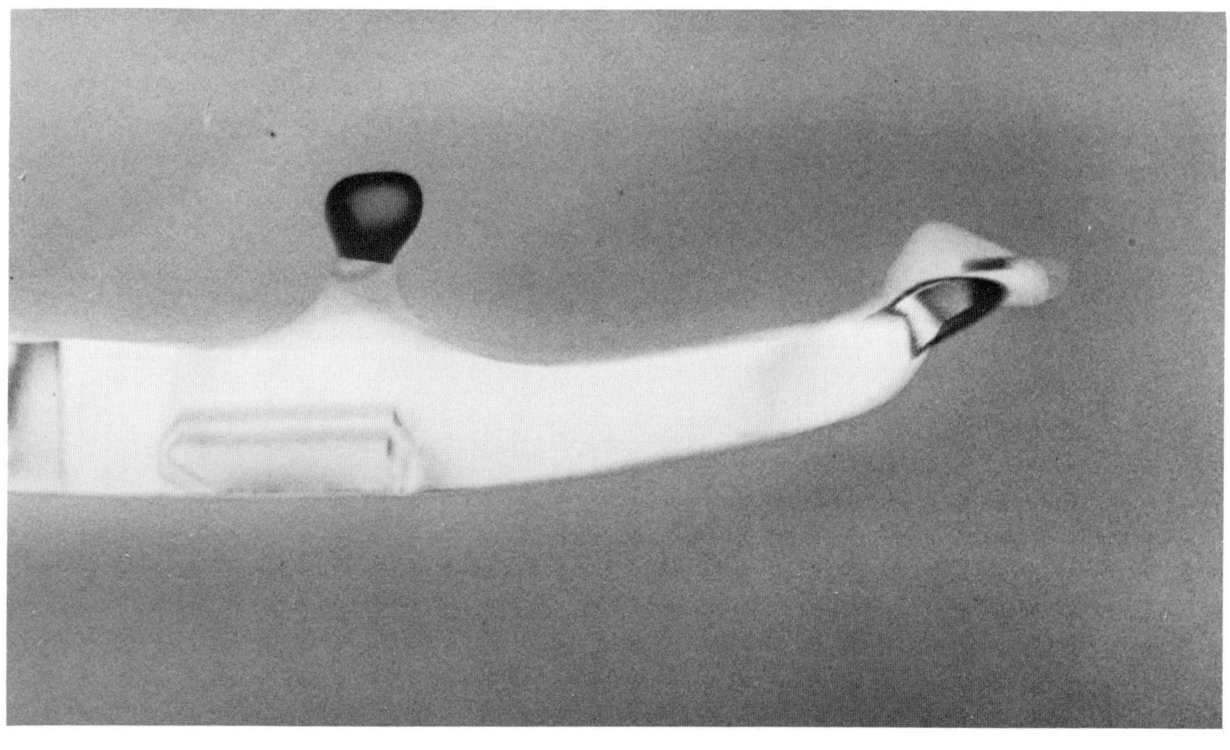
1.0μm

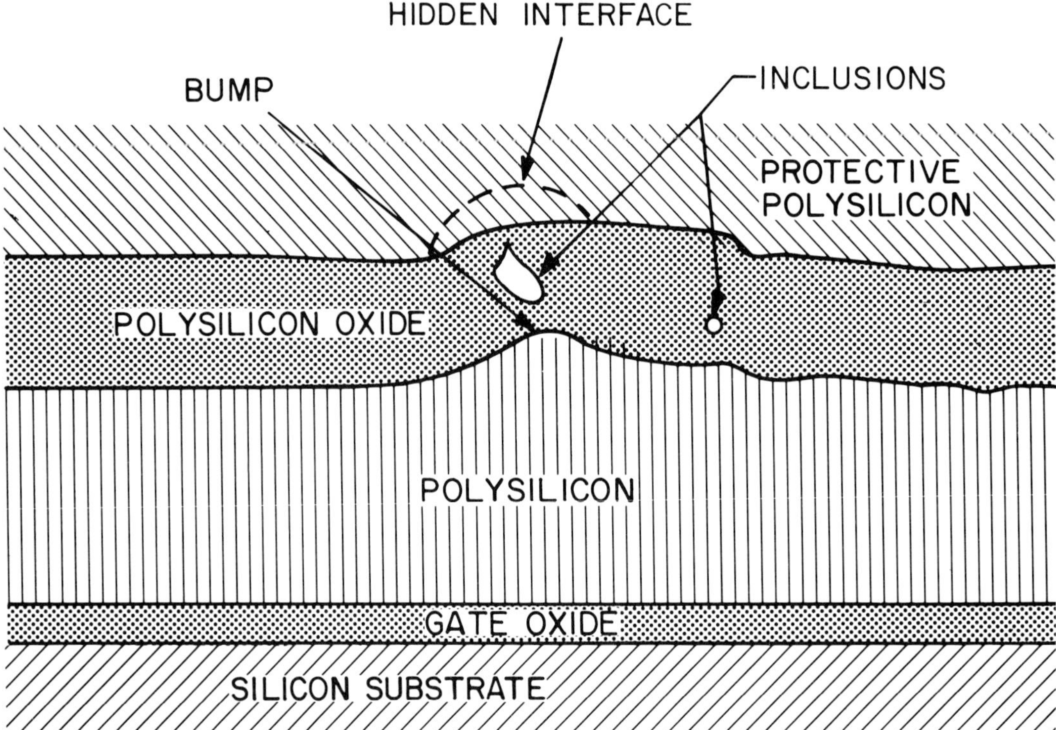

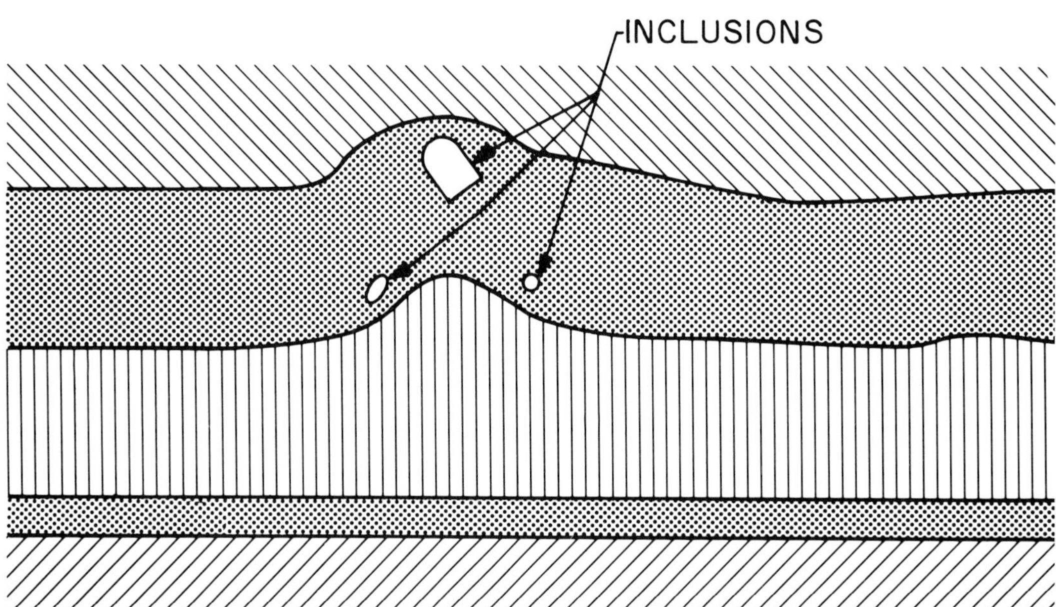

FIGURE 31. Doped polysilicon oxidized at 1050°C in wet oxygen showing bumps caused by anomalous nucleation of polysilicon and inclusions of silicon in the vicinity of the bumps.

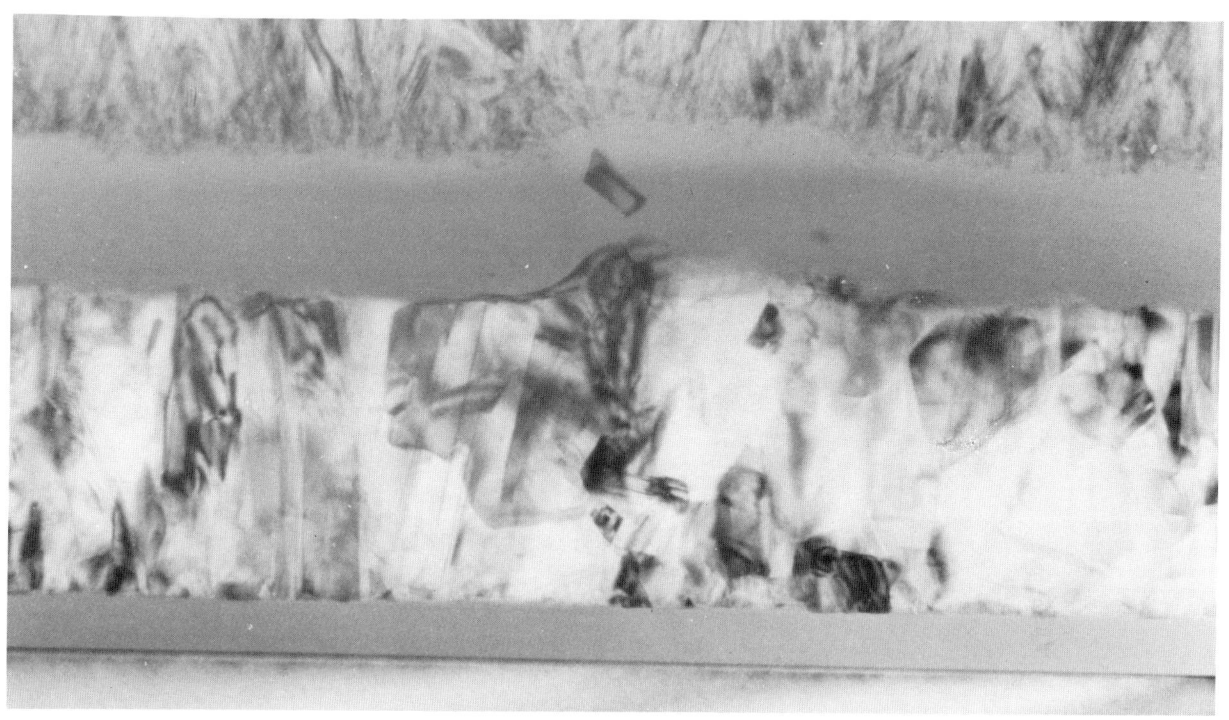

5000Å

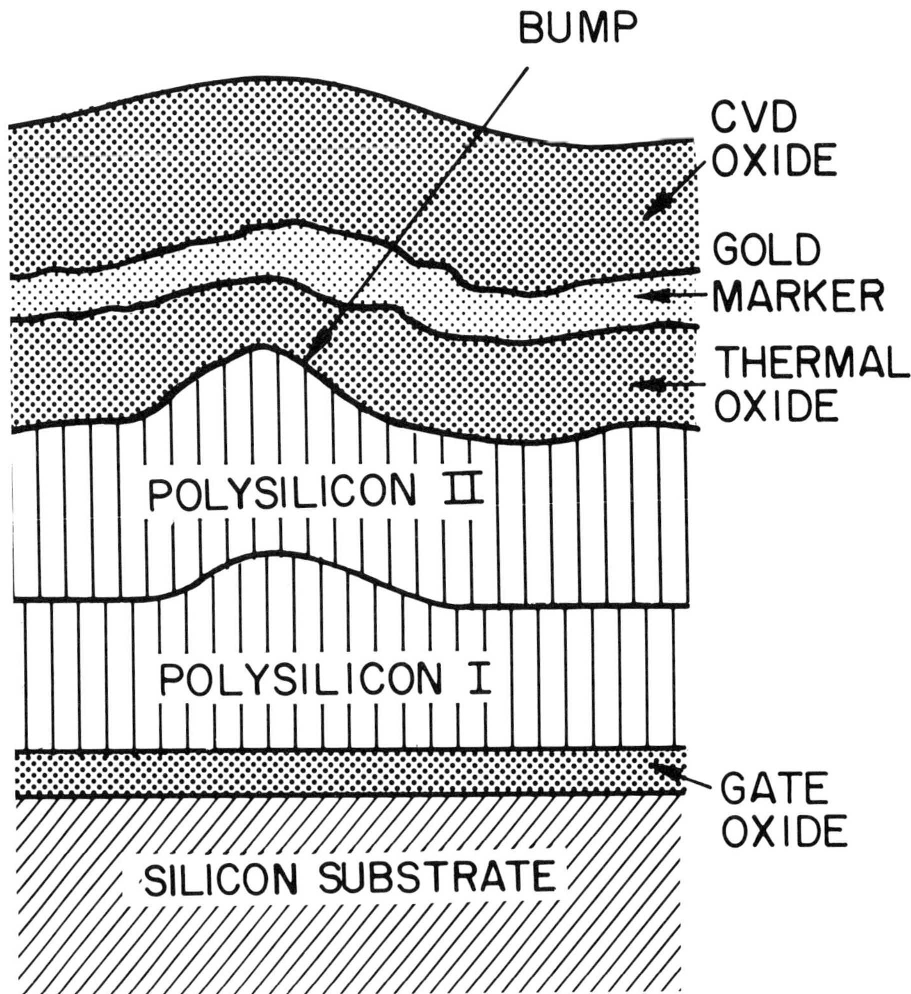

FIGURE 32. Undoped polysilicon after 950°C wet oxidation. The gold layer enhances visibility of the thermal oxide topography, showing that the rough texture of the polysilicon/oxide interface is maintained at the top surface of the oxide.

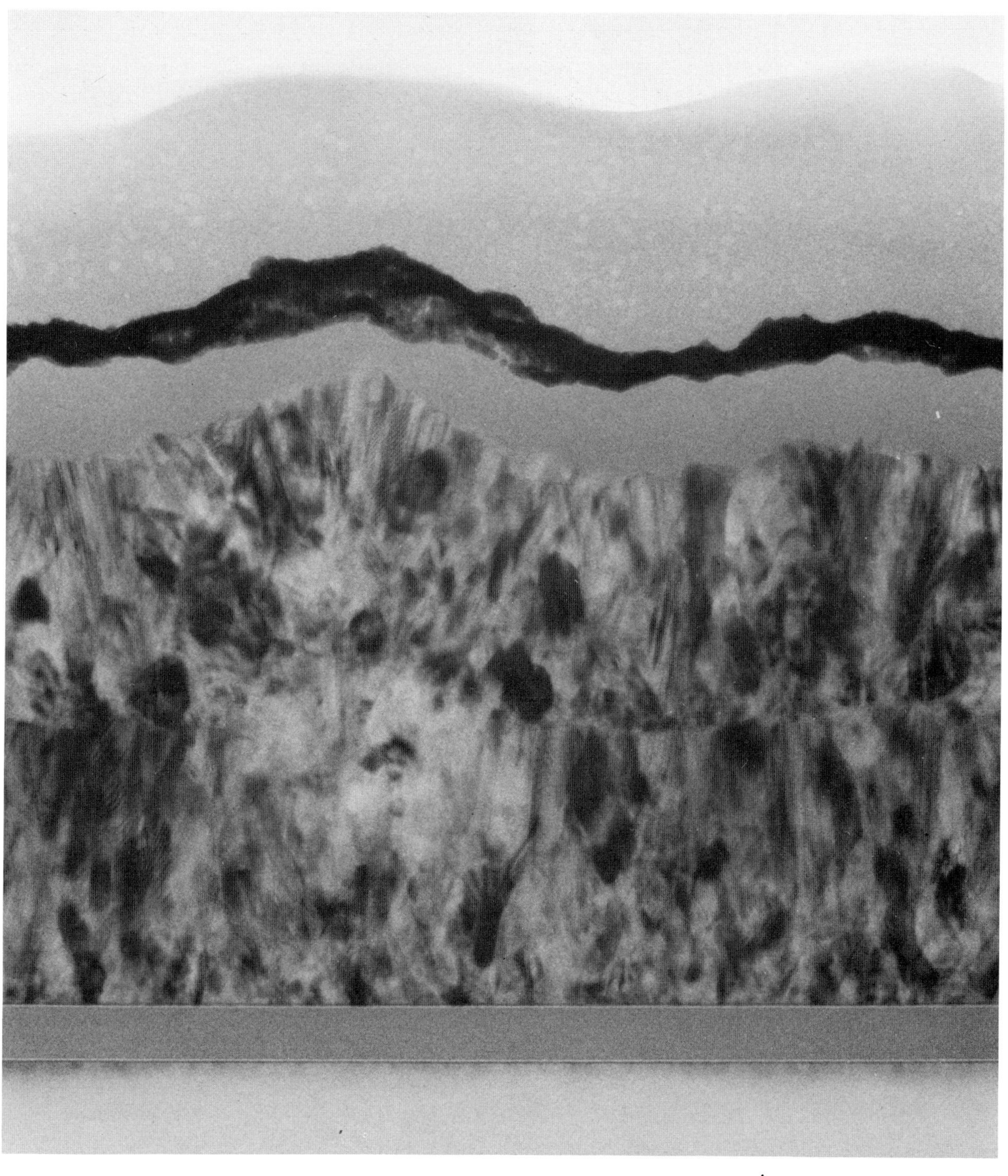
5000Å

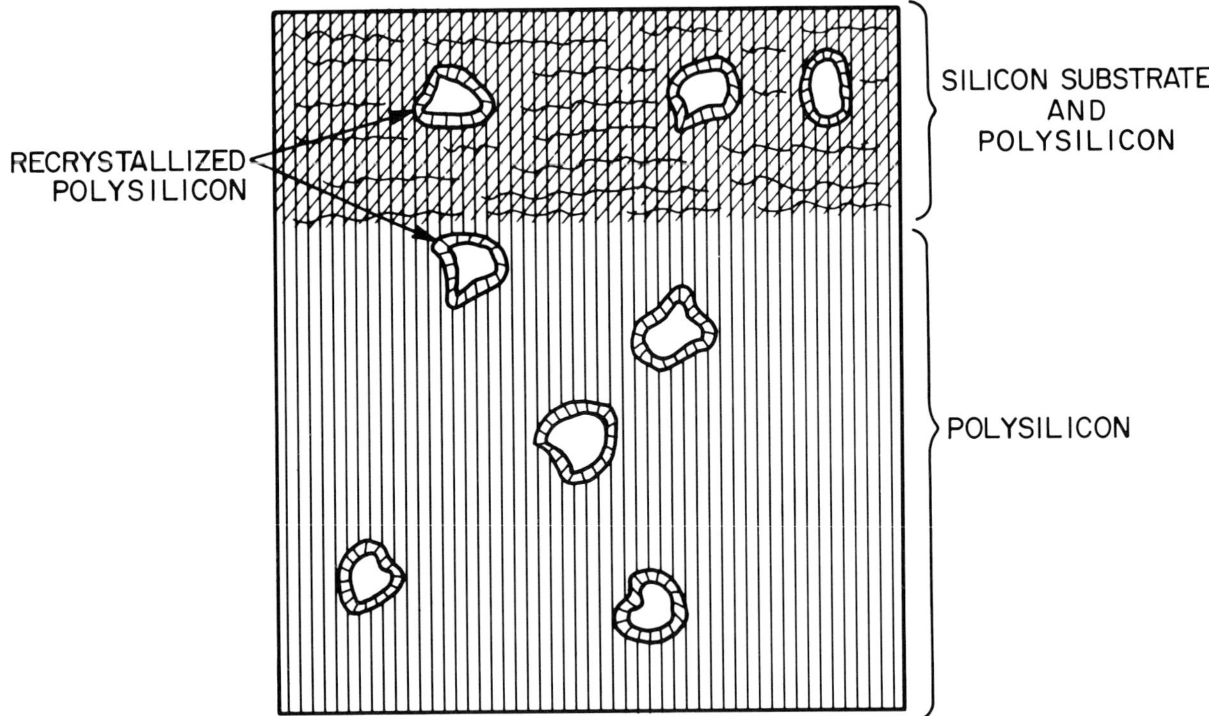

FIGURE 33. Pits forms in an Al-SiO_2-Si capacitor during electrical breakdown. The capacitor field plate consists of 1500-Å aluminum.

Both protuberances and bumps give rise to inclusions of silicon within the oxide (Fig. 31). These inclusions are the third type of textural feature formed by the oxidation of polysilicon. A fourth textural feature is interface roughness, which is caused by the grain structure of the polysilicon (Fig. 32) and is modified by subseqent processing.

Protuberances of the type shown in Fig. 30 minimally disturb the upper oxide surface but locally decrease oxide thickness. Bumps (Fig. 31) and surface roughness (Fig. 32) cause a corresponding texture that is not quite as rough at the upper surface. All four types of features (including inclusions) are usually present and influence the dielectric properties of the oxide. Breakdown is usually symmetric with respect to applied polarity, and higher leakage currents occur when the polysilicon is cathodic.[34]

Self-healing breakdown occurs in silicon capacitors consisting of a thin field plate electrode because of the loss of conducting material by volatilization during breakdown. Figures 33 and 34 show breakdown sites that demonstrate the high temperature created during discharge. The offset of the two circles formed by the intersection of the 5-μm diameter cylindrical pit with the two oxide surfaces (Fig. 34) shows that the pit can form with its axis skew to the wafer surface.

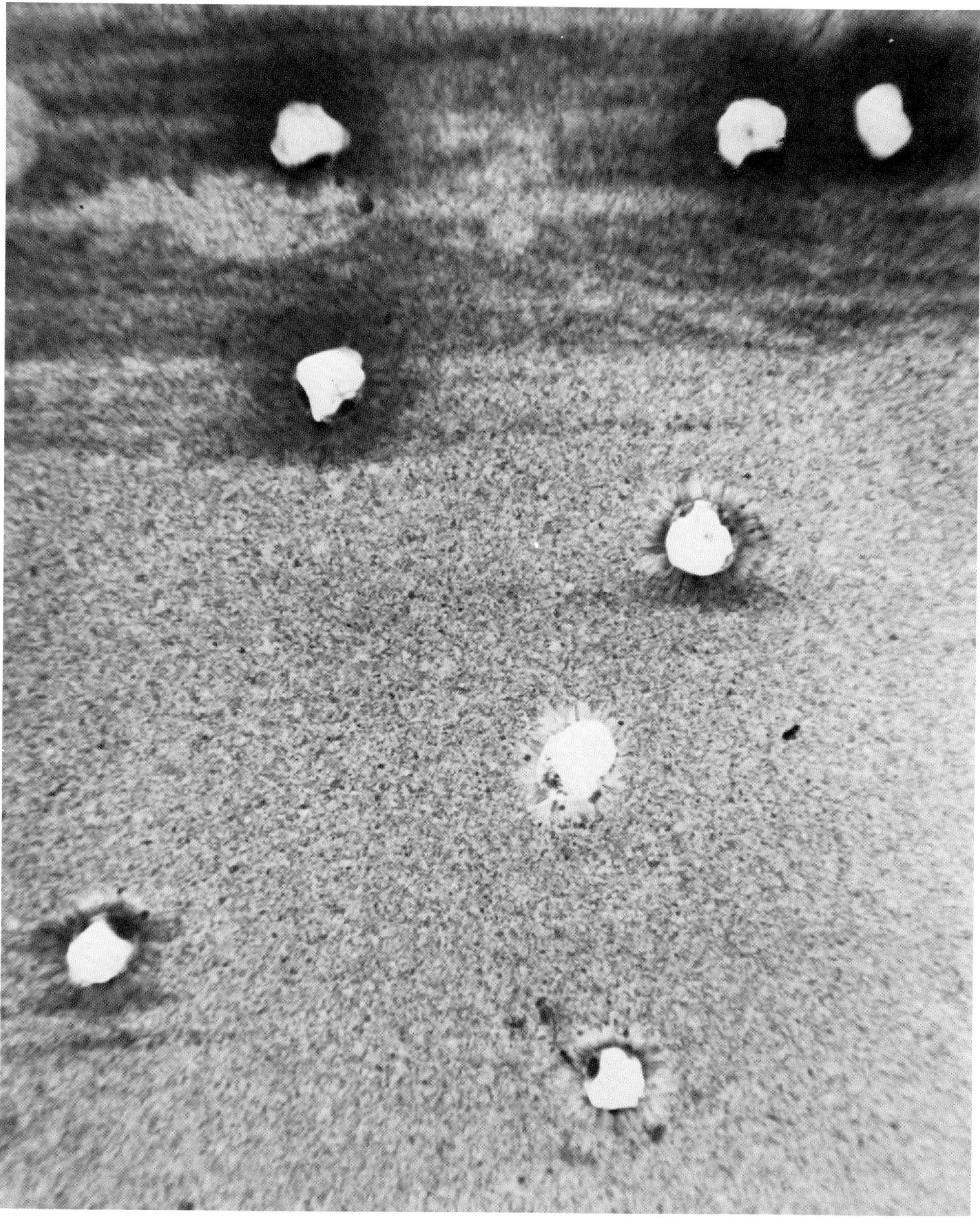
5.0μm

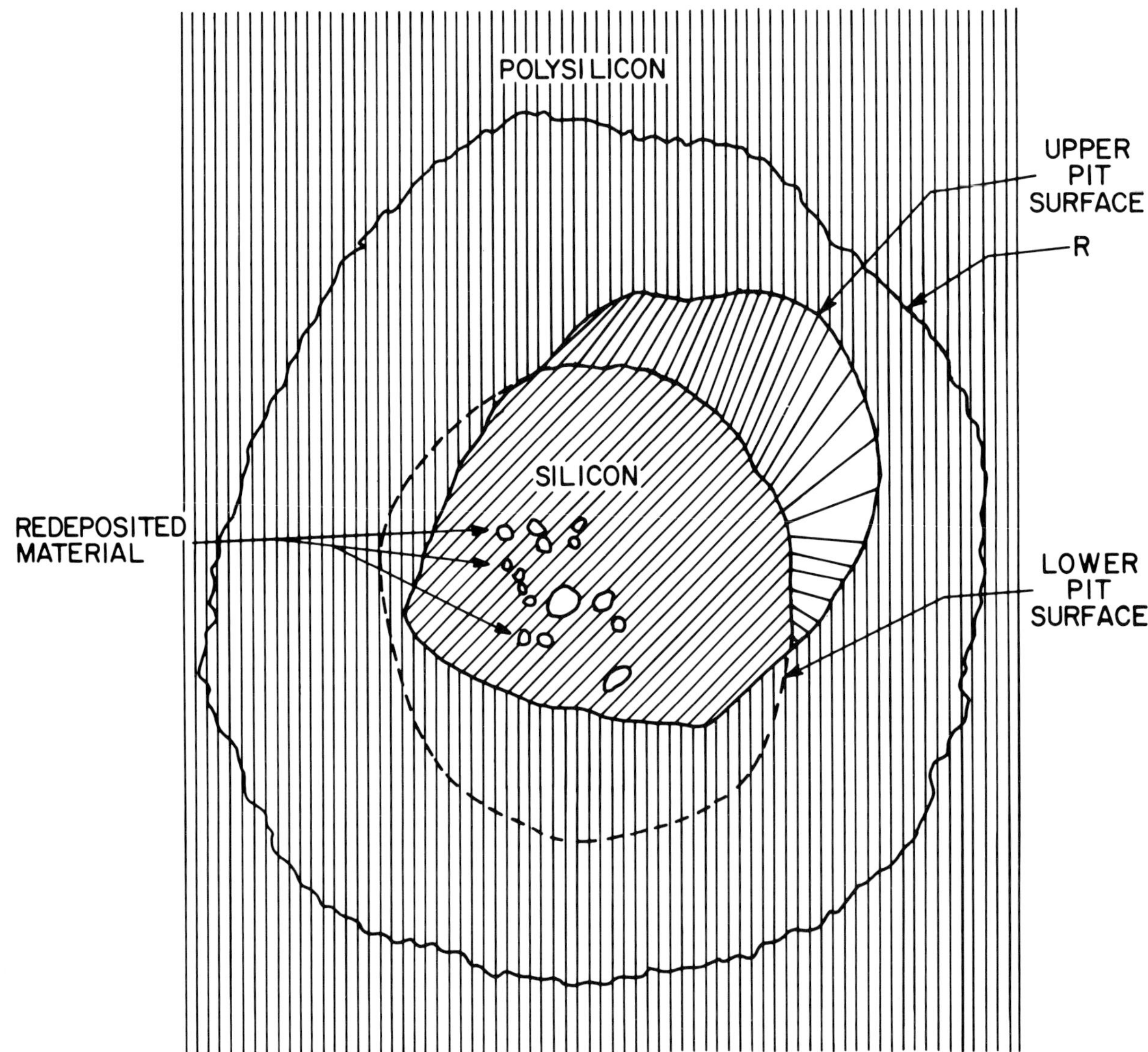

FIGURE 34. One breakdown site in the sample shown in Fig. 33. Local heating during breakdown is sufficient to volatilize material to form the pit and to recrystallize aluminum at the edge of the pit for a distance of 0.5 μm defined by *R*.

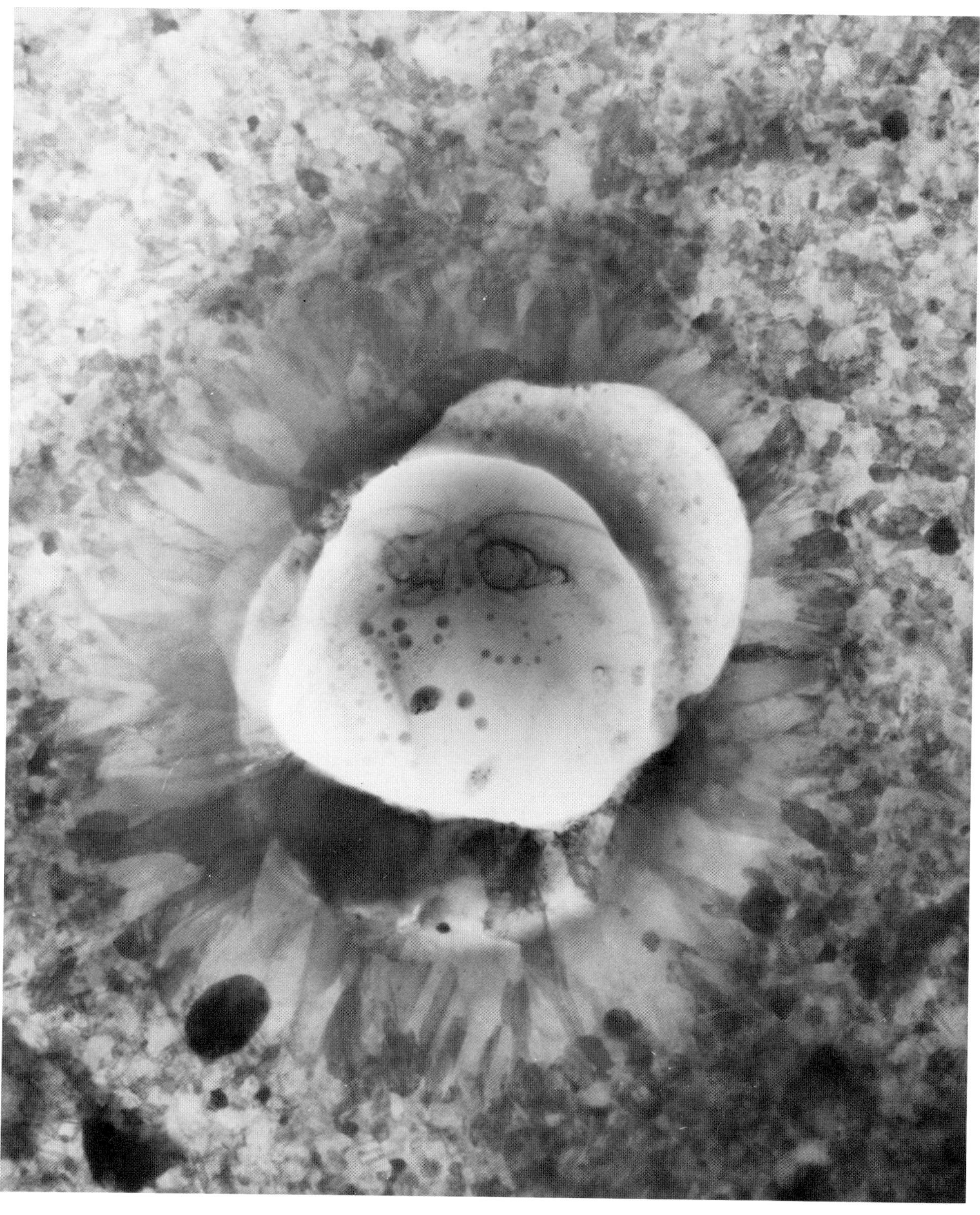

5000Å

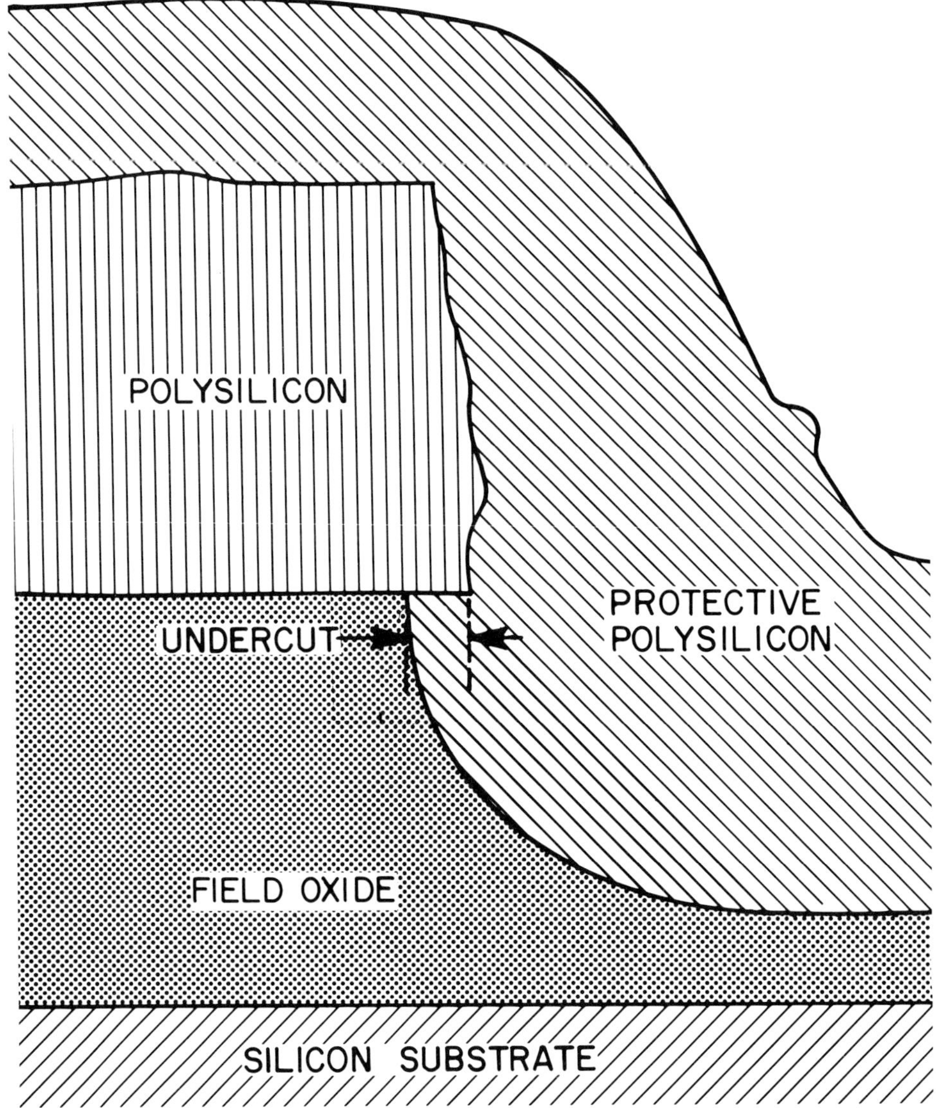

FIGURE 35. Undercutting caused by partial BHF etching of field oxide.

3.3 CHEMICAL VAPOR DEPOSITION AND PLASMA DEPOSITION OF OXIDES

Oxide layers that are needed at a later point in processing often cannot be grown by thermal oxidation, either because free silicon is unavailable, or because silicon is available for oxidation but limitations on junction depth or other diffusion-related properties prohibit using elevated temperatures needed for oxidation. Both chemical vapor deposition and plasma deposition of oxides are used in these cases. Two major topological problems with these oxides are the successful coverage of stepped surfaces, particularly steps with undercuts and the elimination of severe steps on the surfaces of these oxides, so they can be successfully covered by subsequent layers.

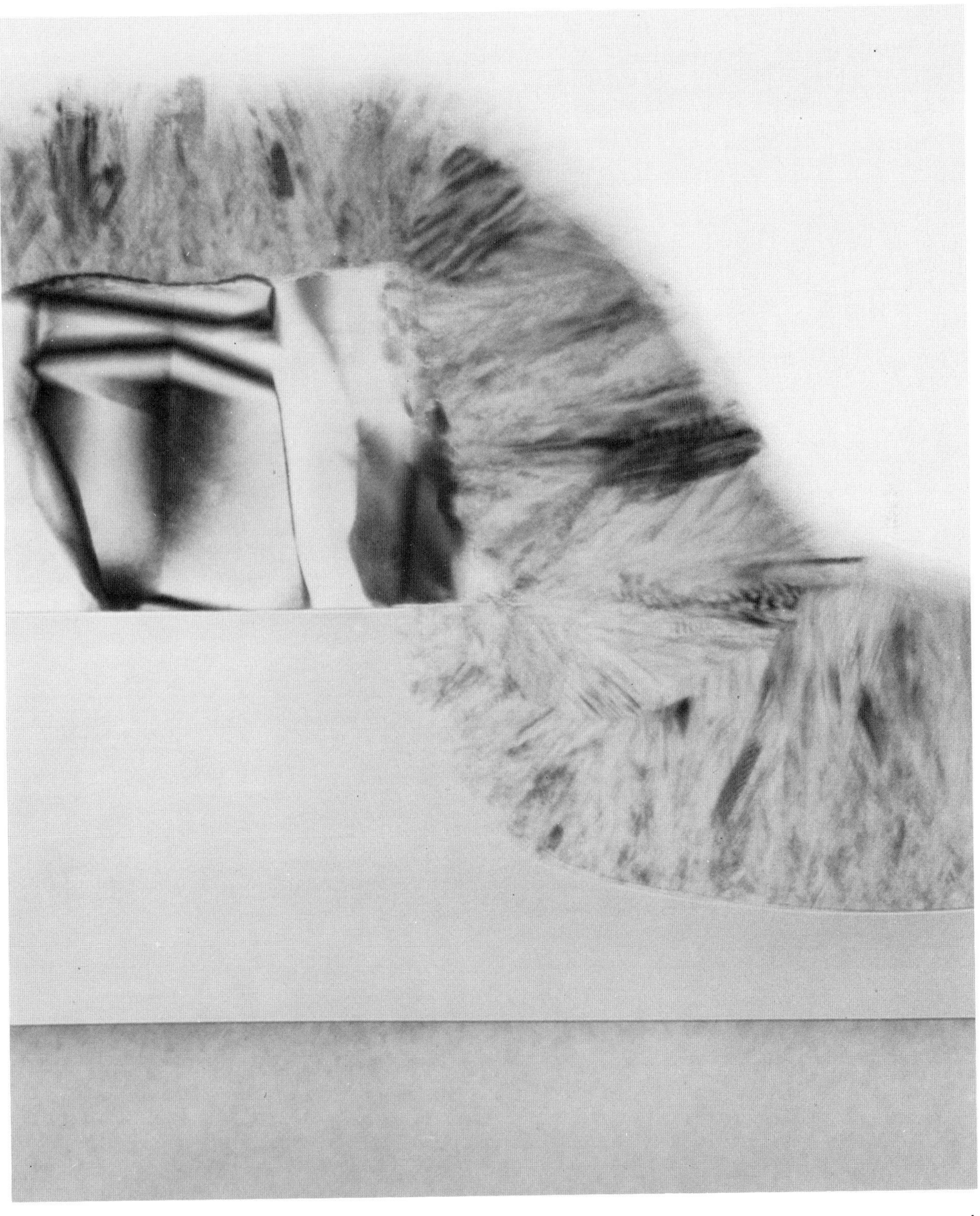
5000Å

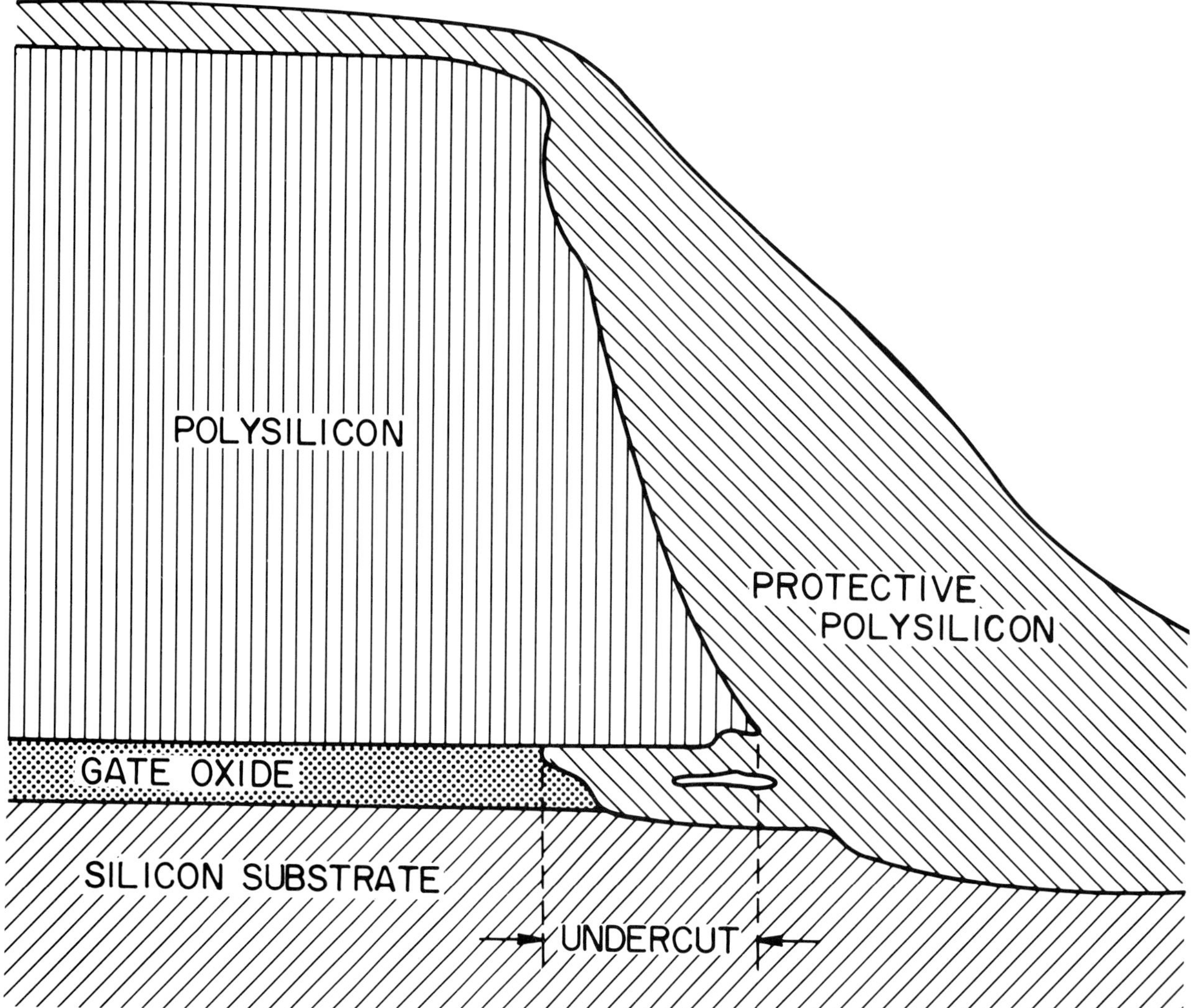

FIGURE 36. Undercutting after complete BHF etching of source–drain region. Note the ability of polysilicon to conform almost totally to the high aspect-ratio region at the undercut.

Steps containing undercuts are the most difficult to cover successfully. Figures 35 and 36 show undercuts created by BHF etching of oxide; undercuts are hard to avoid when using etching procedures that etch isotropically. Note that deposited polysilicon successfully fills these undercuts. Deposited oxides containing dopants, such as phosphorus, flow sufficiently at 1050°C, thus conforming to steps,[35] but undoped deposited oxides have trouble in conforming to

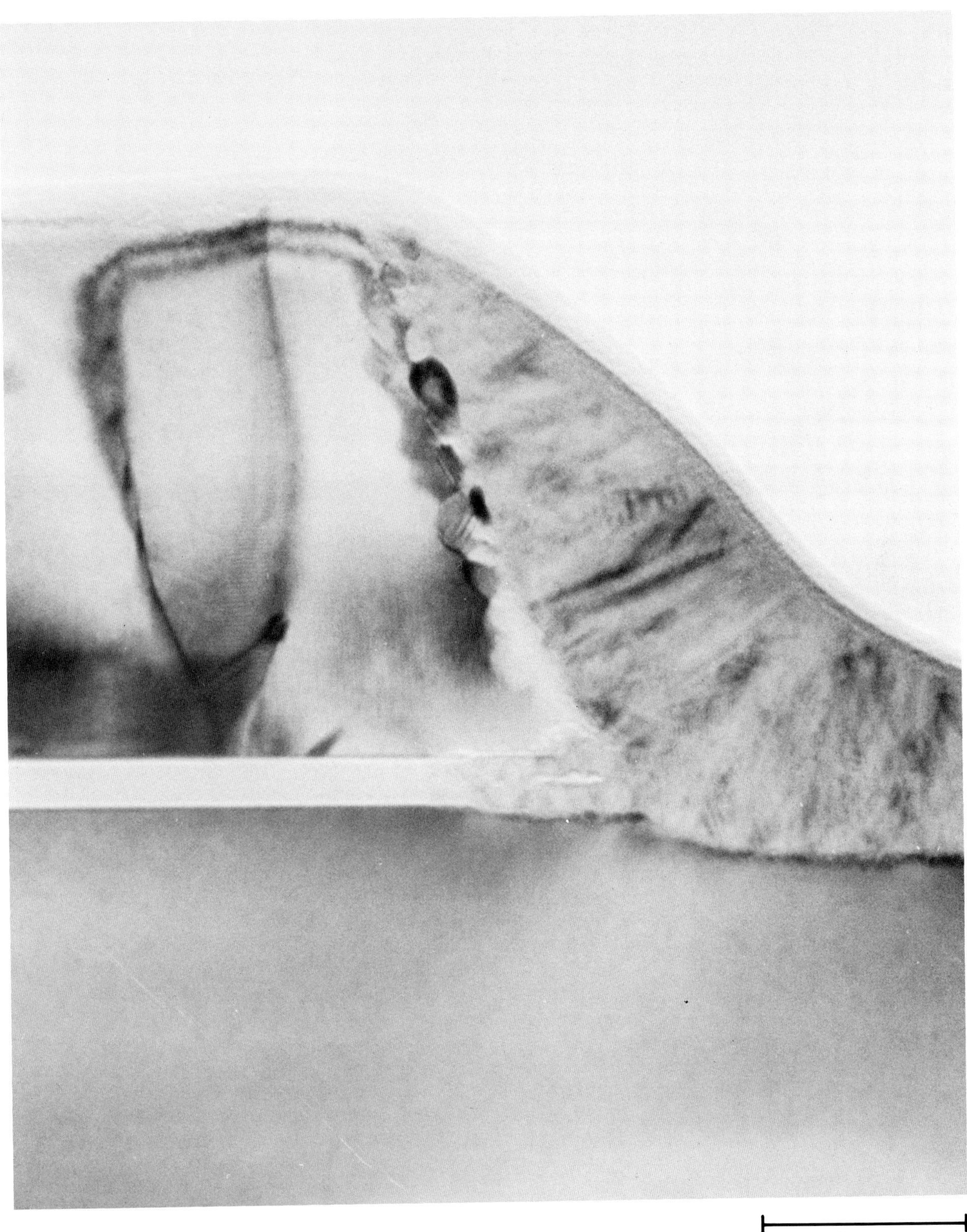
2000Å

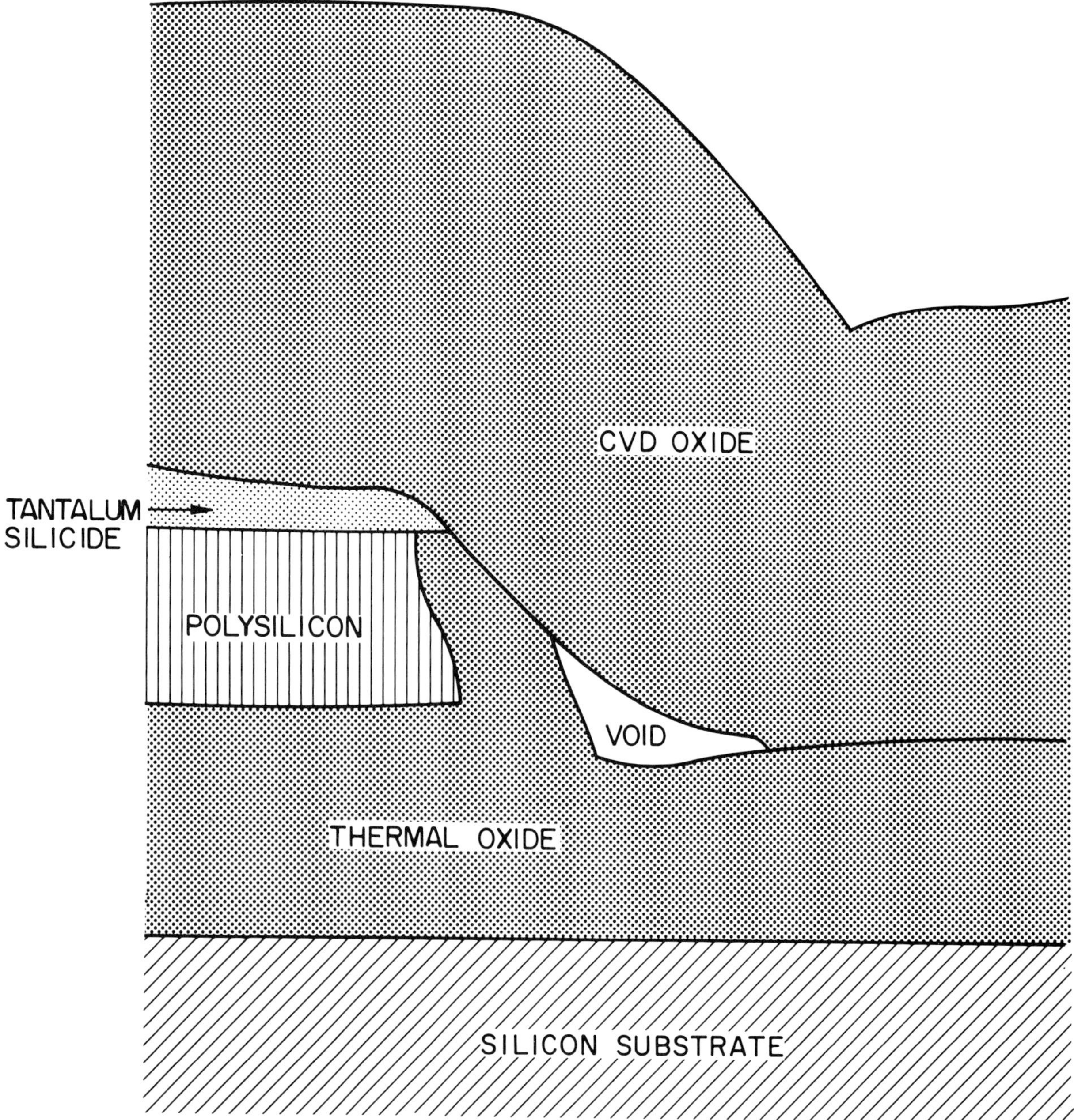

FIGURE 37. Void formation due to poor step coverage of undoped CVD oxide deposited at 450°C.

even simple steps with no undercuts (Fig. 37). In MOS processing, the clearing of oxide from a source–drain region is usually followed by source–drain doping, a thermal drive-in process that may be an oxidation, and CVD deposition of an intermediate oxide (Fig. 8). When the gate oxide is severely undercut as shown in Fig. 36, it is not successfully covered and leaves a void even after

5000Å

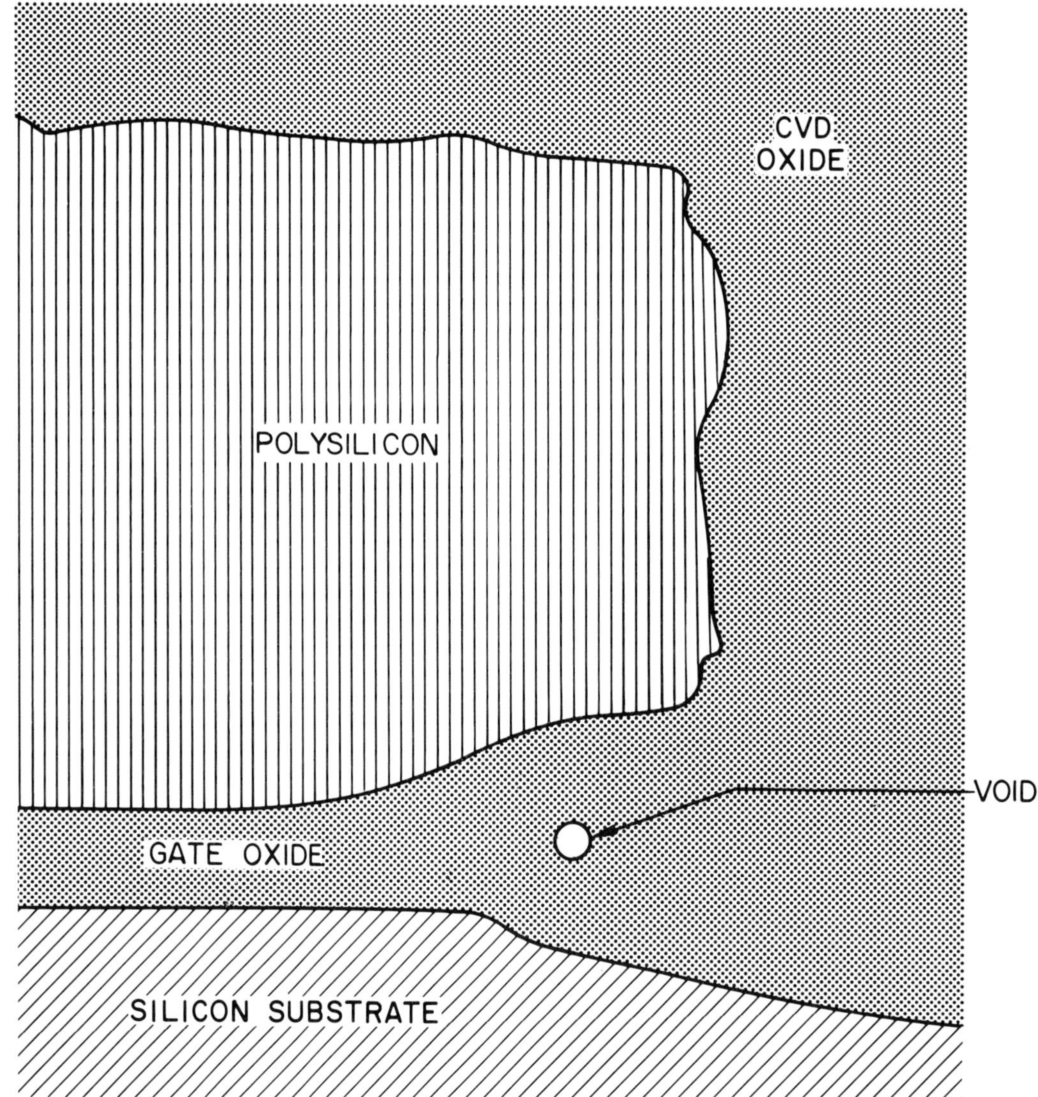

FIGURE 38. Void formation caused by undercutting of polysilicon during oxide etch (see Fig. 36). After oxide etch, 1200-Å oxide was thermally grown, and phosphorus-doped CVD oxide was deposited, leaving a void. The void was made slightly larger to improve TEM viewing by briefly treating the sample with BHF.

phosphorus-doped CVD oxide flows over at 1100°C. The visibility of the void in Fig. 38 has been enhanced by treating the TEM sample with BHF. Note the linear void, which runs the length of the polysilicon edge; such a void constitutes a potential failure mode.

2000Å

CHAPTER FOUR

METALLIZATION

4.1 INTRODUCTION

Components on a VLSI chip can be interconnected by using low resistivity metallization or highly conductive, diffused regions within the silicon. Processing simplicity requires that as few different metallizations as possible be used for interconnections and field plates for Schottky barriers and MOS structures. Furthermore, the metallization must not degrade under device operation and must be compatible with processing. That is, it must be unreactive with contiguous solid-phase materials in the device and with chemical processing steps used in device fabrication. Also the metallization should not interfere with device performance. At the same time, the material must be patternable. Both aluminum and highly doped polycrystalline silicon (polysilicon) films have met most of these requirements for a number of years, and various silicides are now being studied as possible metallization material. Polysilicon has the added advantage of forming an oxide that is a useful dielectric.

As transistor dimensions and design rules become smaller, the fraction of chip area occupied by metallization tends to remain the same. For example, the BELLMAC*-32 microprocessor, a 32-bit microprocessor chip made with 3.5-μm design rules, occupies 1.46 cm^2 surface area, contains 1×10^5 transistors, and has 0.68 cm^2 of aluminum metallization (46.6% coverage). An improved version of the chip made with 2.5-μm design rules occupies 1.00 cm^2 surface area, has 1.4×10^5 transistors, and has 0.47 cm^2 of aluminum metallization (47.0% coverage). Deep UV, X-ray, and electron-beam lithography can decrease metallization linewidths to 1 μm or less; but metallization path-length resis-

*Trademark of Western Electric, Inc.

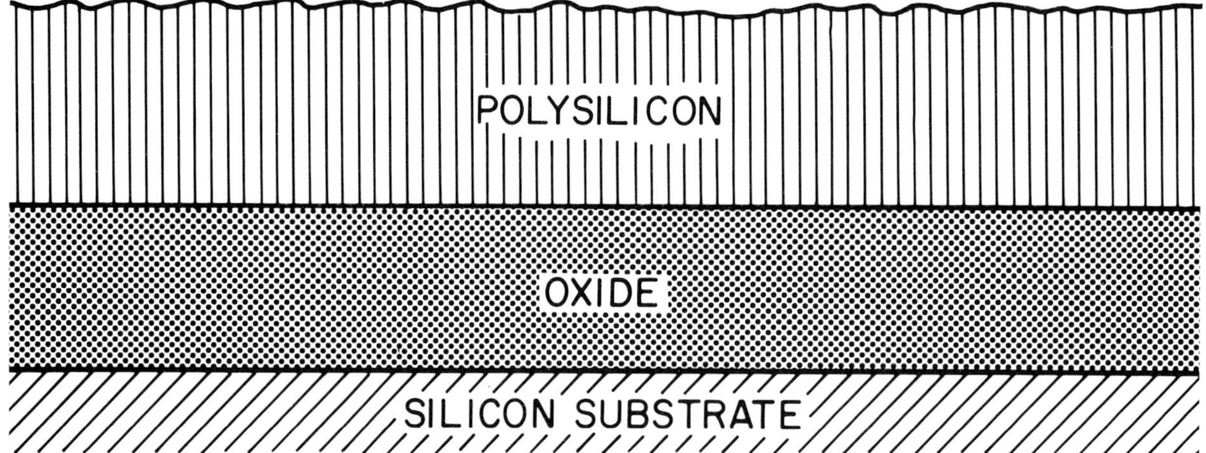

FIGURE 39. (upper & middle)

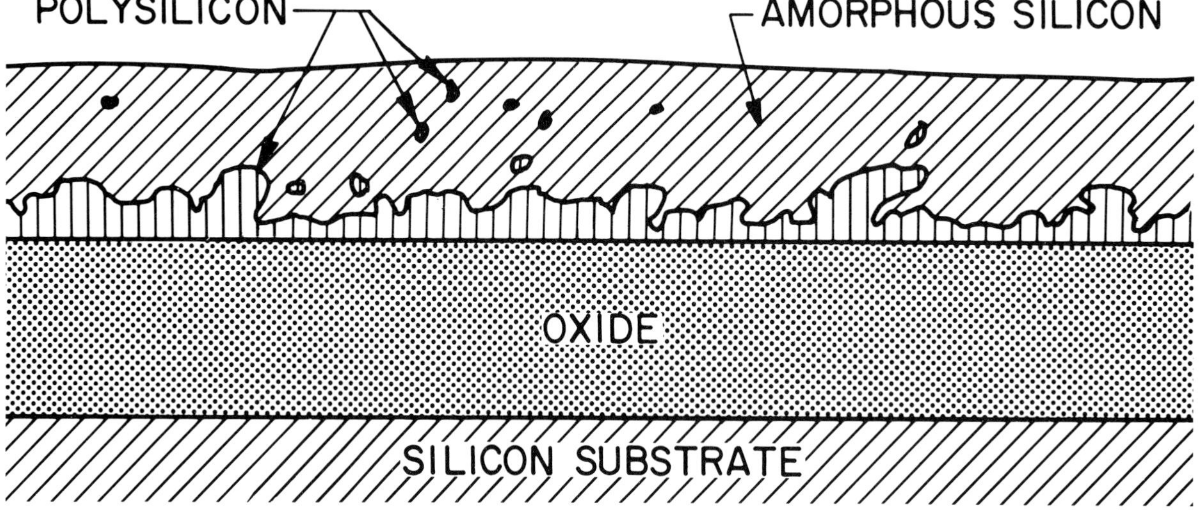

FIGURE 39. (lower)
Polysilicon morphology can vary from columnar (upper) to small grained (middle) when deposited in different reactors under nominally the same conditions, or to partly amorphous (lower) when deposited at 625°C. Deposition temperatures lower than 600°C produce an amorphous film.

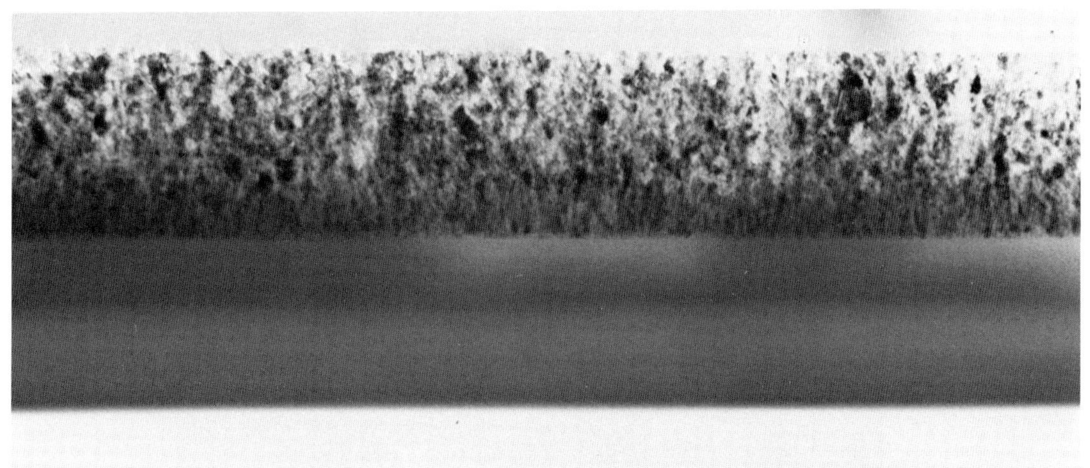

5000Å

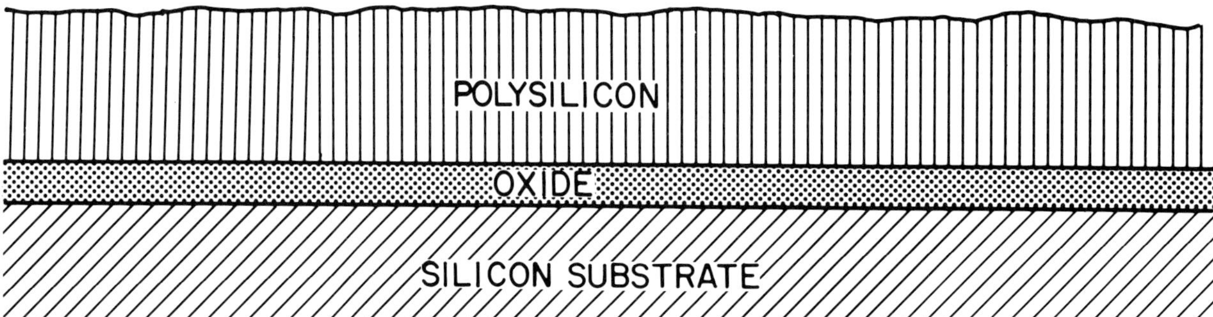

FIGURE 40. Grain growth occurs after doping polysilicon with phosphorus (from PBr_3) at 950°C. The degree of grain growth depends on the thickness of the polysilicon layer.

tance increases as linewidth decreases, increasing the *RC* time constant and the delay time of circuit elements. Small device dimensions also place greater requirements on metallization step coverage. Sloped walls begin to occupy an increasingly significant fraction of device dimensions as gate lengths and window diameters fall below 2 μm, and vertical walls produced by anisotropic etching methods are not easily covered by a uniform thickness of certain metallizations, particularly aluminum.

To solve these problems, processing schemes have been devised that use various combinations of metal silicide, polysilicon, and/or aluminum films at each metallization level in the device. These metallizations are described in the next three sections.

4.2 POLYSILICON

Polysilicon films are often grown from the pyrolysis of SiH_4 at 650°C. Since the dielectric properties of thermal oxides grown from polysilicon partly depend on polysilicon grain structure (Section 3.2), an analysis of grain structure dependence on deposition parameters becomes relevant to device performance.

Figure 39 (top and middle) shows micrographs of polysilicon deposited at the same temperature (640°C) in two different reactors under nominally the same conditions. The grain texture is mostly columnar and has surface undulations with a period of about 0.1 μm in the upper micrograph; grain size is much smaller in the center micrograph. Electron and X-ray diffraction studies show that both films are polycrystalline with very little evidence for preferred orientation. Columns in columnar films grow normal to the local surface, and at an inside step the surfaces of these columns meet at 45 degrees. When the deposition temperature is lowered to 625°C, the film becomes mostly amorphous (Fig. 39, lower), and lowering the temperature further produces a completely amorphous film. As previously discussed in Chapter 3, anomalies occurring during polysilicon deposition sometimes result in the appearance of localized regions of thicker deposits (bumps) that add to the texture of the polysilicon film surface and are maintained during subsequent oxidation (Figs. 31 and 32).

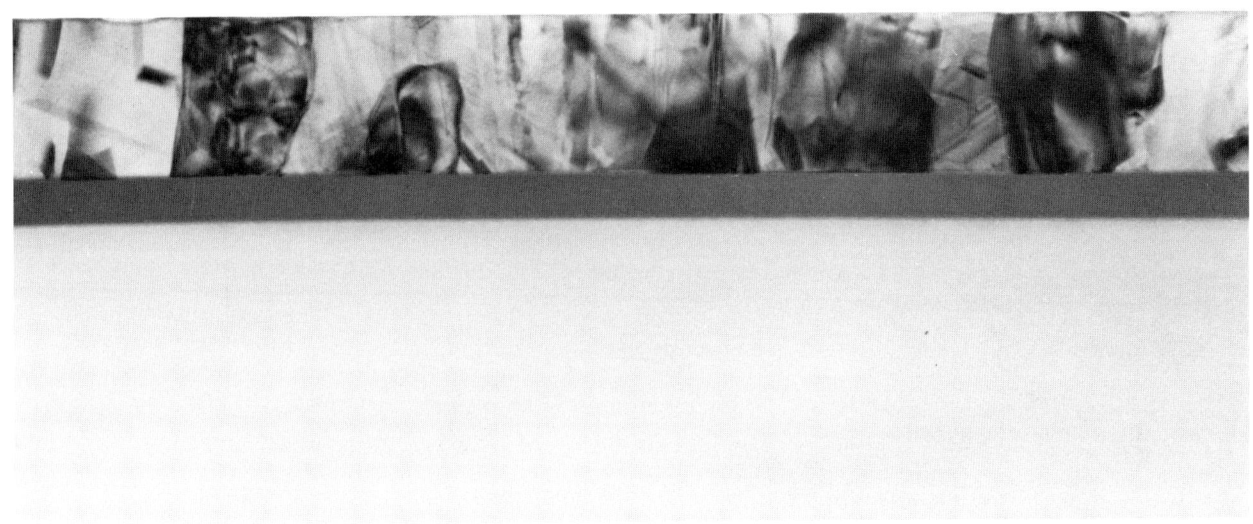

5000Å

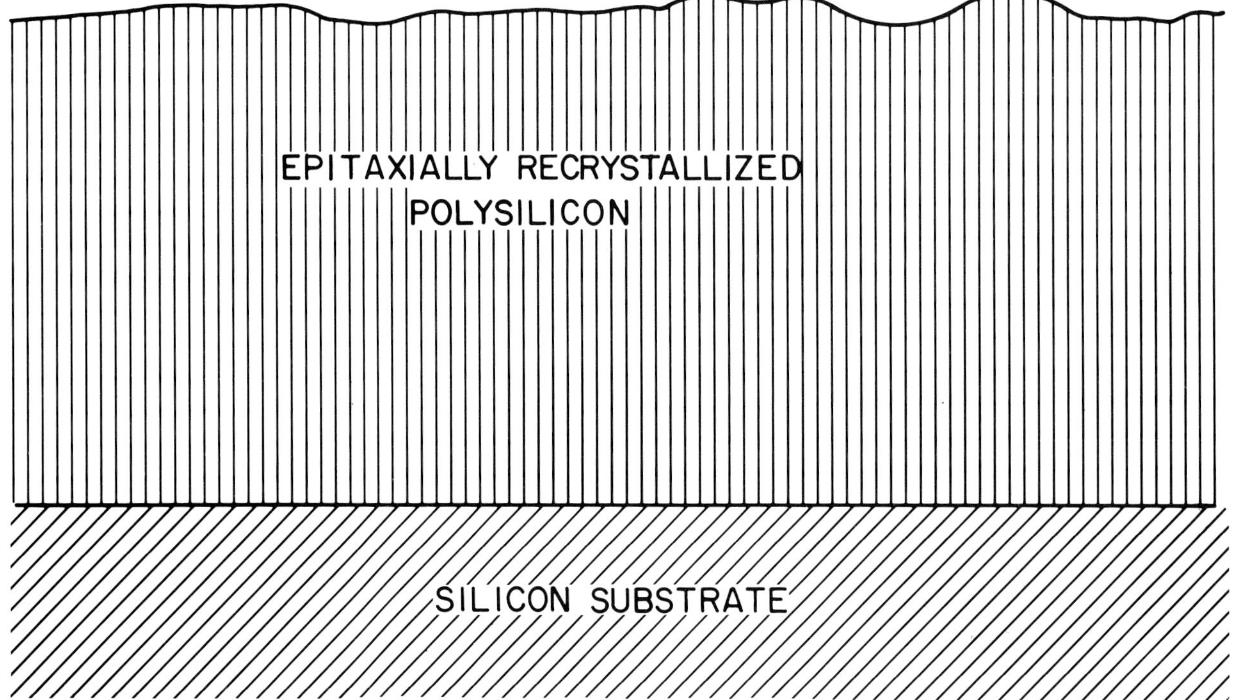

FIGURE 41. Epitaxial recrystallization of polysilicon films deposited on (100) silicon following 1000°C PBr_3 treatment for 2500-Å film (upper) and 5000-Å film (lower).

Grain growth of polysilicon occurs during gas phase doping above 900°C. The degree of growth depends on the temperature, anneal time, and film thickness. Figure 40 shows three polysilicon films after PBr_3 doping at 950°C and shows that grain growth increases with decreasing film thickness. Polysilicon films deposited on well-cleaned silicon {100} substrates convert to epitaxial films following 1000°C PBr_3 treatment[36] (Fig. 41). These epitaxial films contain a high defect density that decreases with increasing temperature of PBr_3 treatment and with decreasing film thickness (compare Fig. 41, upper with Fig. 41, lower).

Slight additional grain growth occurs after oxidation of doped (and diffused) polysilicon films, but the major effect of oxidation on grain morphology is on the texture of the polysilicon/oxide interface. (Certain features of this texture were discussed in Section 3.2.) Silicon protuberances form within the oxide (Fig. 30) and crystallites at the ends of the protuberances eventually become isolated and appear as silicon inclusions (Fig. 31). Except for protuberances and bumps, the polysilicon surface texture during oxidation is largely governed by the grain size determined by prior processing. The major influence of oxidation on surface texture is through variation in oxidation rates of differently oriented grains; a minor influence seems to be the presence of emerging grain boundary sites.

2000Å

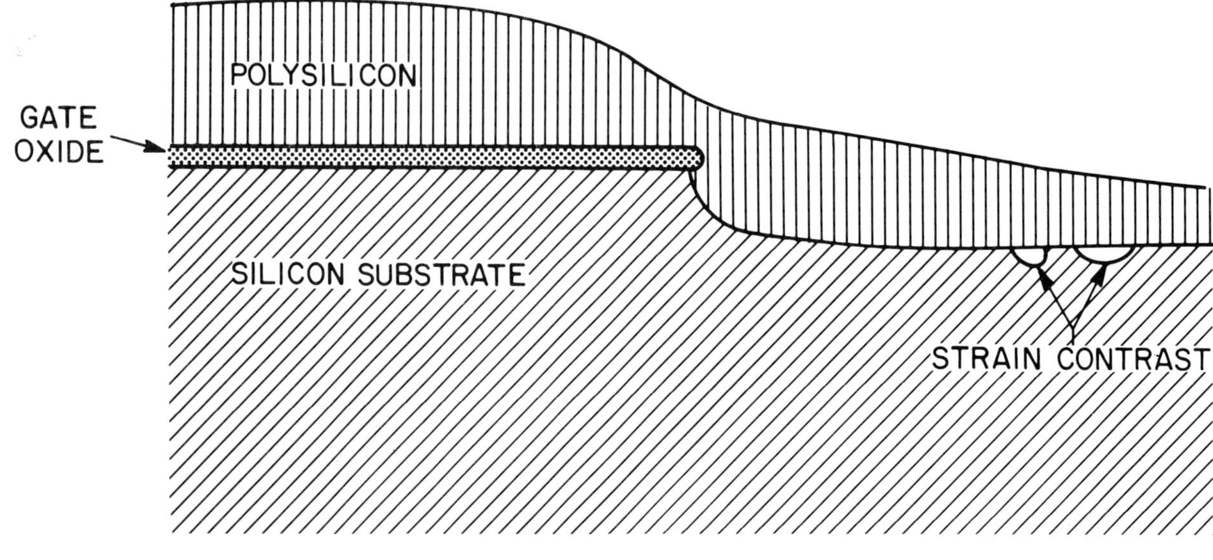

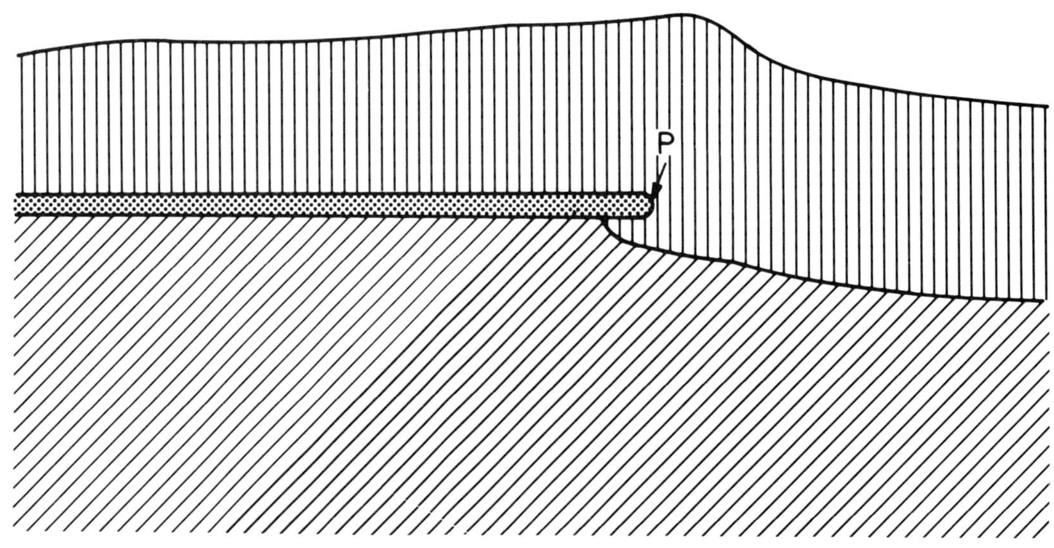

FIGURE 42. Polysilicon contacts to the silicon substrate at the edge of gate oxide. Local strain in the substrate is caused by the presence of polysilicon grain boundaries at the interface. The apparent penetration of gate oxide into the polysilicon at P is a projection of the slanted edge of the oxide onto the plane of the page.

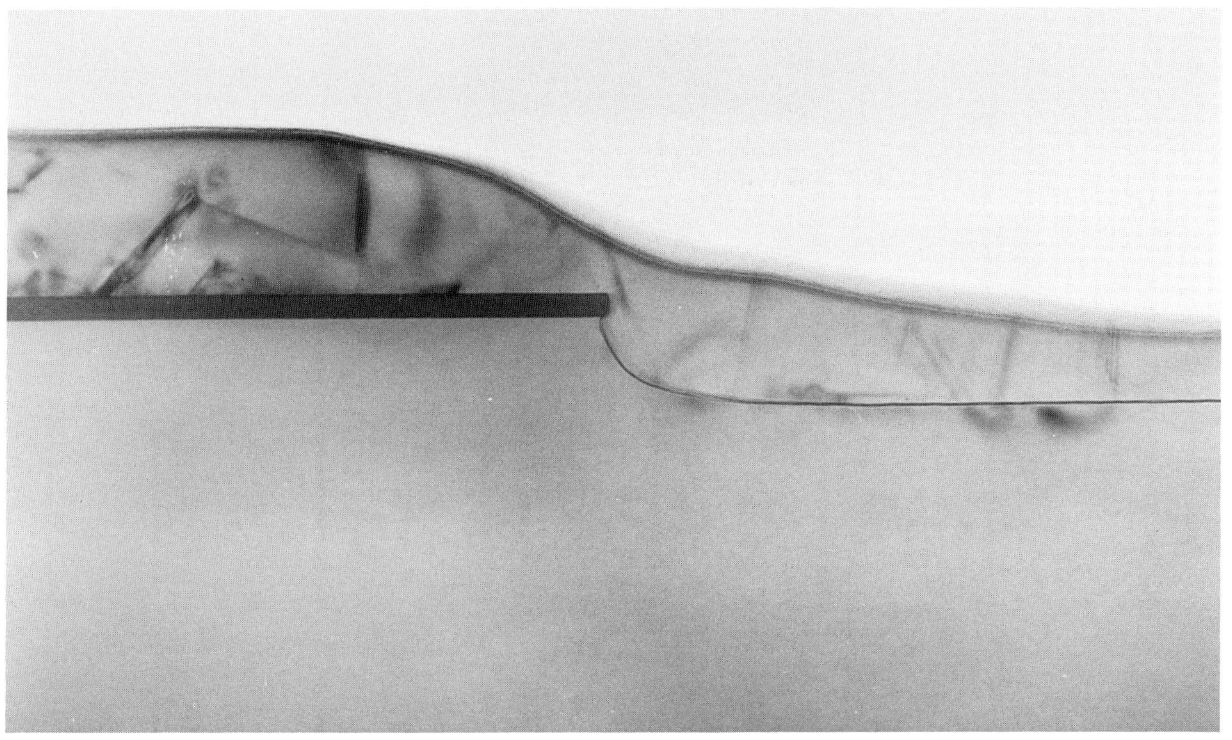

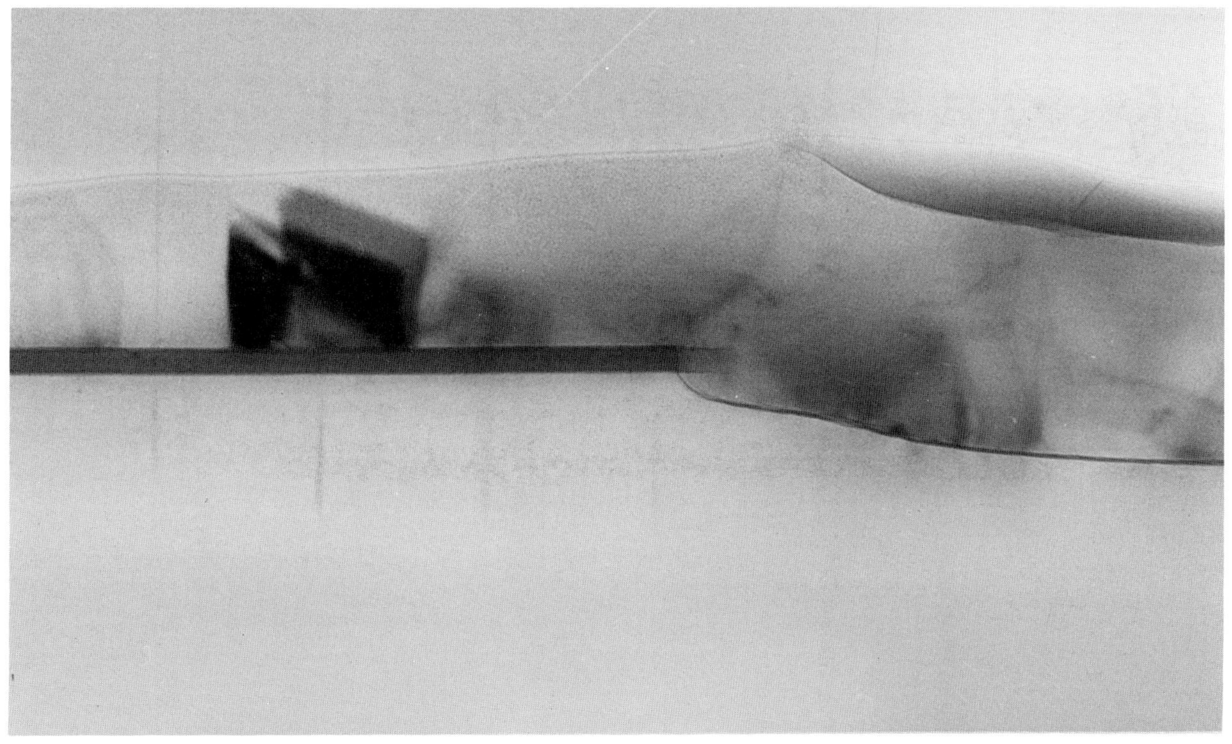

5000Å

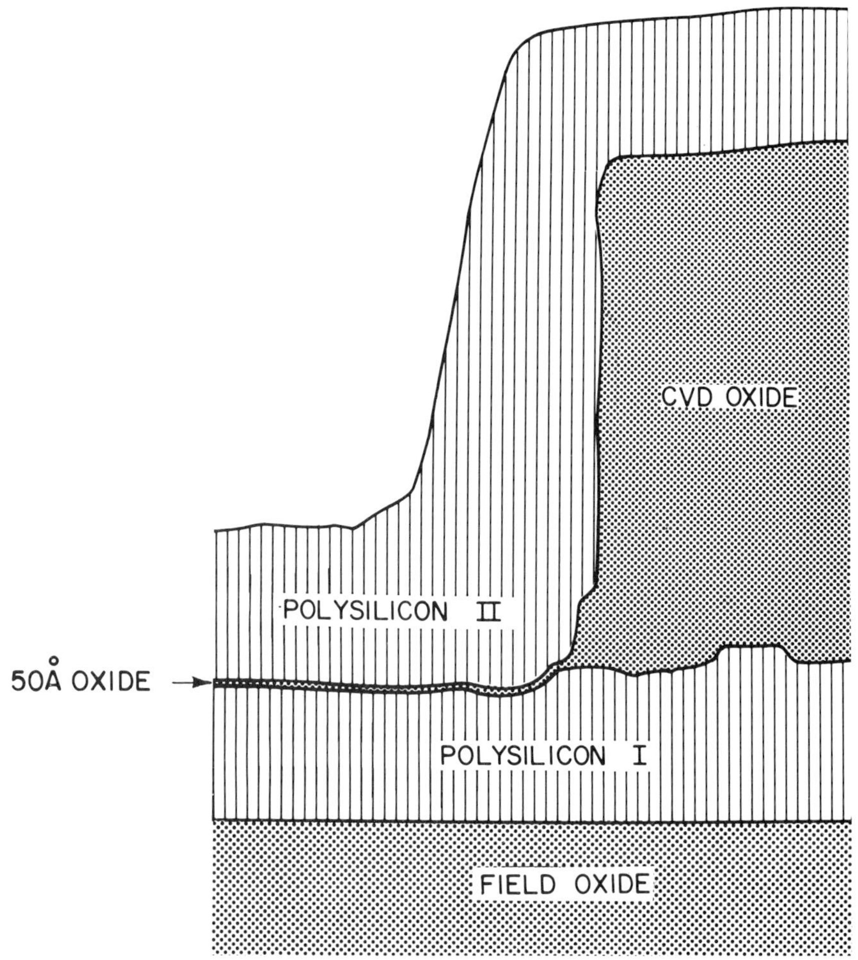

FIGURE 43. Highly resistive contact of polysilicon II to polysilicon I caused by the presence of 50-Å interfacial oxide.

Polysilicon contacts to the silicon substrate or to underlying polysilicon metallization are made by cleaning the substrate in BHF to remove unwanted oxide, and then depositing doped and undoped polysilicon onto this surface. Low resistivity ohmic contacts are readily formed (Fig. 42). Sites of emerging polycrystalline grain boundaries at the interface are sufficiently stressed to produce local changes in contrast in the substrate under two-beam or multiple-beam viewing conditions (Fig. 42, upper). Such strain contrast is fairly easy to see at good contact regions but becomes impossible to produce when an oxide film thicker than about 50 Å exists at the boundary (Fig. 43). The 50-Å oxide layer shown is caused by insufficient cleaning of polysilicon I prior to polysilicon II deposition. Detecting a thin (50 Å) layer is difficult when the sample is too thick, for reasons discussed in Section 2.2, and the absence or presence of a

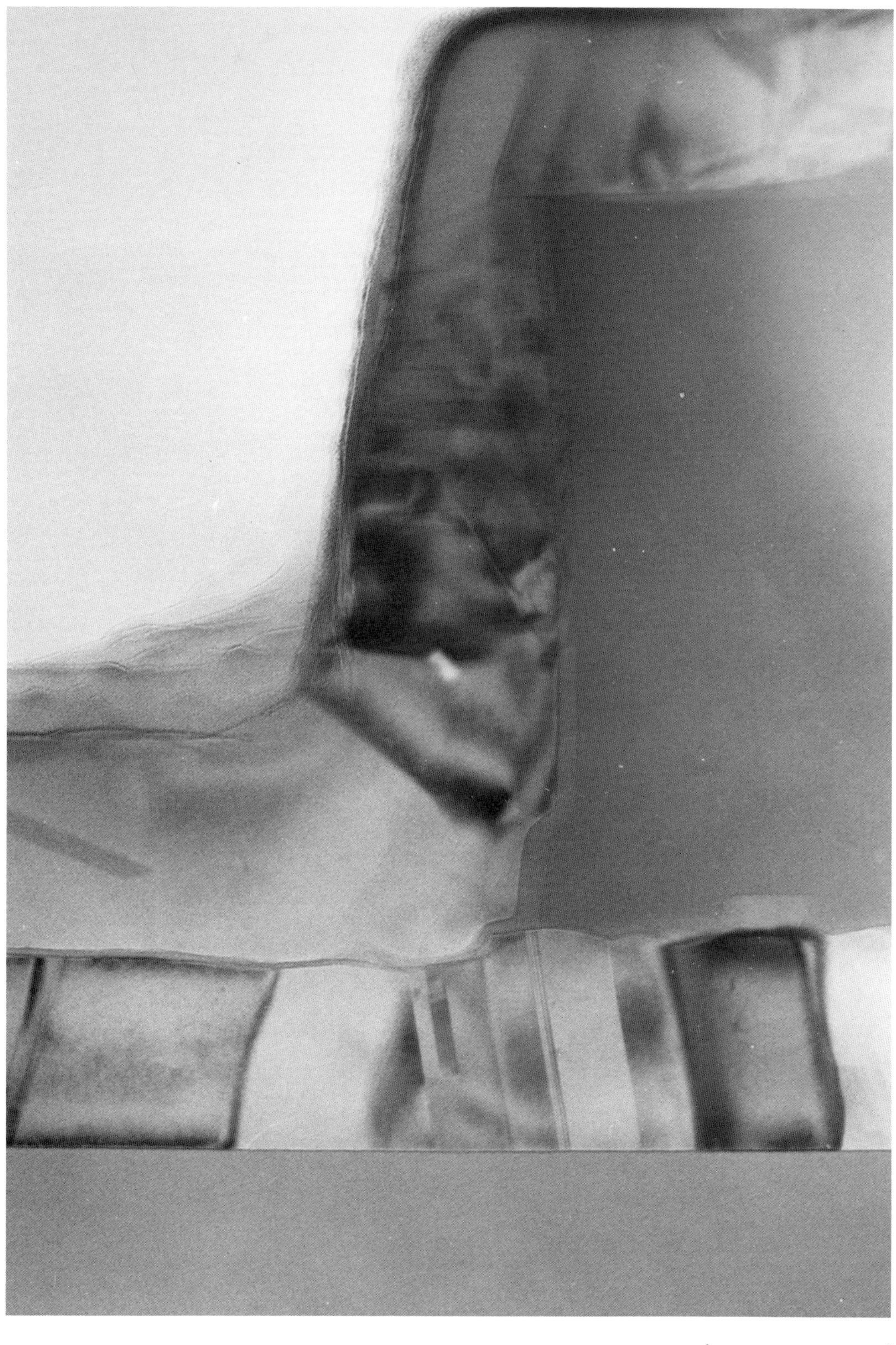

2000Å

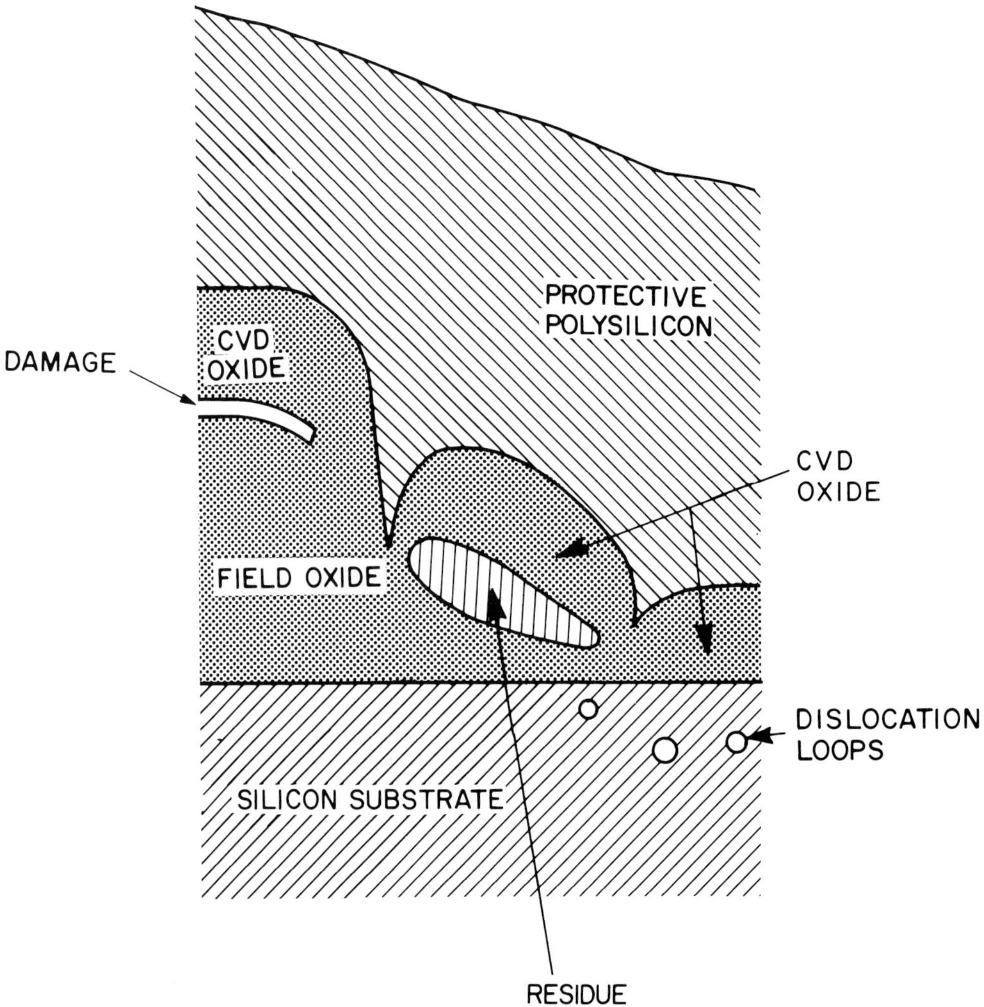

FIGURE 44. Region at the edge of field oxide after arsenic implantation and annealing. Unetched polysilicon remains as a residue. The implanted oxide layer is visible as a region of damage, and dislocation loops are found in the source–drain region.

substrate strain field at the polysilicon grain boundaries is a better indicator of the cleanliness of the contact.

Improving and tightening processing specifications sometimes introduce new processing problems. Vertical walls produced by reactive ion etching of oxide are conformably coated with polysilicon, but a vertical segment of polysilicon at a wall's edge is usually significantly larger than the mean thickness of the film. Therefore, polysilicon pattern etching must be carried out long enough to remove this vertical element and thus prevent the situation shown in Fig. 44 from occurring. A polysilicon residue is clearly visible; other features in this figure are described more completely in Chapter 6.

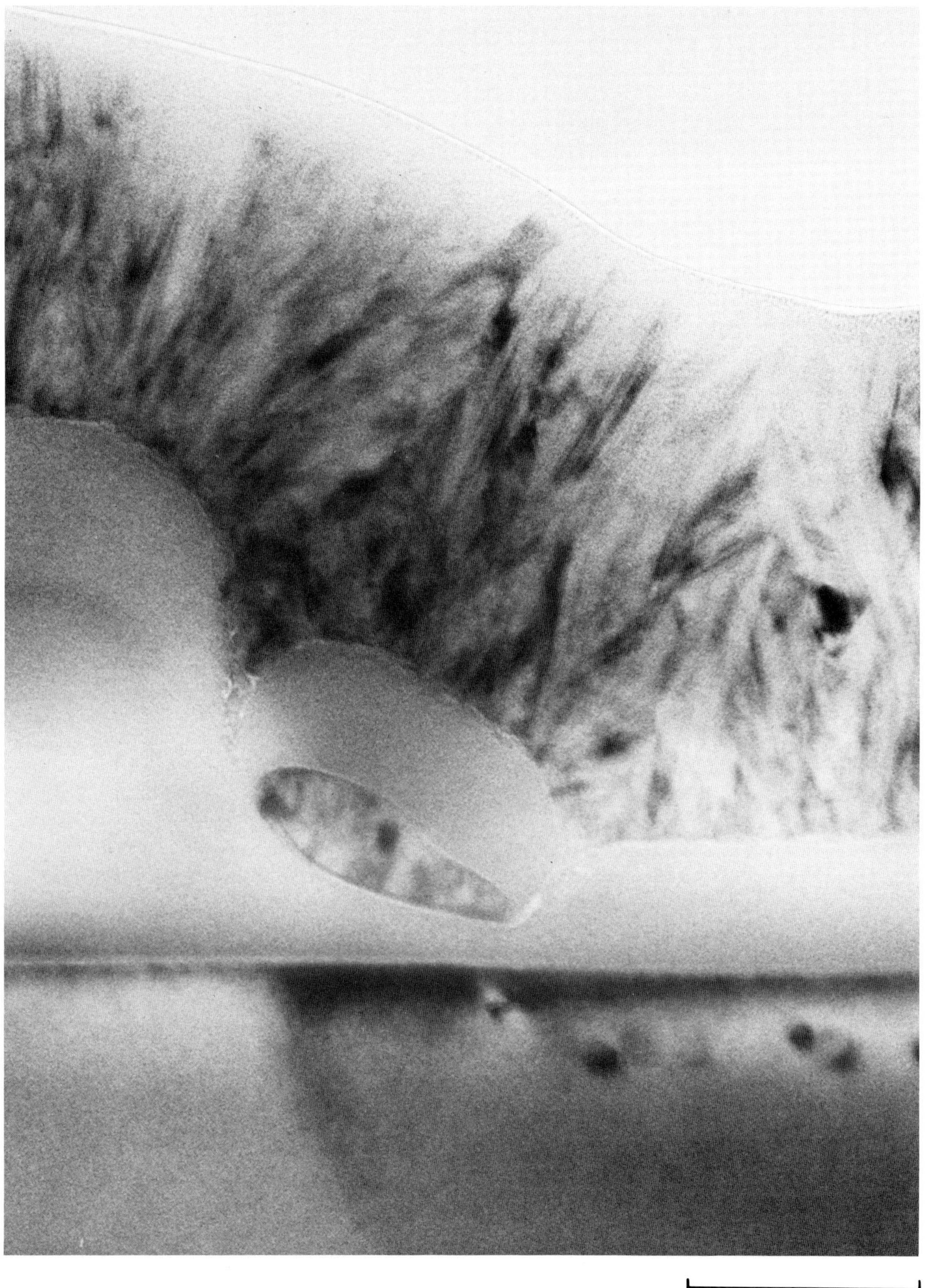
2000Å

4.3 ALUMINUM

Aluminum has been used as a metallization material since the beginning of IC technology. This metal is highly conductive, adheres well to oxides, and can be both deposited and patterned with relative ease. The main difficulties with aluminum are poor step coverage, electromigration, and substrate interactions at contact sites.

Aluminum grain size depends on a number of factors, including deposition conditions, film thickness, and thermal treatment. Electron-beam deposited films with average thicknesses of 0.5 μm (Fig. 45, upper) and 1.6 μm (Fig. 45, lower) have mean grain sizes of 0.2 and 2 μm, respectively; following 450°C heat treatment, these values increase to 2 and 20 μm, respectively.[37] Perhaps the most striking impact of grain size is on the electromigration resistance of aluminum conductors. Doping aluminum with a variety of elements has been proposed as a method for inhibiting electromigration,[38] but the degree of inhibition of a particular alloy composition also depends strongly on grain size, which in turn partly depends on deposition conditions. Thus, 2.6-μm aluminum metallization with 0.5% copper and a 4.5-μm mean grain size deposited by electron-beam evaporation gives an extrapolated lifetime (under certain accelerated test conditions) of 12 hr, while metallization consisting of inductively evaporated alloy of the same composition with a 0.8-μm grain size has a mean lifetime of less than 1 hr.[39] When the linewidth is decreased to 1.0 μm, the lifetime of the electron-beam evaporated material increases to more than 80 hr. An important parameter that correlates with the increase in lifetime is the absence of "triple points"—regions where three grains meet within the metallization. Figure 46 shows a contact pad and part of a 1.0-μm meander line; the grain size ranges from 1.5–15 μm. In Fig. 47 (upper) the narrow meandor consists of a linear arrangement of grains and has no triple points. In Fig. 47 (lower) the 3.0-μm line contains triple points and shows a corresponding decrease in lifetime.[39]

Aluminum, unlike polysilicon, does not grow with uniform thickness over a stepped surface. Metallization thinning at steps (Fig. 48) is related to the substrate temperature and angle of incidence of the metal vapor to the local surface.[40] A diffused source, such as a sputter source, produces more uniform step coverage than a point source, which produces different coverage at steps on wafers that are covered simultaneously (compare Fig. 49, upper, with Fig. 49, lower). Elevating the substrate temperature during aluminum deposition can increase step-coverage uniformity. More satisfactory solutions to step-coverage problems lie in minimizing substrate topography by using recessed oxides (Section 3.2) and planarization methods,[12] or by changing the metallization. The metallization can be changed by either replacing the aluminum or adding a second conducting film of polysilicon to the metallization.[41]

FIGURE 45. Grain-size difference in aluminum films 0.5 μm thick (upper) and 1.6 μm thick (lower). Grain size is related to film thickness.

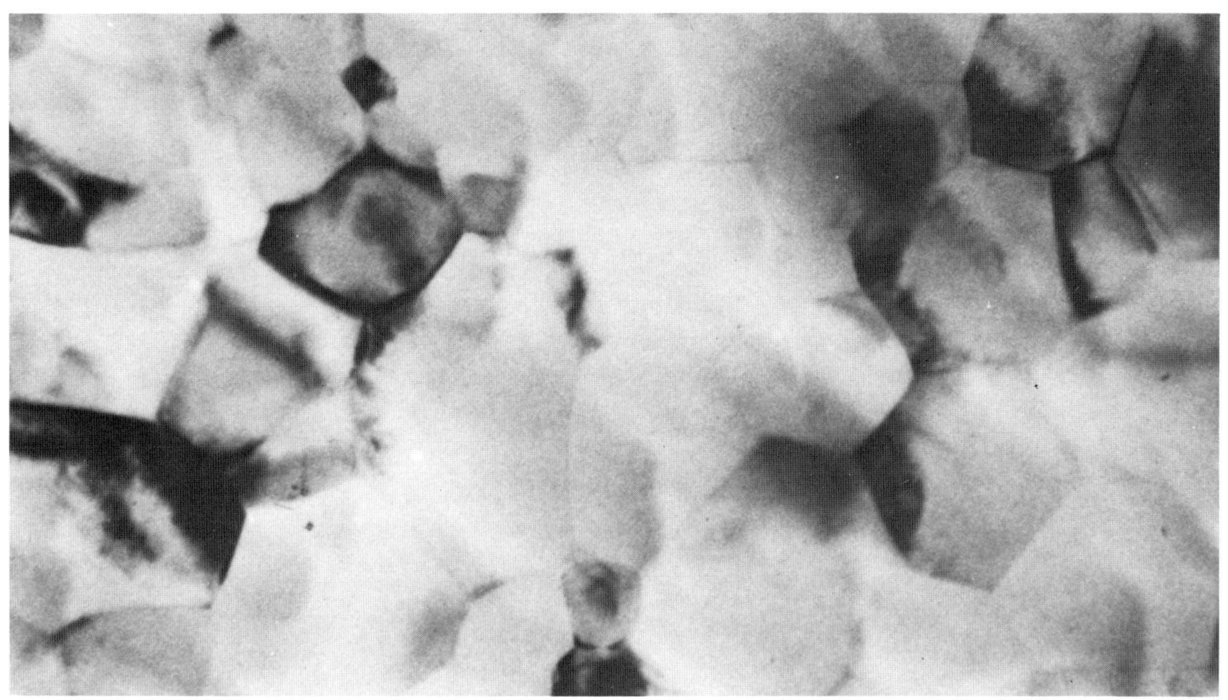

2000Å

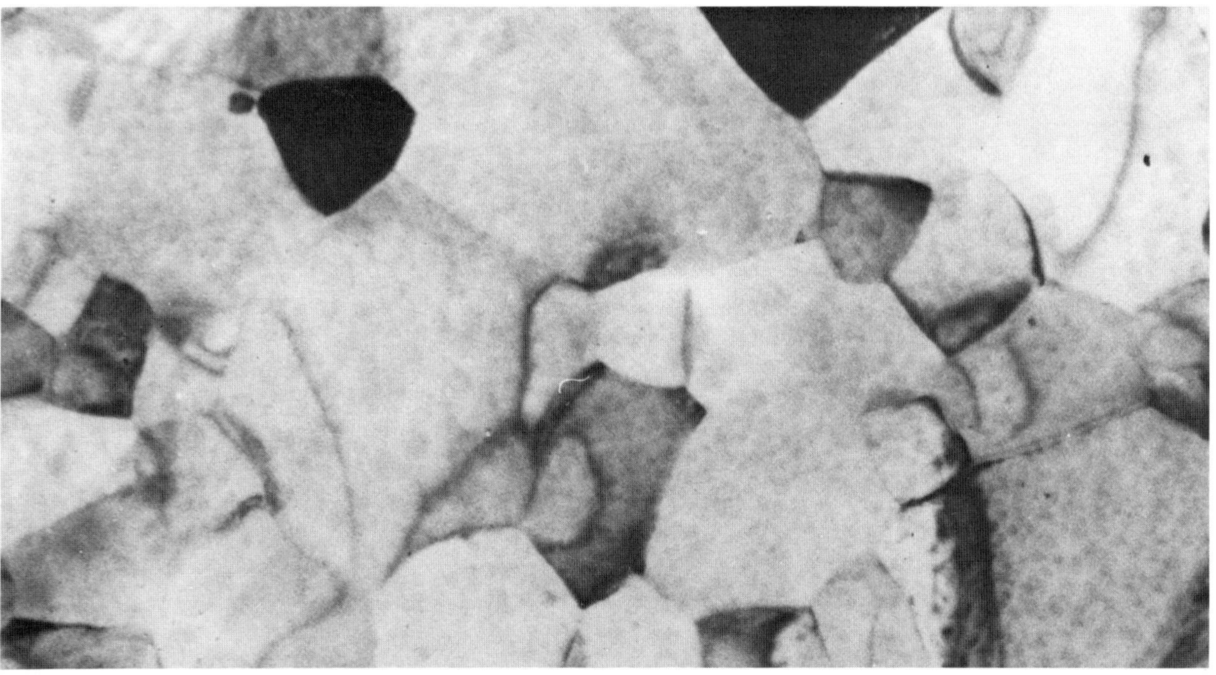
2.0 μm

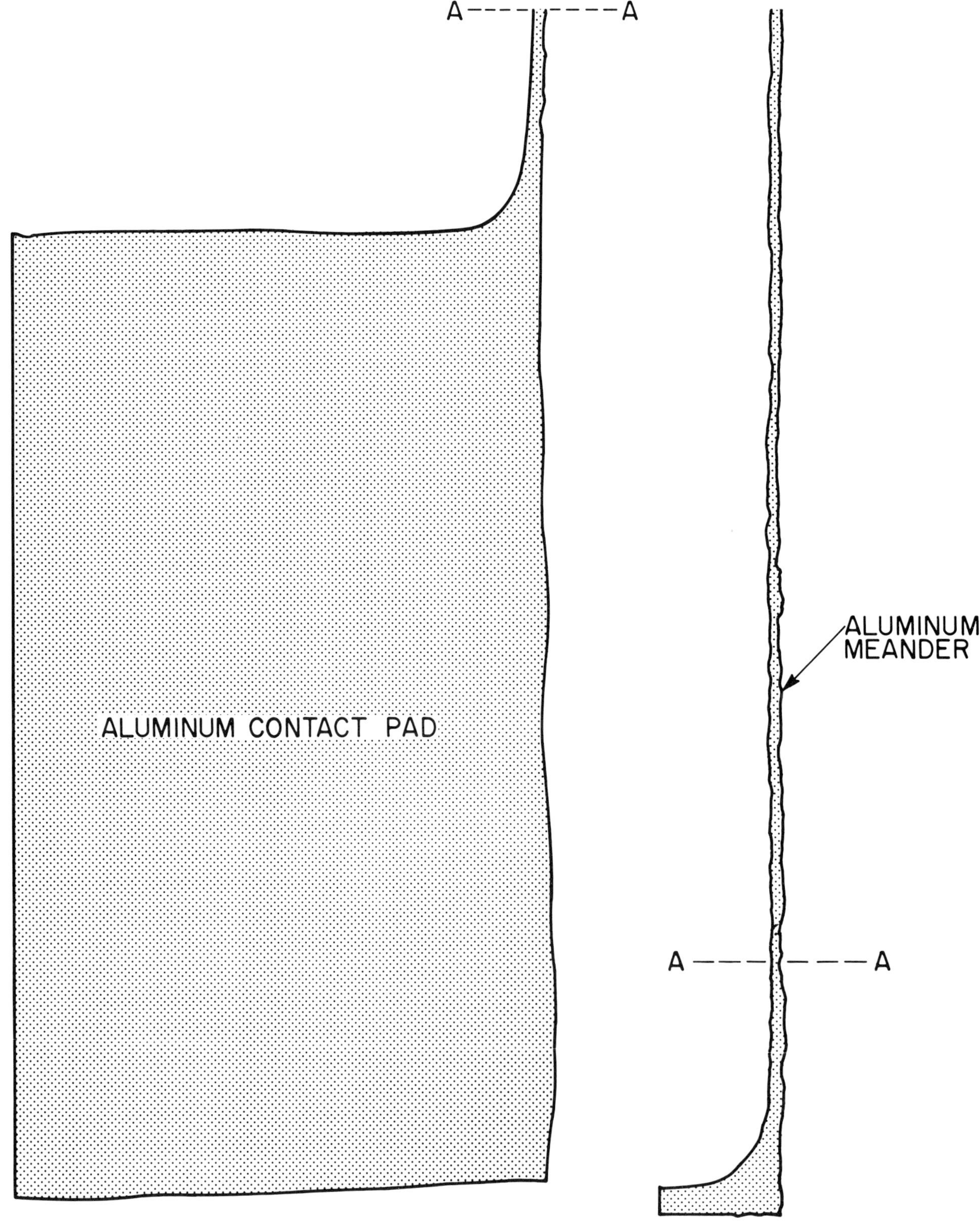

FIGURE 46. Part of an aluminum contact pad and a 1.0-μm meander line. Grain size ranges from 1.5 to 15 μm.

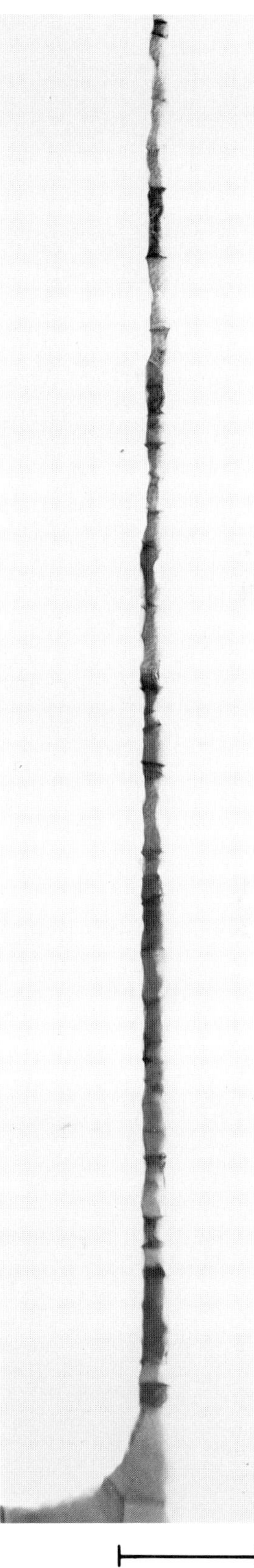
10.0 μm

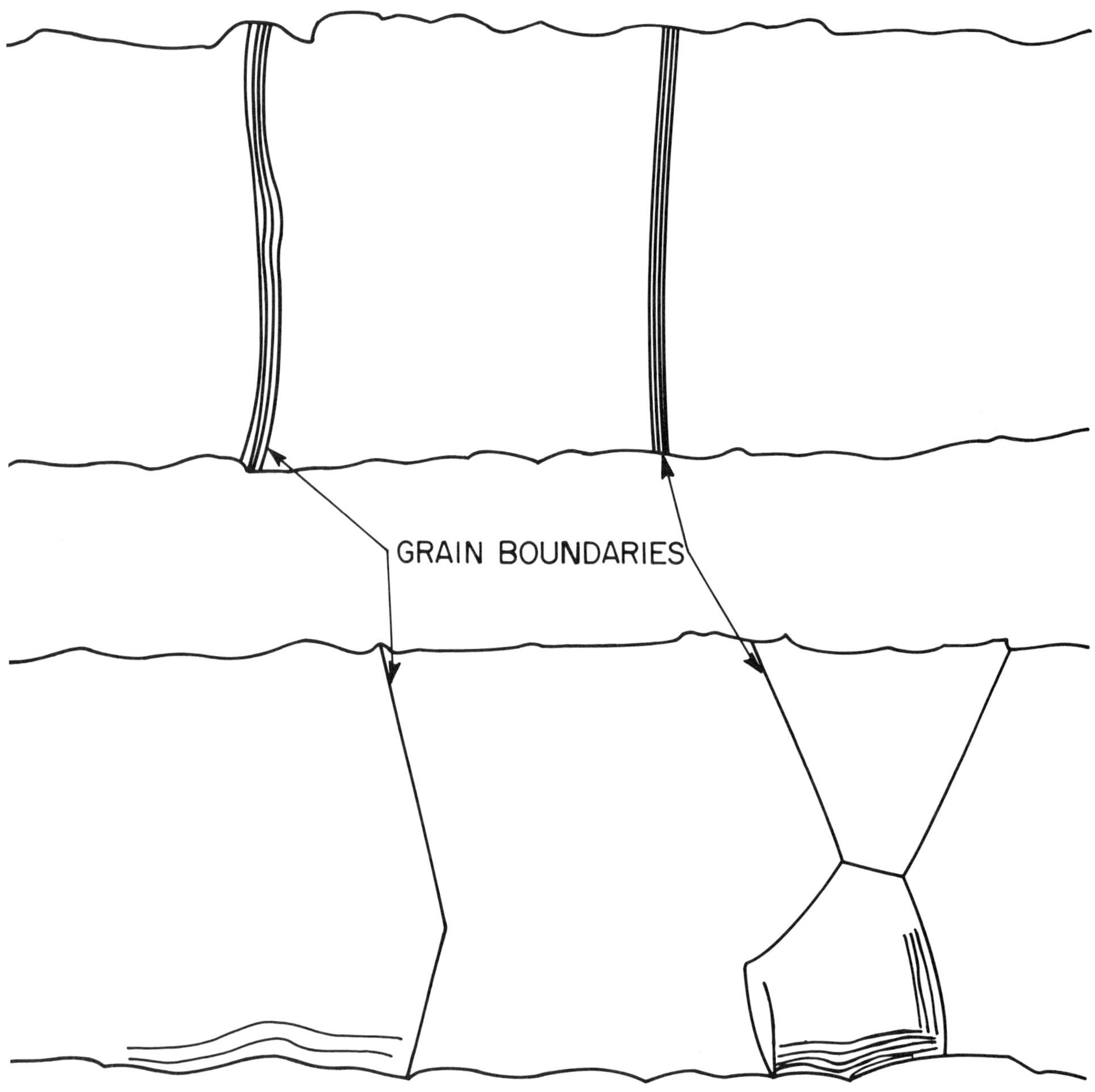

FIGURE 47. Aluminum lines of width smaller than the aluminum grain size contain only transverse grain boundaries (upper). When the linewidth becomes larger, longitudinal grain boundaries and triple points appear (lower).

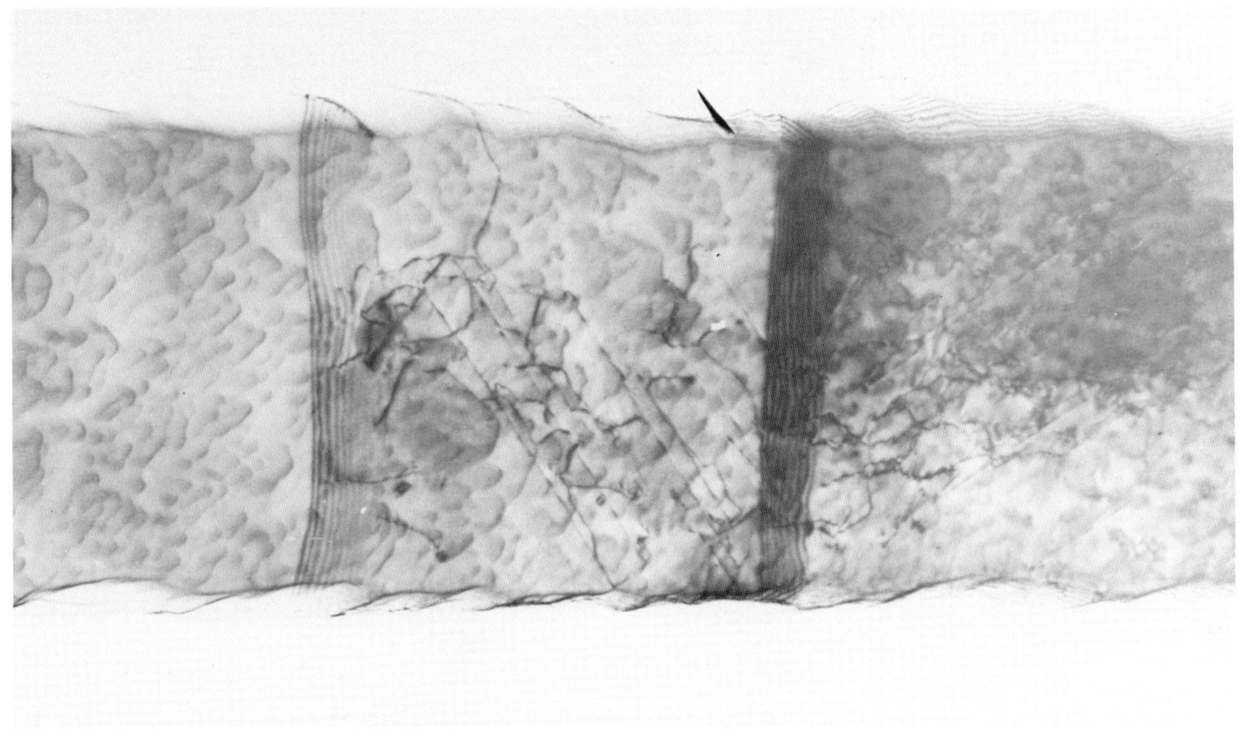
5000Å

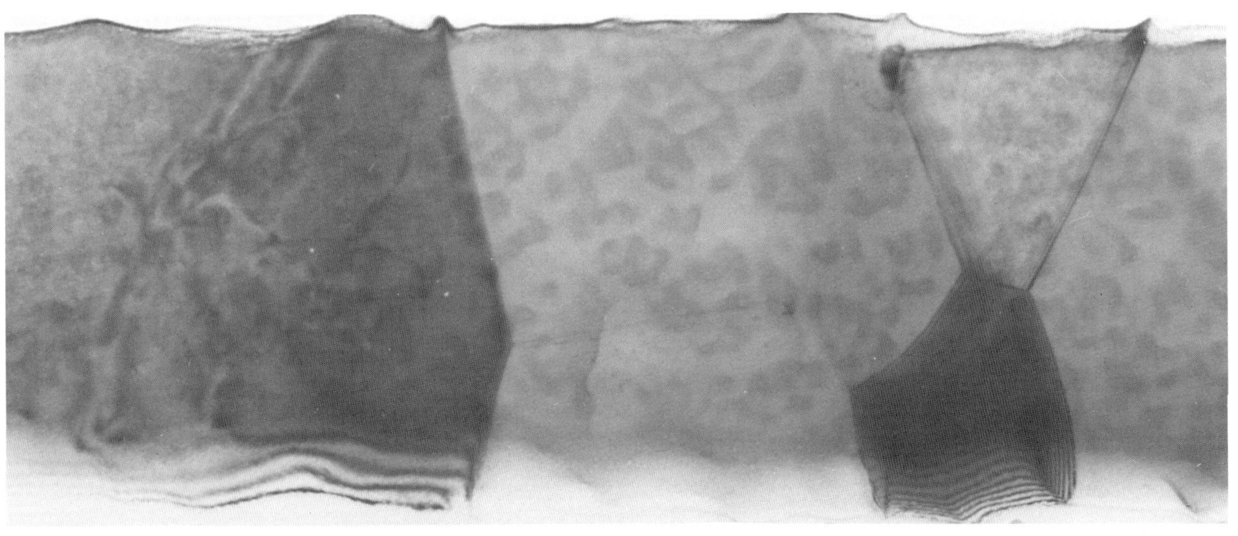
2.0μm

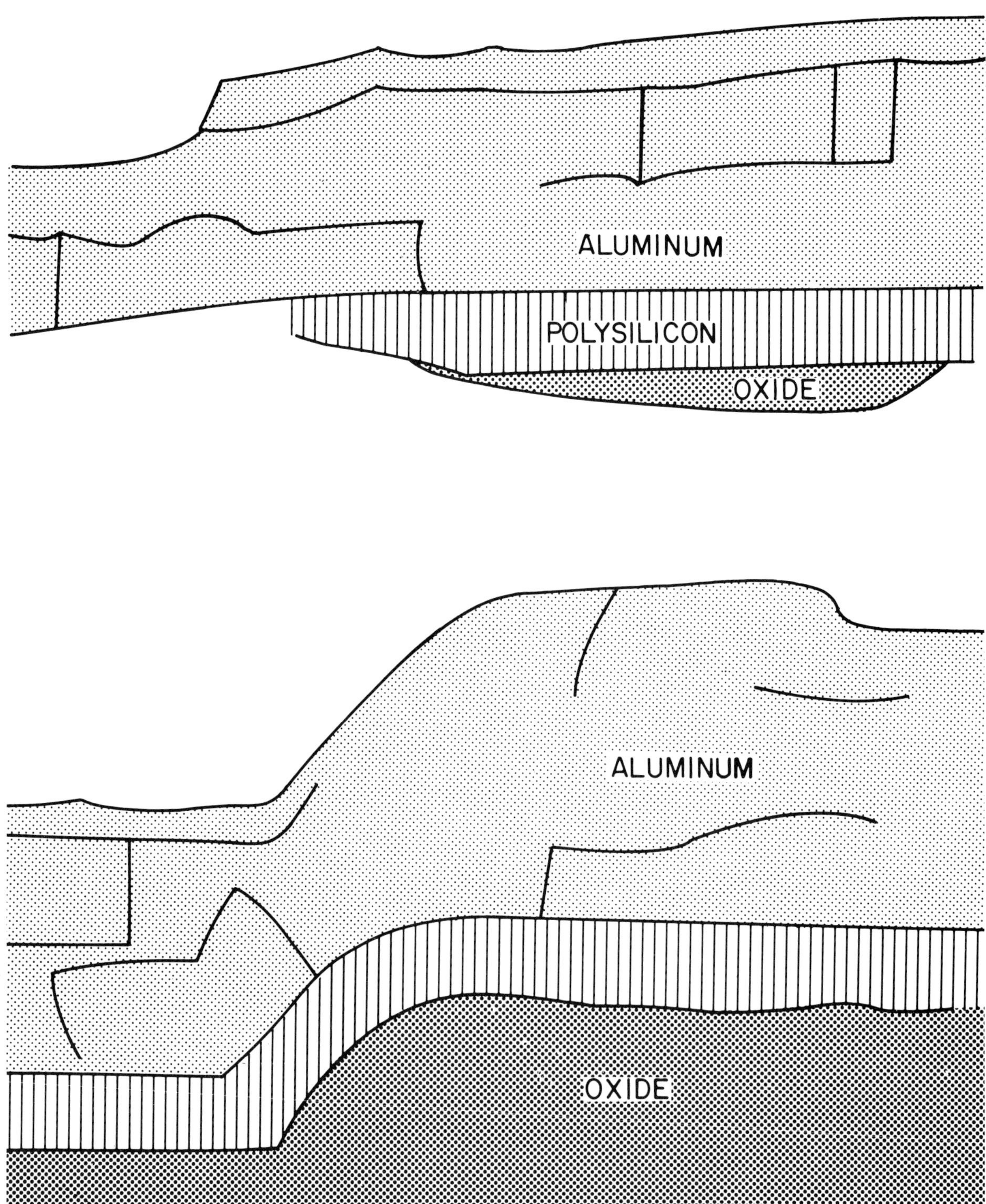

FIGURE 48. Aluminum grains often have the shapes of platelets with the short dimension perpendicular to the local substrate. A few grains are indicated in these two photographs. Platelet morphology is disturbed at a step and thinning occurs (lower).

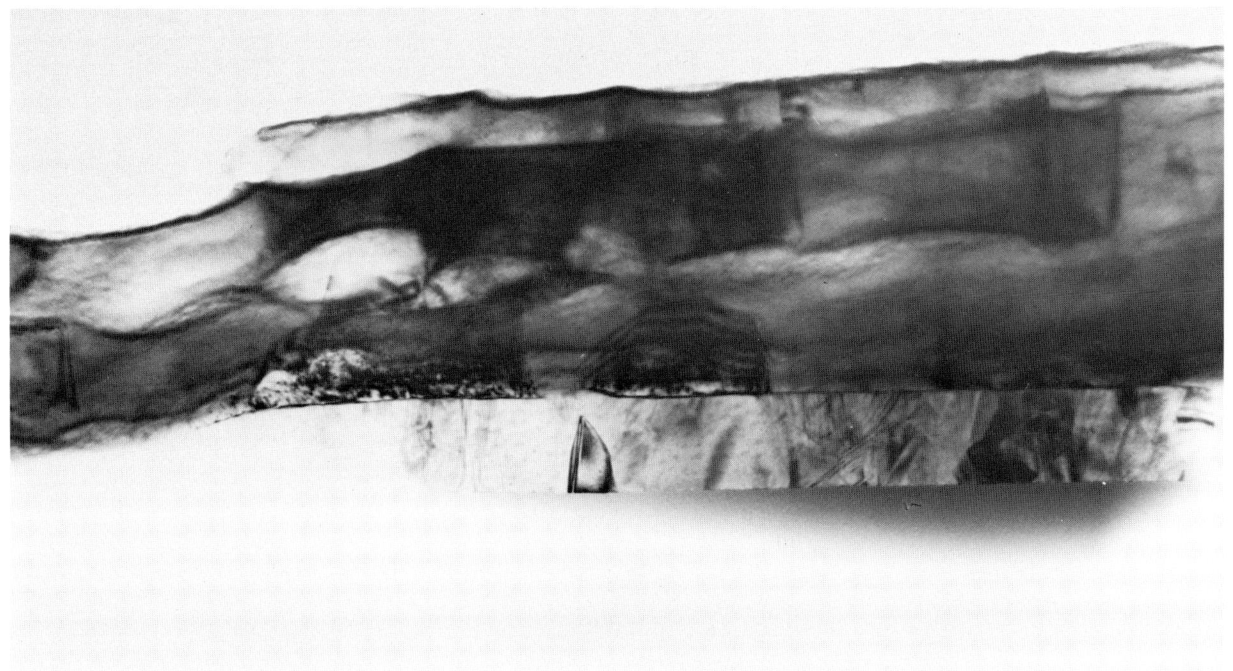

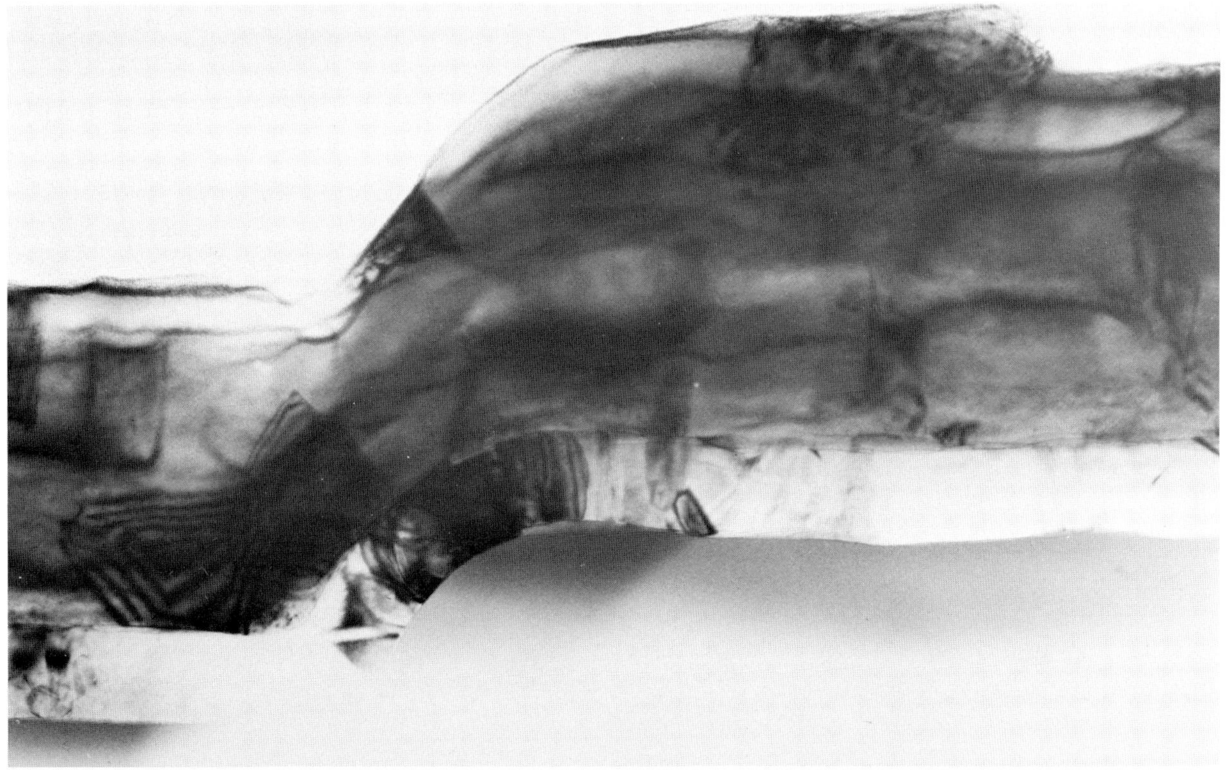
5000Å

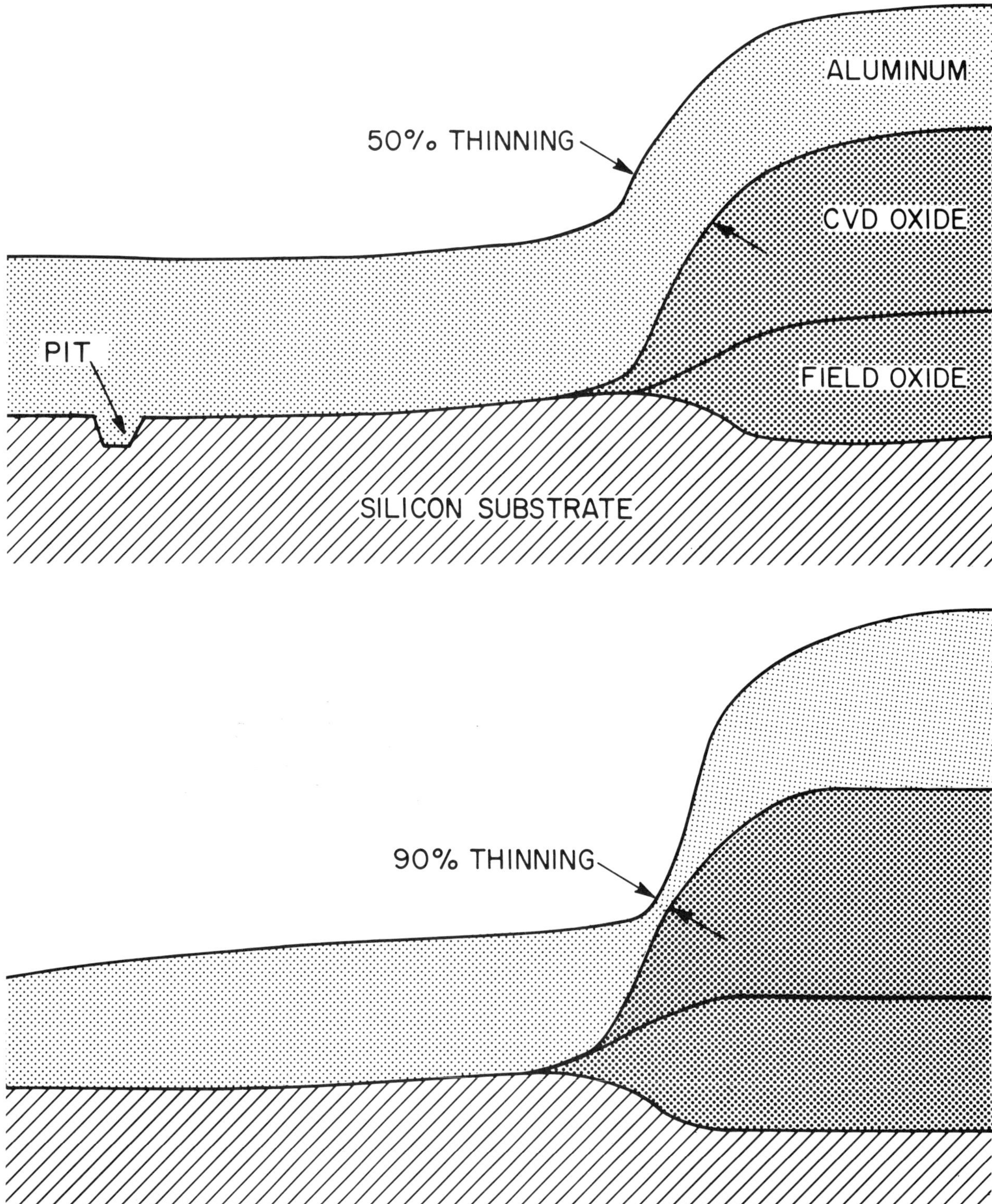

FIGURE 49. The variation in step coverage of aluminum over the semirecessed field oxide step in two wafers simultaneously coated, caused by the difference in the mean angle of incidence of the aluminum source. An aluminum-filled pit is also shown.

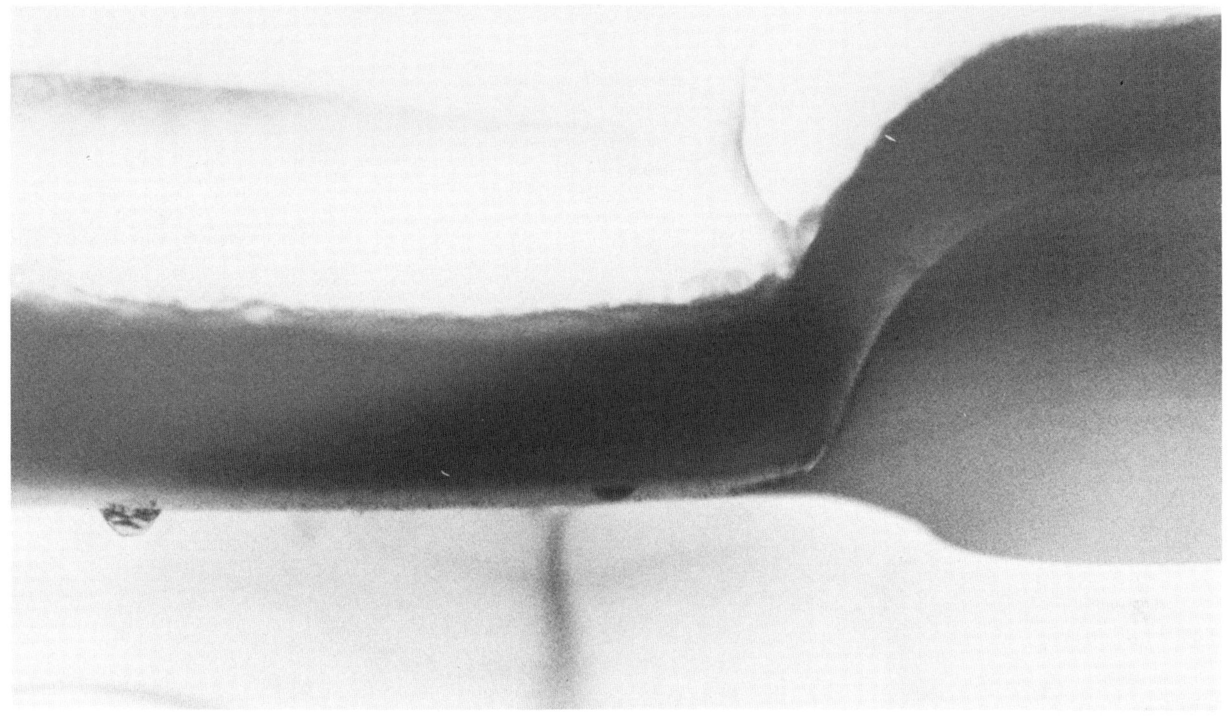

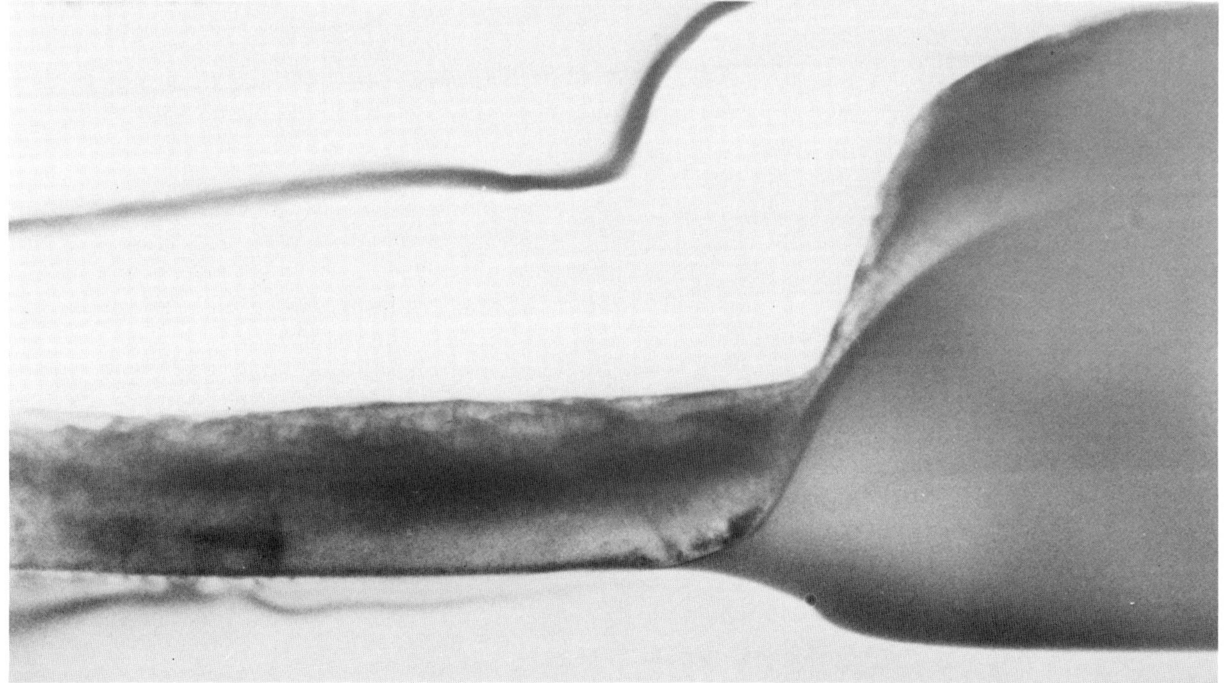

1.0 μm

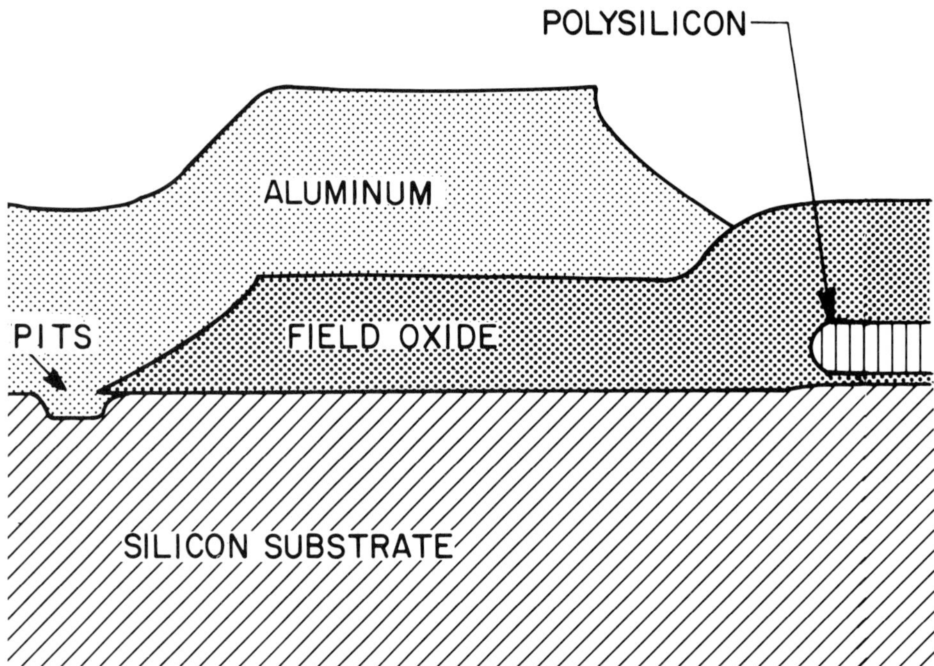

FIGURE 50. Interaction between the metal and substrate causes aluminum regions to form in silicon at the contacts. This interaction causes silicon to diffuse into the aluminum, leaving aluminum-filled pits.

A moderate-temperature (450°C) heat treatment decreases aluminum contact resistance to polysilicon and to the substrate and causes some alloy formation at the interface. The high solubility of silicon in aluminum causes silicon to diffuse into the metal, leaving spikes or pits in the silicon. Figures 49 through 51 show that such regions of interaction are clearly visible for the case of contacts to the silicon substrate. In Fig. 50 the pits have a maximum depth of approximately 0.2 μm, which is typical. In Fig. 51 a replica of a substrate contact region (made after removing the aluminum) shows a higher pit density at the side of the contact that is closest to the metallization runner where a semi-infinite sink is available for silicon dissolution. Such pit regions constitute potential failure sites, and the problem becomes more acute for shallow junction devices and for devices with small windows, since the pit size correlates inversely with window sizes.[42] This interaction can be eliminated by using aluminum metallization already saturated with silicon or by adding a doped polysilicon layer under the aluminum.

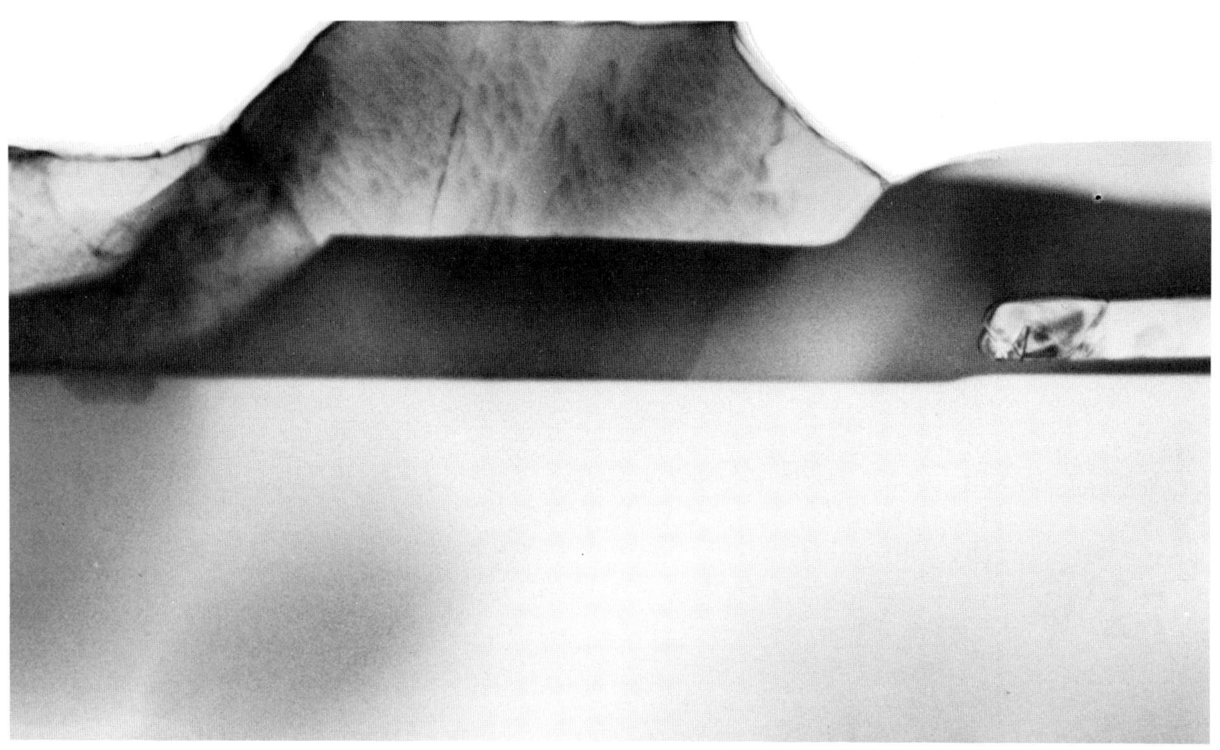
1.0μm

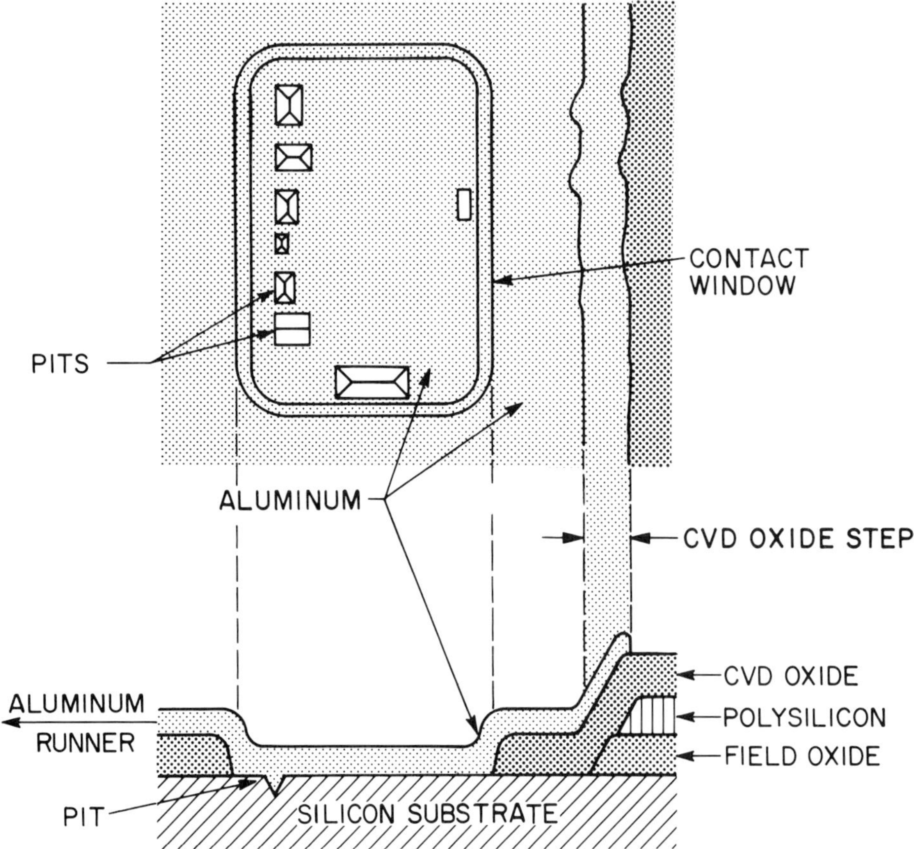

FIGURE 51. Replica of aluminum–silicon contact (upper) showing pits along the edge of the contact area, concentrating on the side nearest the aluminum runner. A cross section of this region is diagramed in the lower figure; the aluminum runner continues off to the left.

4.4 SILICIDES

Increasing resistance as device dimensions shrink and high oxidizability of polysilicon and aluminum are problems that can be solved by using other metallizations. Various metal silicides have been used, and relevant properties and processing parameters described.[43] The silicides of tantalum, molybdenum, tungsten, cobalt, and platinum are five of the more important ones.

Silicides are formed either by depositing the metal onto silicon followed by sintering or by codepositing the metal and silicon followed by sintering. The choice of method is governed partly by processing problems (e.g., metal over oxide does not form a silicide and might more easily be removed than a codeposited alloy) and partly by essential differences in final properties. For ex-

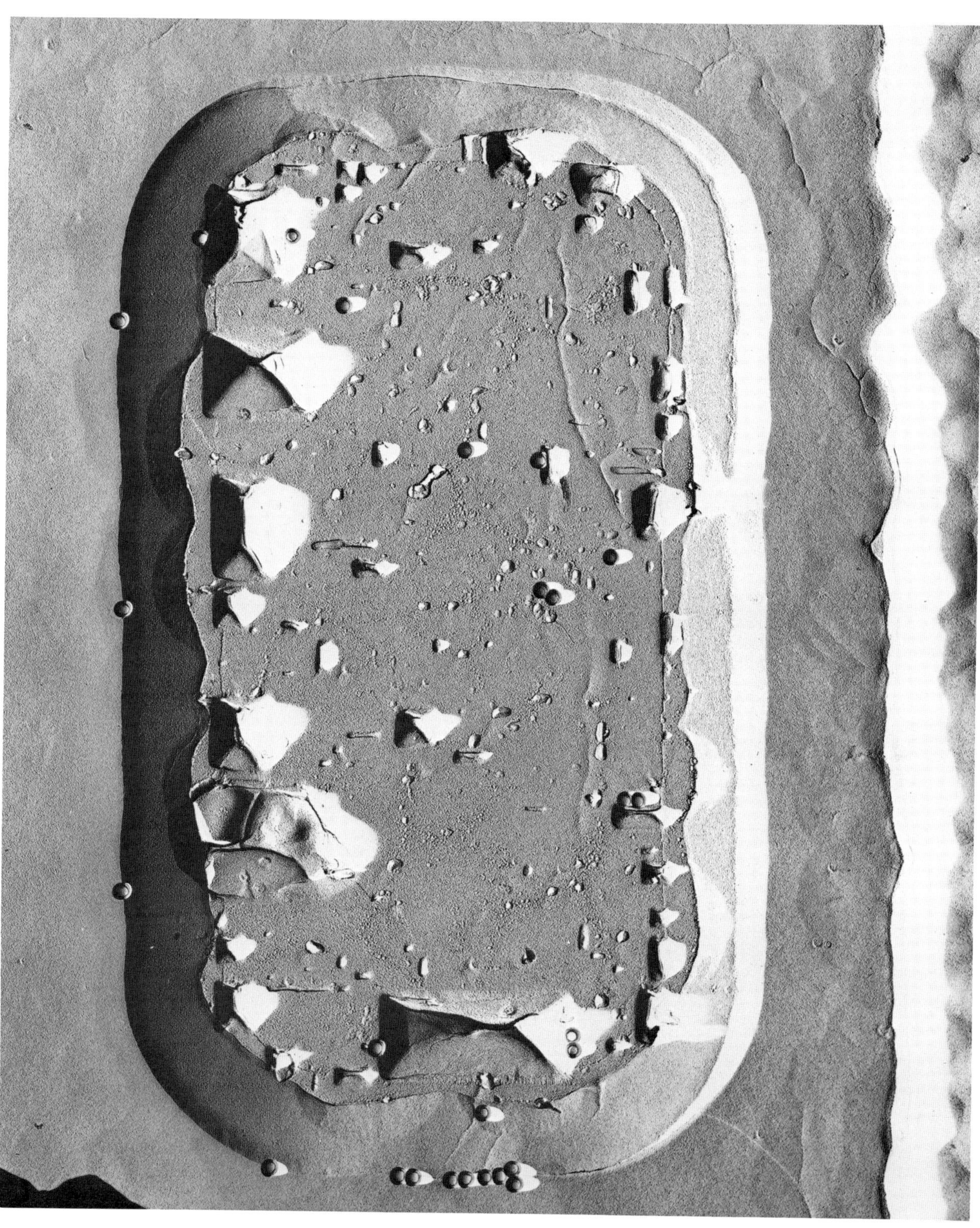
2.0μm

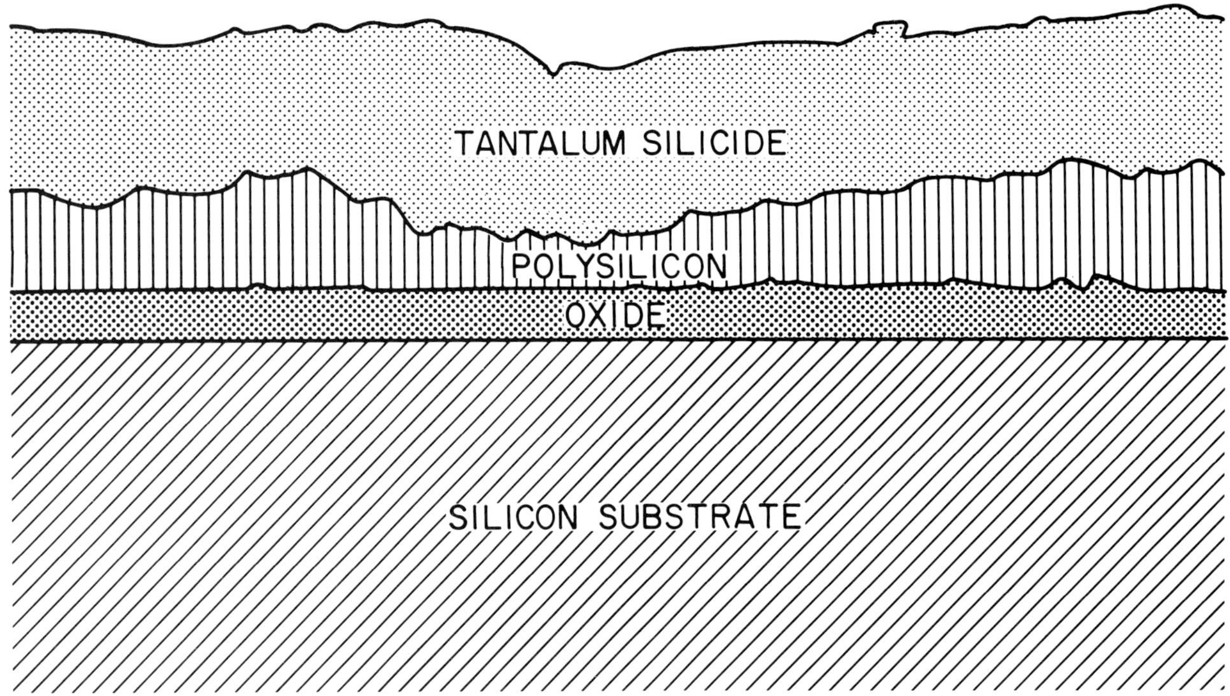

FIGURE 52. Tantalum silicide films formed by tantalum sputtered onto polysilicon (upper), and by cosputtering tantalum with silicon (Si/Ta = 2.1) onto polysilicon (lower). In both cases deposition was followed by sintering at 1000°C for 30 min in hydrogen.

ample, codeposited material is often amorphous, and produces a smaller volume change and smoother surface upon sintering. Tantalum deposited onto polysilicon forms a film of irregular thickness after 30-min sintering at 1000°C in hydrogen (Fig. 52, upper), while codeposited material forms a smoother surface (Fig. 52, lower).[44] Higher temperature sintering of codeposited tantalum and silicon often forms voids (Fig. 53).

Oxidation behavior differs from the various silicides when they are in contact with SiO_2, but in all cases, oxidation of silicides in contact with silicon leads to the growth of SiO_2.[43] Figure 54 shows sintered $TaSi_2$ over polysilicon before (upper) and after (lower) 1050°C wet oxidation. These photographs show that oxidation consumes the polysilicon with essentially no change in the morphology of the silicide. Once the supply of silicon is exhausted, the silicide begins to oxidize. Thus, the processing technology that requires high temperature (wet) oxidation is compatible with the presence of silicide if a supply of silicon is available.

Both metal deposition and cosputtered metal-silicon deposition produce films with a thickness uniformity that depends on the angle of inclination of the substrate. Platinum deposited onto a gently sloping polysilicon surface forms an even film of silicide after sintering (Fig. 55). A film deposited on a wall of steeper slope, such as at the edge of a gate, produces a slightly thinner

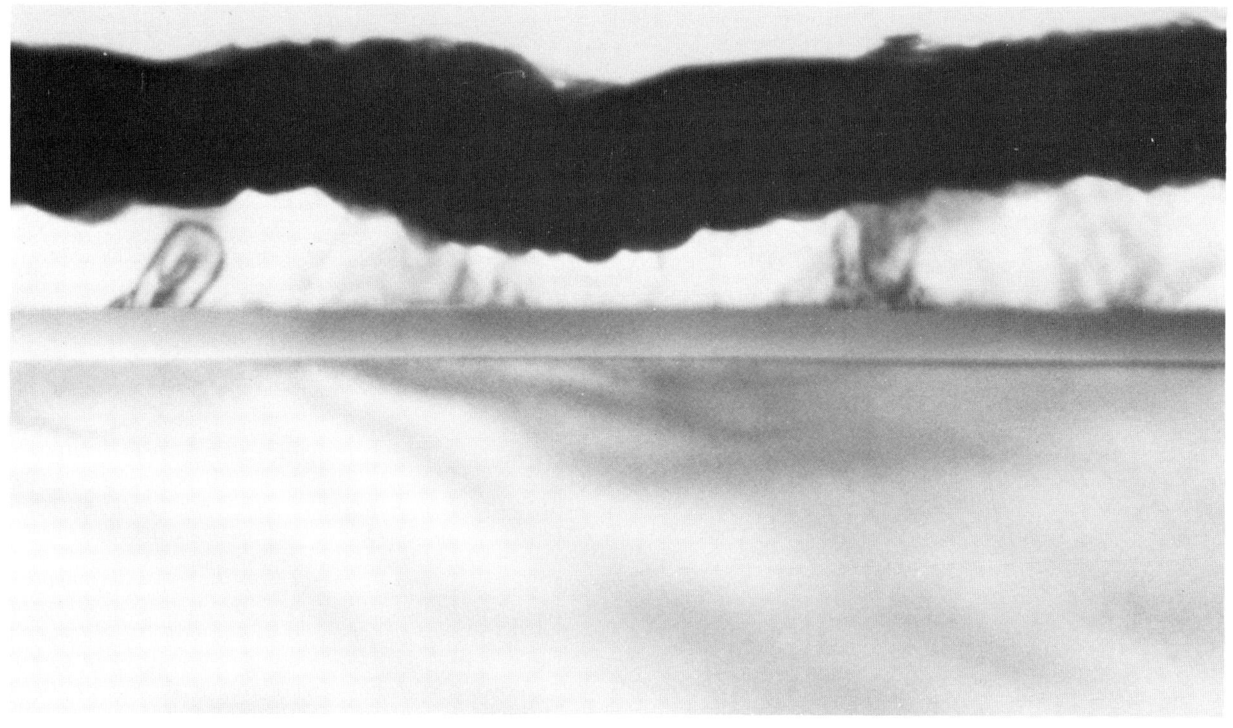

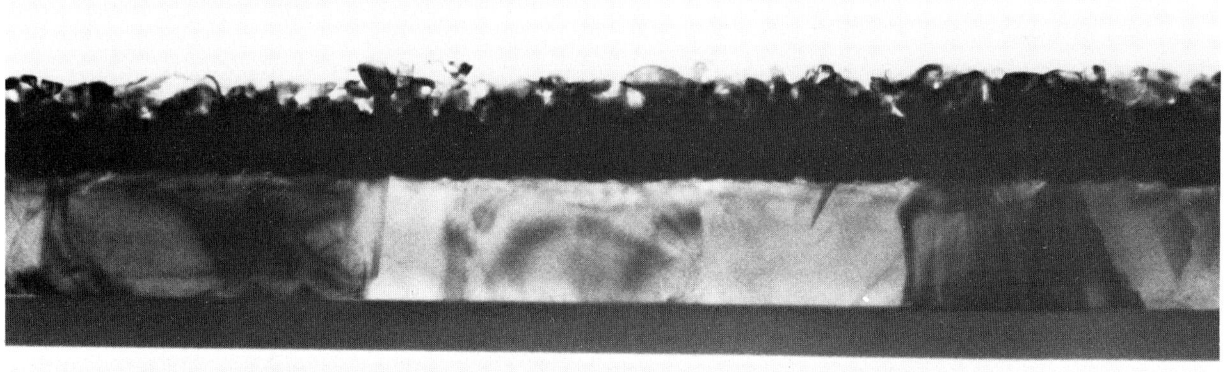

5000Å

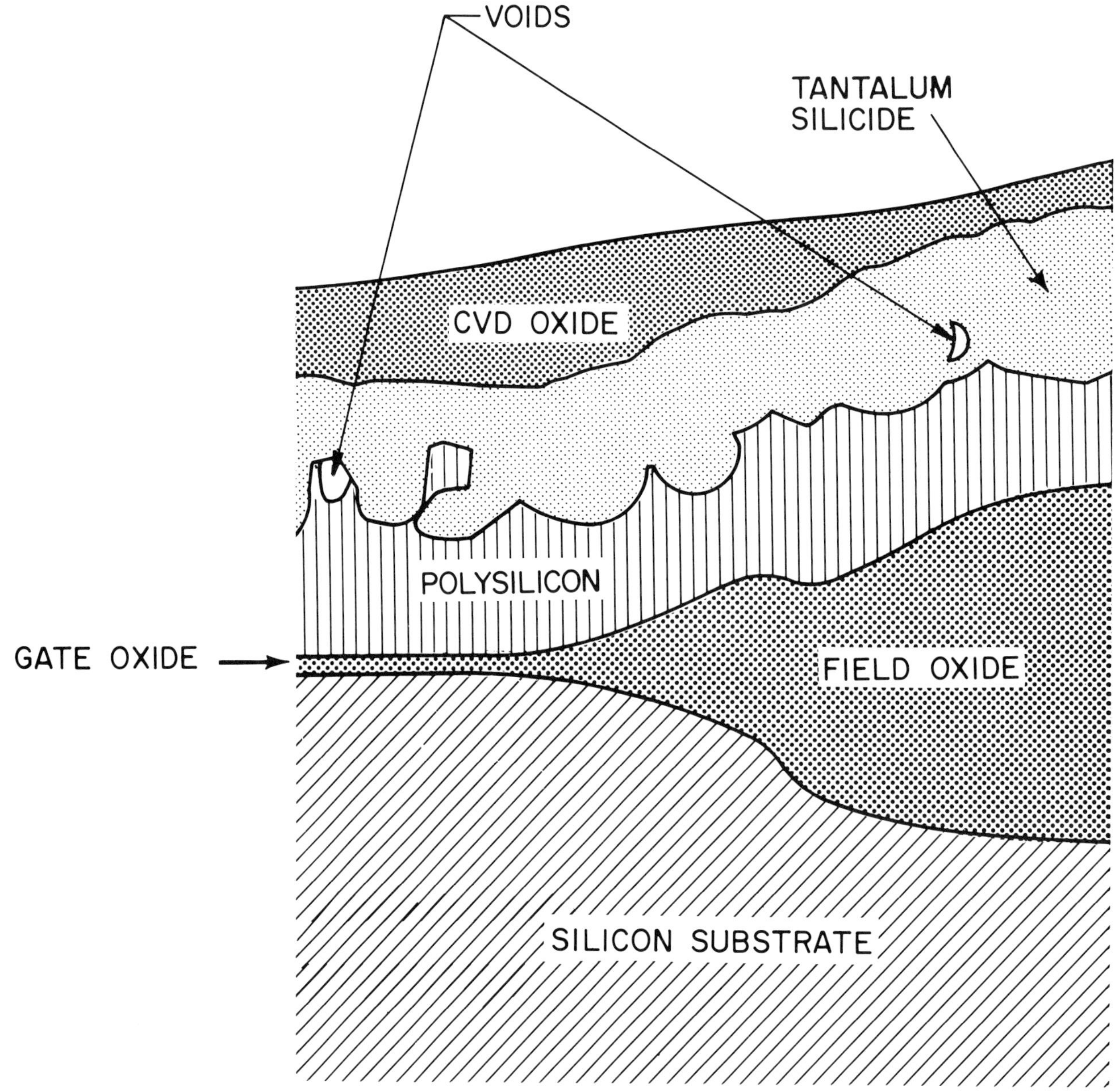

FIGURE 53. Tantalum silicide metallization showing rough silicide–polysilicon interface texture and the formation of voids. This silicide was formed by sintering codeposited material at a higher temperature than that used for the sample shown in Fig. 52 (lower).

3000Å

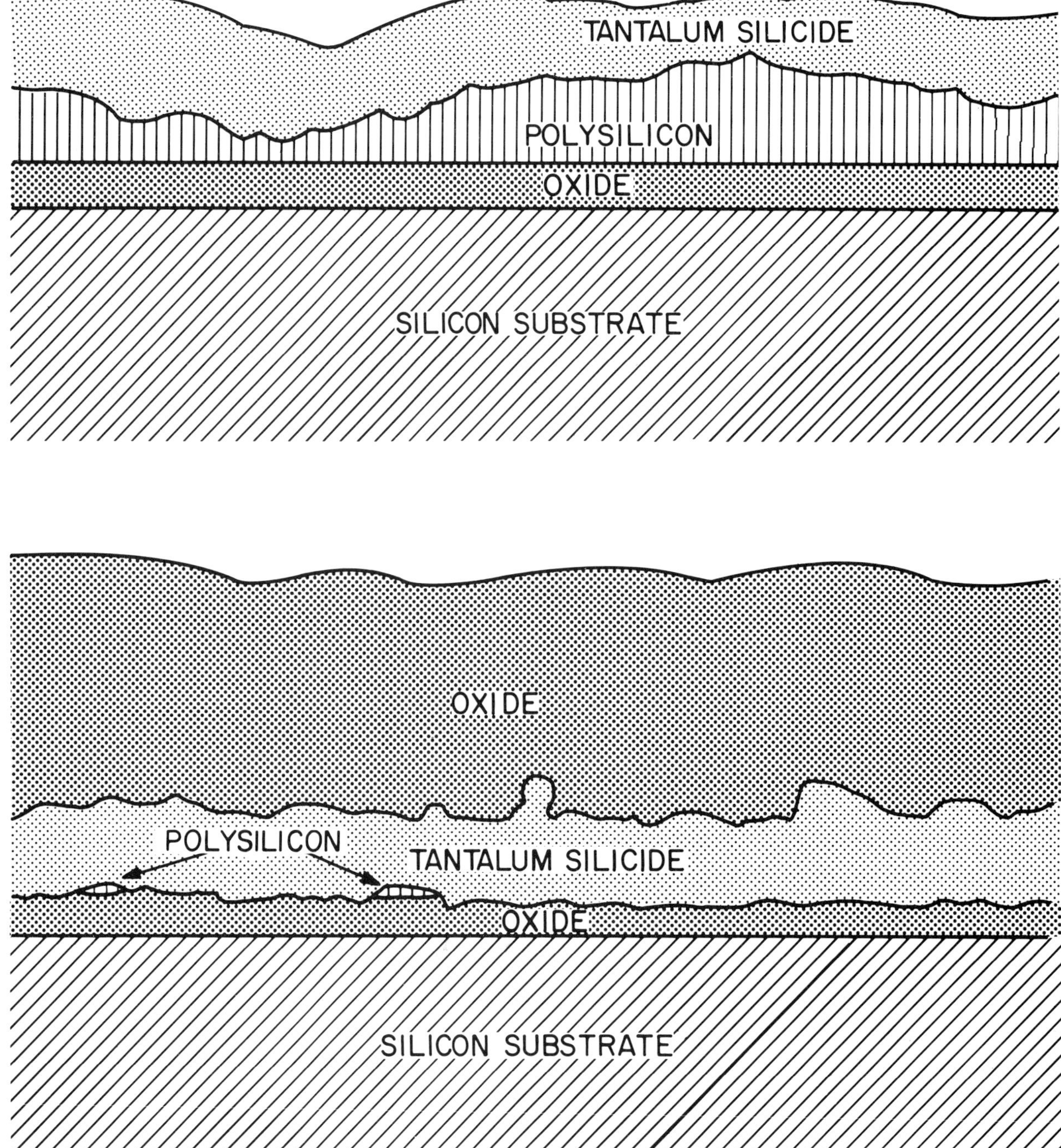

FIGURE 54. Sintered $TaSi_2$ on polysilicon before oxidation (upper) and after 90 min wet oxidation at 1050°C (lower) showing almost total consumption of the underlying polysilicon.

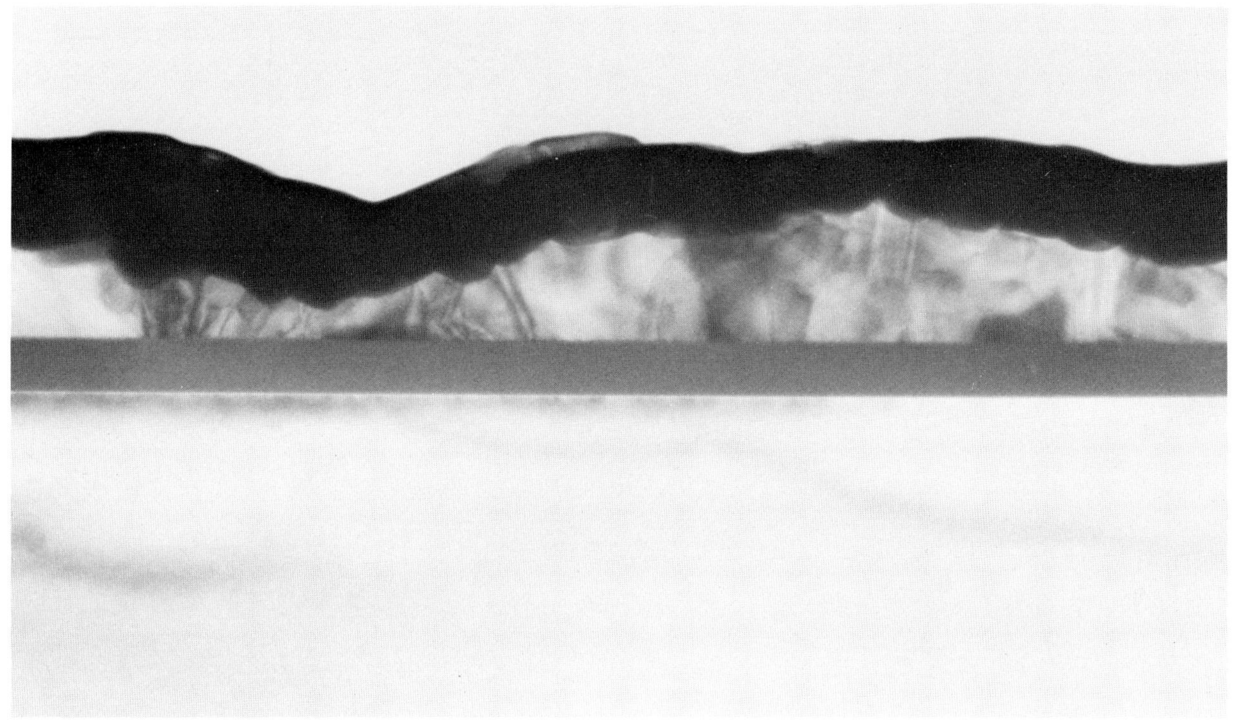

5000Å

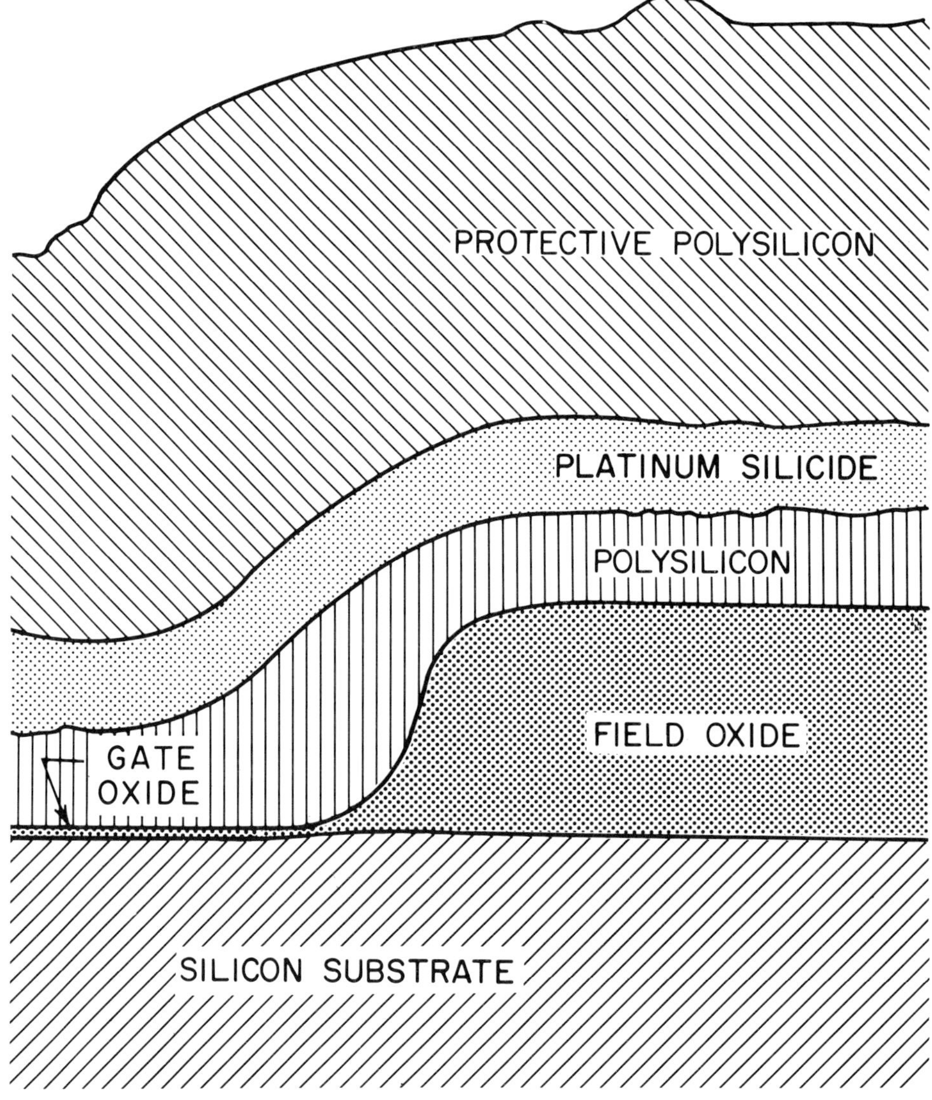

FIGURE 55. Platinum silicide formed by sintering platinum deposited over a gradually sloping polysilicon surface at an oxide step. Platinum covered the entire surface and an even coating of silicide resulted.

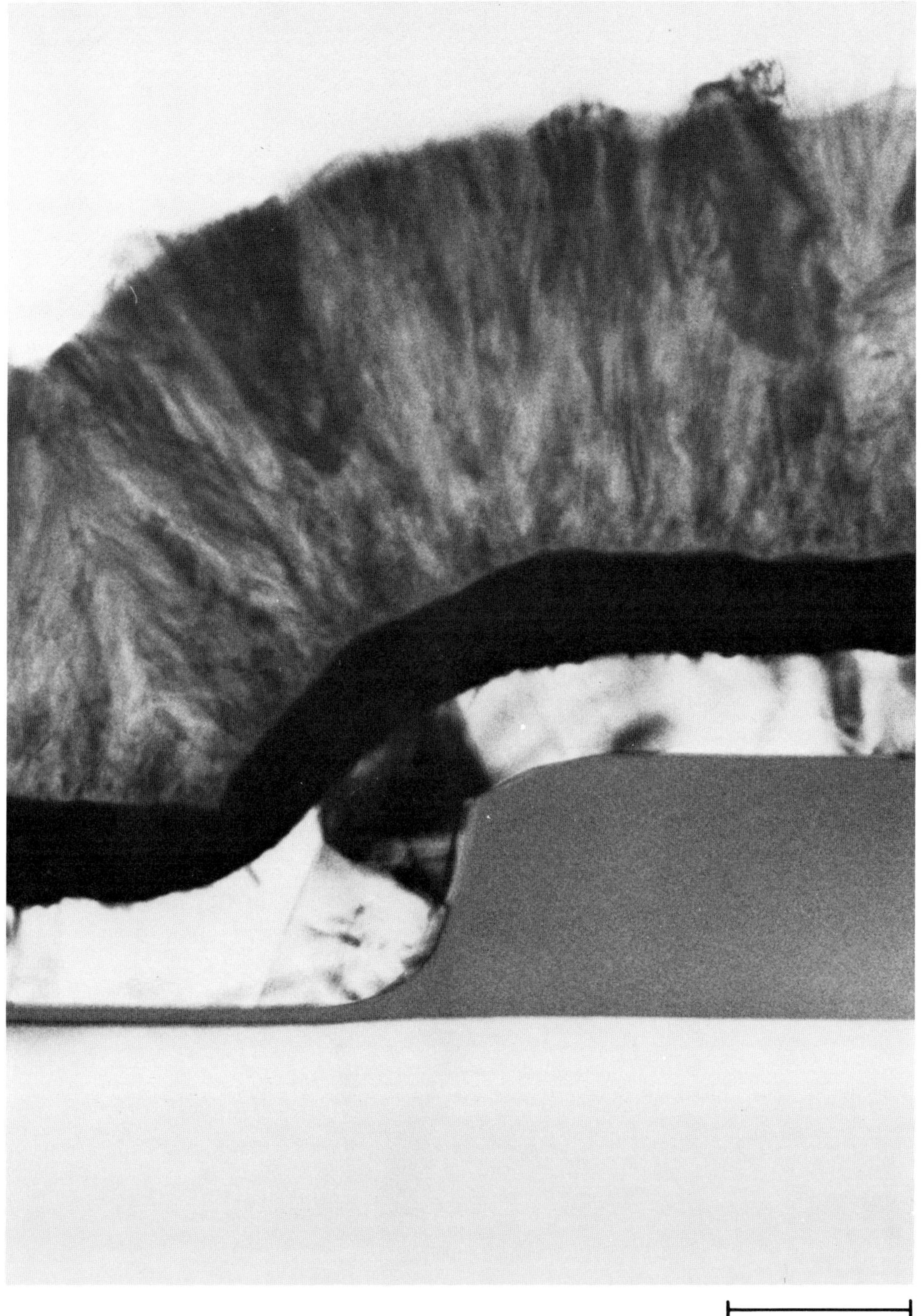
2000Å

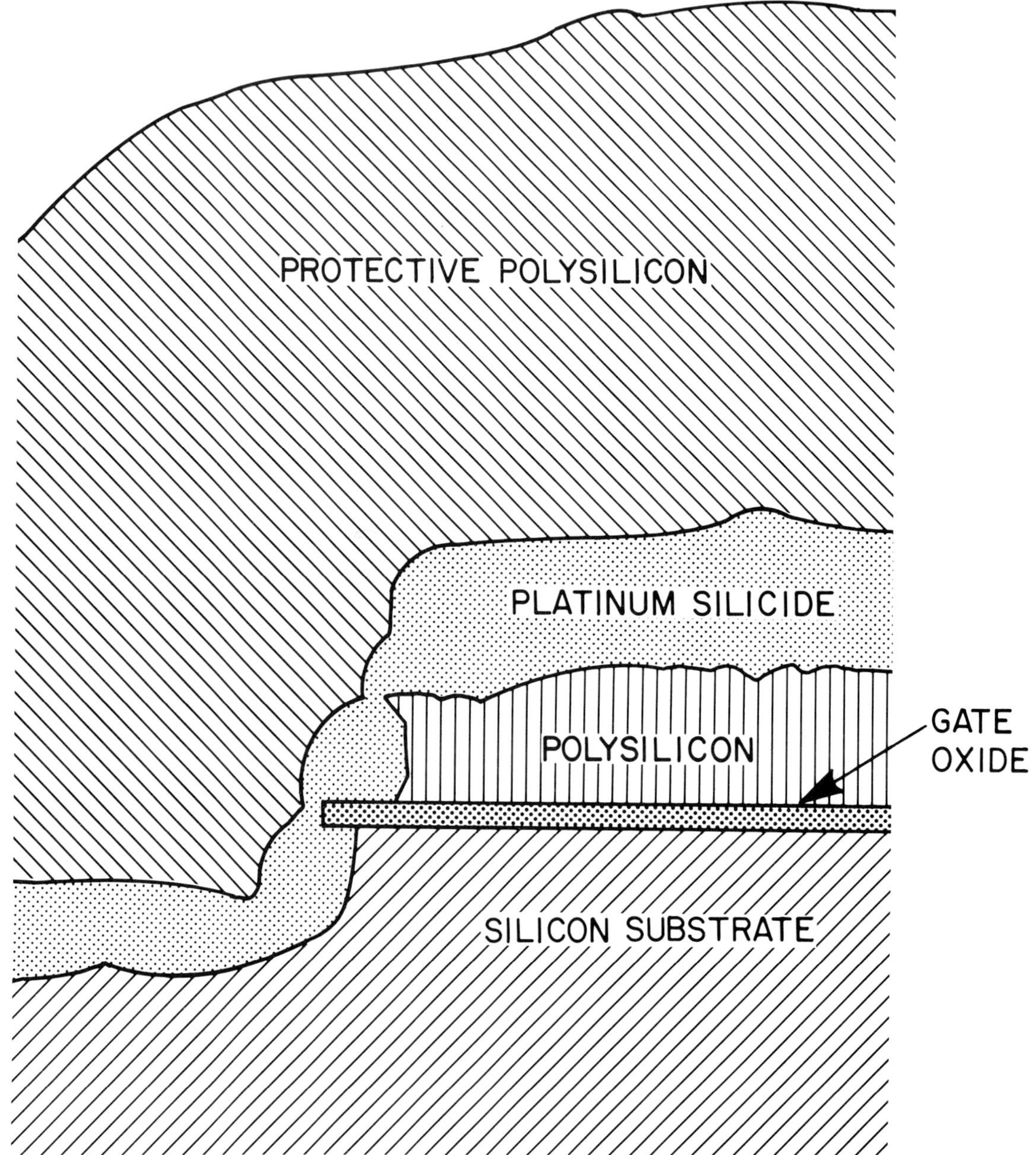

FIGURE 56. Platinum silicide formed by sintering platinum deposited at the edge of a gate bordering on the source–drain region. A continuous film of silicide at the edge of the gate causes shorting between the gate and the source–drain region.

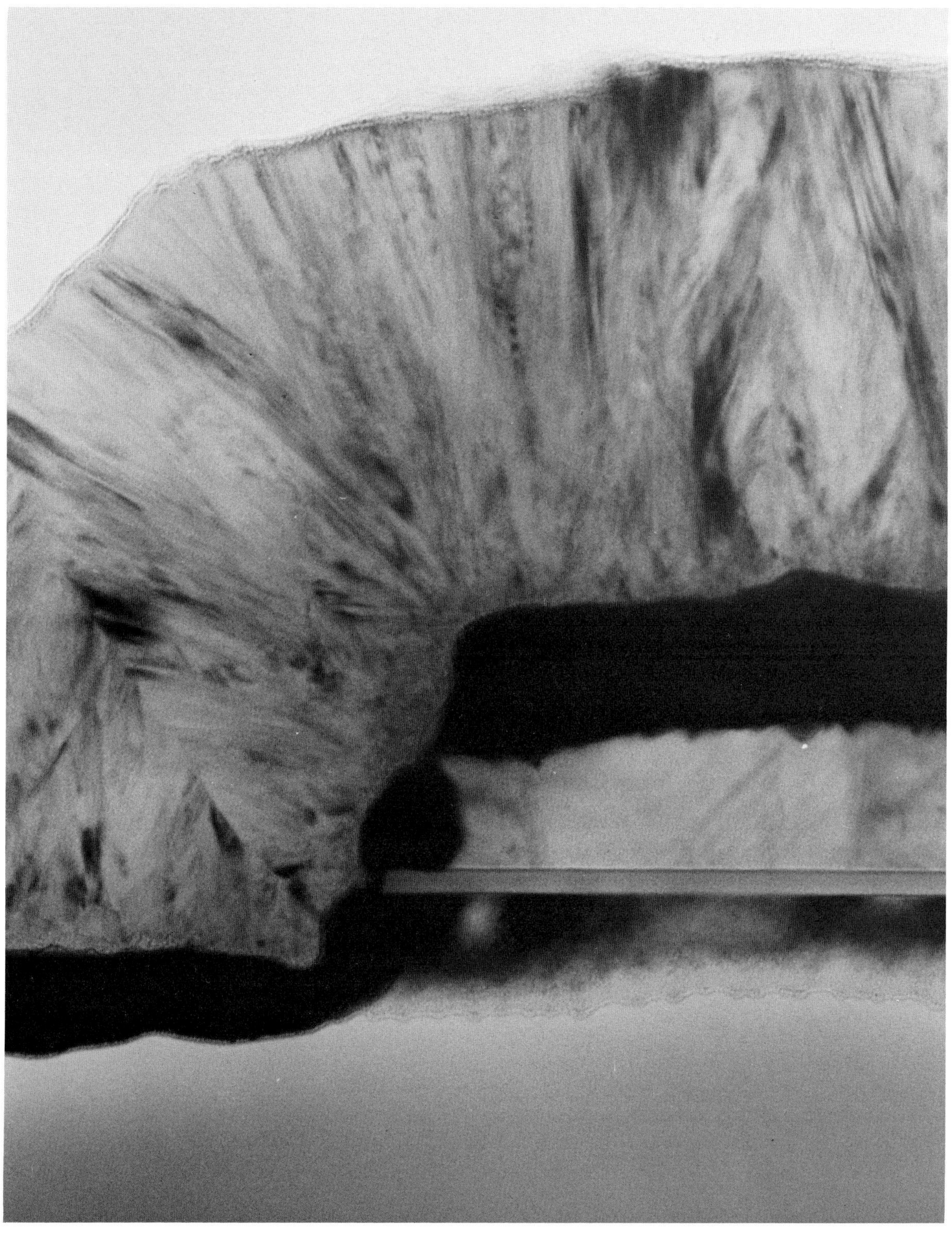
5000Å

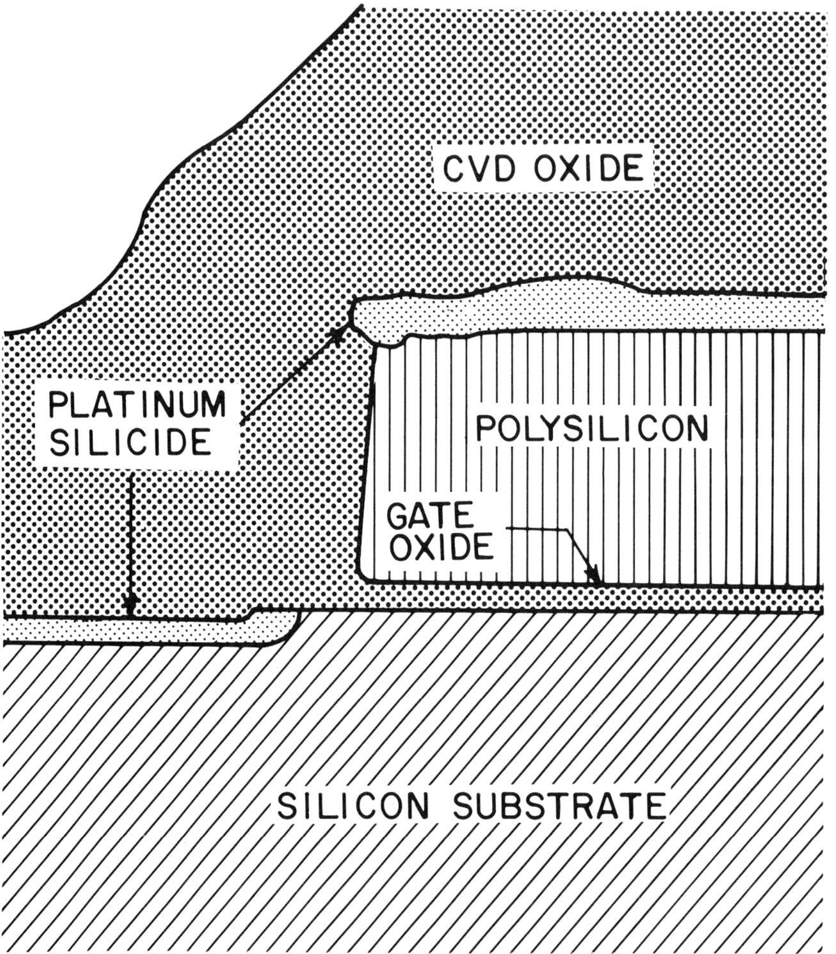

FIGURE 57. Platinum silicide formed in a manner that leaves no silicide on the wall of the polysilicon gate.

deposit that forms a silicide after sintering (Fig. 56). The presence of silicide on the gate-side wall (Fig. 56) constitutes a failure site. Complete isolation of the gate from the substrate requires a different procedure for silicide formation. In one variation, useful for making Schottky-barrier contacts to the source–drain regions,[45] a thin CVD oxide is deposited after polysilicon patterning and is subsequently removed from the horizontal surfaces; oxide is left protecting the side wall. Subsequent platinum deposition and sintering is followed by removal of unreacted metal from the side walls, which leaves the structure shown in Fig. 57.

Silicide topography depends partly on stoichiometry, which also controls film stress. Molybdenum silicide films show a marked decrease in stress corresponding to a composition with the ratio of Mo to Si equal to 2.[46] This stress relief is not well understood but is thought to be due to void formation (Fig. 58) along with the formation of the intermetallics $MoSi_2$, Mo_3Si_2, and Mo_3Si.

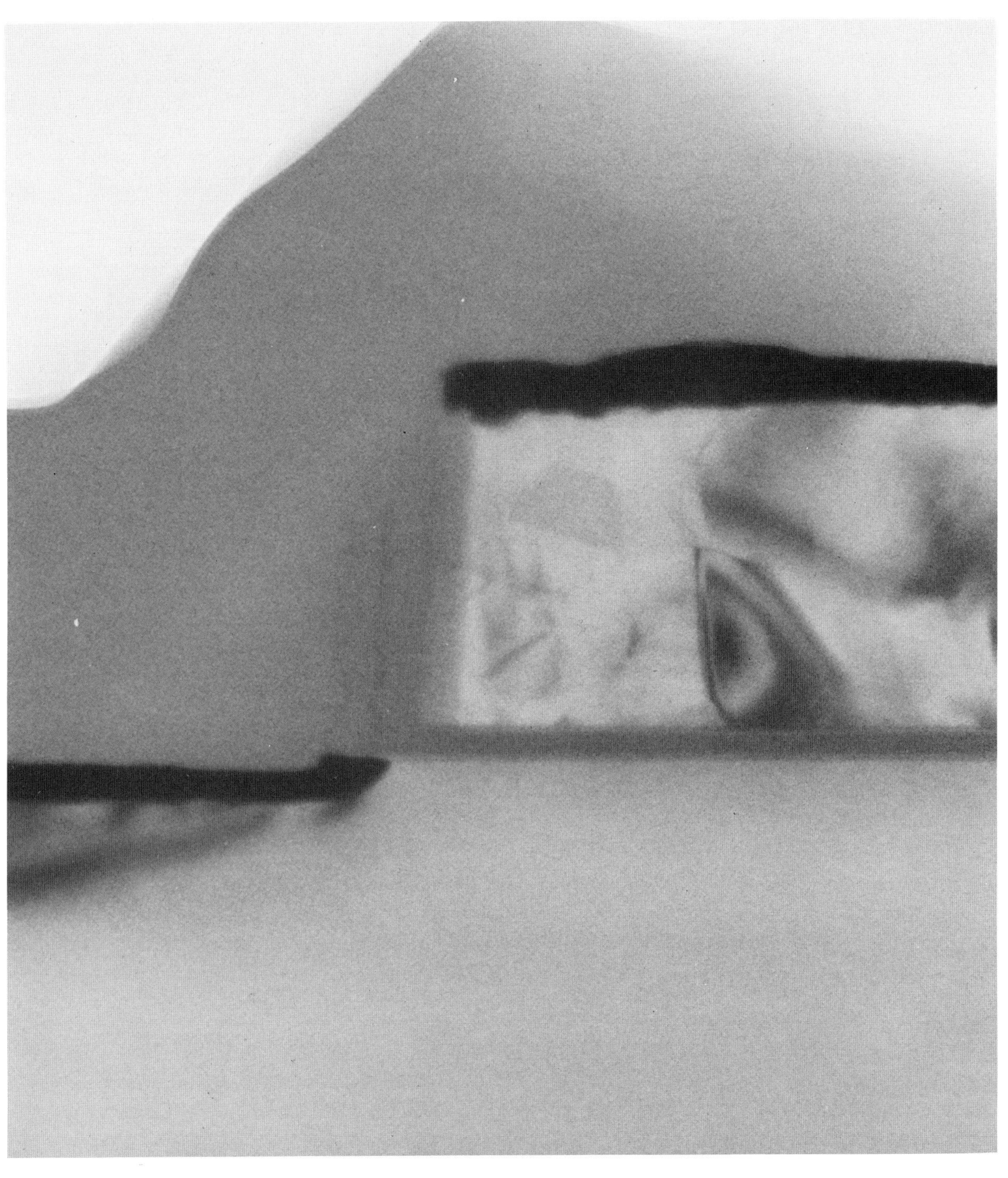

2000Å

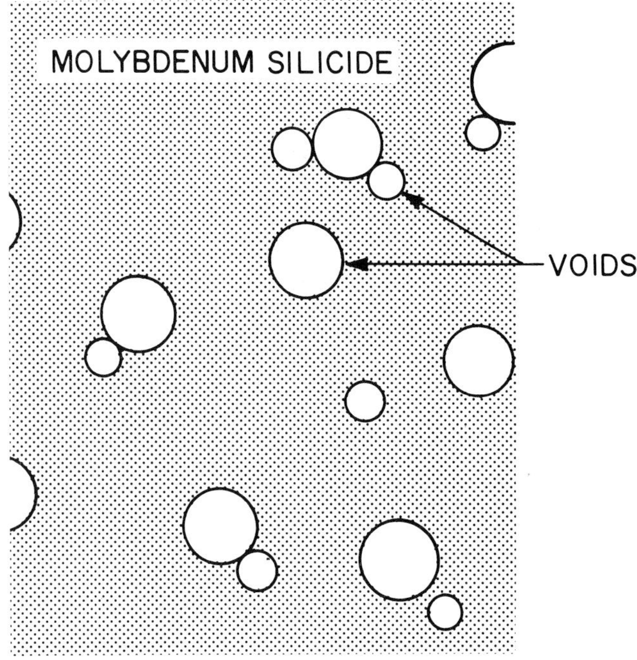

FIGURE 58. Molybdenum silicide sintered at 950°C in hydrogen for 1 hr (Mo/Si = 2.0). Stress relief occurs at this stoichiometry, probably because microvoids form.

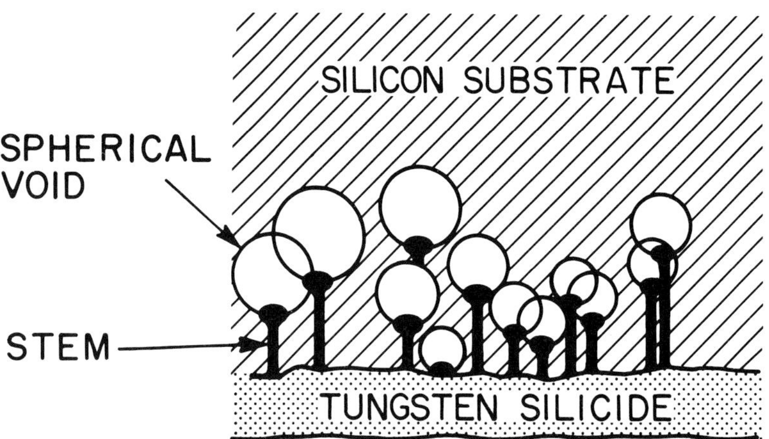

FIGURE 59. Tungsten silicide formed by laser annealing a 480-Å film of tungsten on silicon. The stem consists of WSi_2 needles ending on spherical voids.

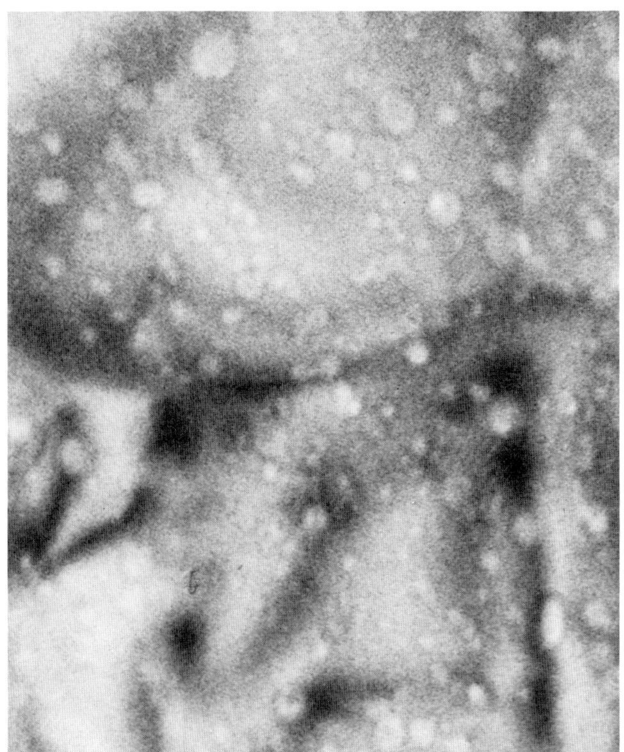

1000Å

2000Å

Bulk silicide texture also results from phase changes occurring during thermal stressing. Figure 59 shows the formation of WSi_2 needles and spherical voids in laser annealed samples of 480-Å tungsten on silicon. The 30-nsec laser pulse is thought to result in constitutional supercooling that causes the formation of WSi_2 needles and vapor bubbles in the molten silicon, which leave spherical voids.[47]

CHAPTER FIVE

JUNCTION DELINEATION

Pn junctions are formed by introducing a *p*- or *n*-type dopant into a substrate of opposite conductivity and annealing the sample. Annealing activates the implanted dopant and produces the desired function depth by diffusing the dopant. The final junction depth can be determined by a number of electrical measurements of the junction or device that contains the junction, or by certain physical measurements. Of the various physical methods available, the most useful ones are spreading-resistance probing, depth profiling of chemical species, and staining followed by optical or sometimes scanning electron microscopy. These three methods are limited in depth resolution, and none can accurately measure the lateral extent of the shallow junction under a feature (such as the edge of a gate electrode).

The depth resolution of spreading-resistance probing for shallow-angle lapped junctions[48] is limited by the probe tip size to about 1000 Å. Depth-profiling methods are excellent sources of information on junction depth and are limited only by the sensitivity limit to detection of the dopant. For example, sensitivity limits for the detection of arsenic by Rutherford backscattering spectroscopy,[49] Auger electron spectroscopy,[50] and secondary ion-mass spectroscopy[51] are 3×10^{18}, 5×10^{18}, and 5×10^{14} atoms cm^{-3}, respectively. The rapid fall-off of the doping profile of arsenic in silicon for shallow (<0.25 μm) junctions at concentrations less than 10^{19} atoms cm^{-3} makes the difference between the junction depth and the delineation depth small, typically 200–300Å.

The traditional method of angle lapping, staining, and microscope examination with interference optics has a depth resolution limited by the spacing of interference lines (0.255 μm for xenon light). Substituting an SEM for an optical microscope and using a cleaved face instead of an angle-lapped surface

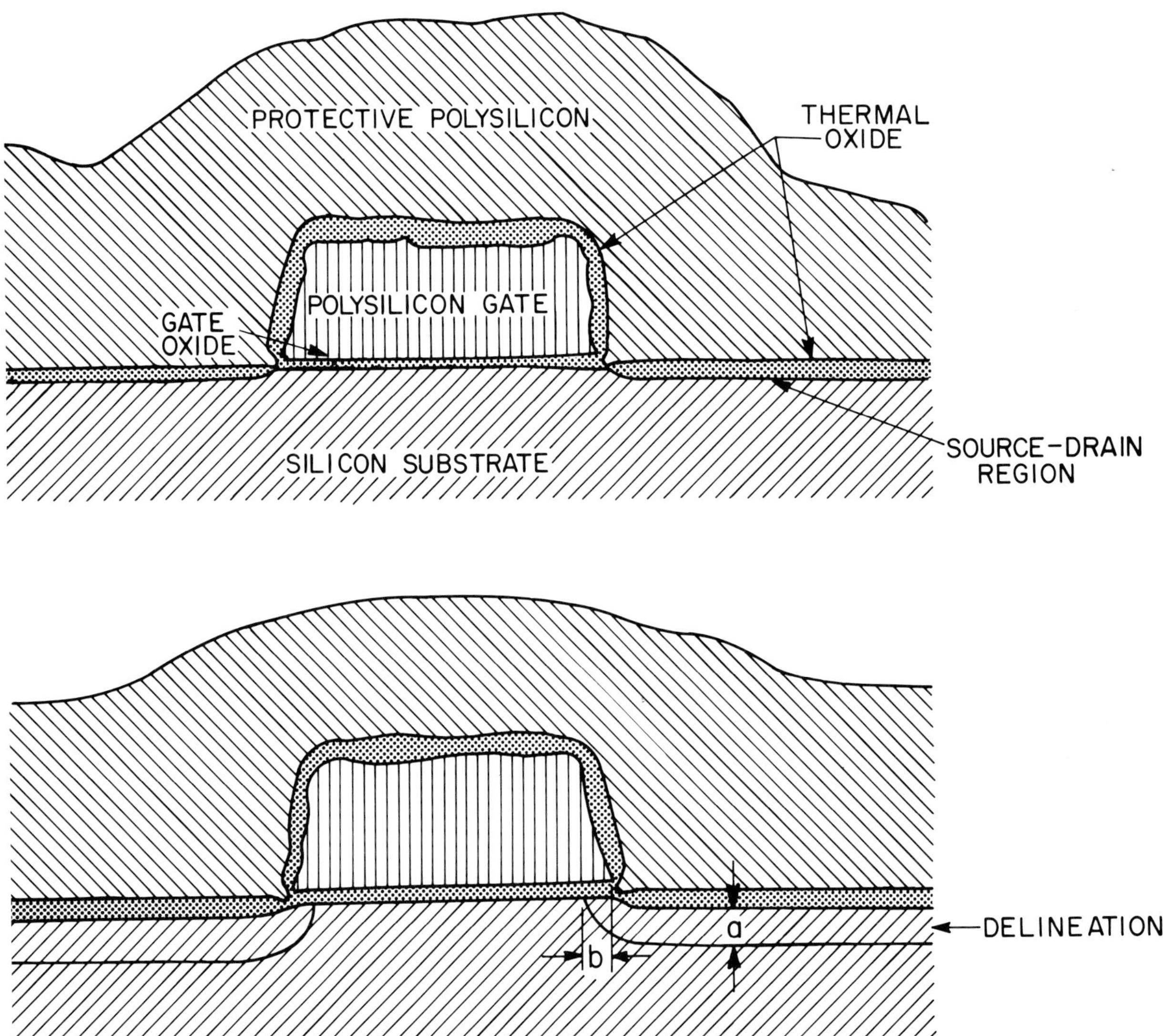

FIGURE 60. Polysilicon gate and source–drain region before (above) and after (below) n^+p junction delineations. The ratio of the lateral extent of the junction to the junction depth (b/a) can easily be measured. The junction was formed by arsenic implantation and diffusion.

offers some improvement in resolution. The highest resolution, however, is obtained by TEM study of thin sections of chemically treated junctions.

Phosphorus- and arsenic-doped n^+p junction samples treated with 0.5% HF in HNO_3 reveal the junction during TEM study [18] (Section 2.5). The solution preferentially etches the n^+ region and produces a delineation at a depth corresponding to an arsenic concentration less than 1×10^{19} atoms cm^{-3}. Differences in junction depth at the source–drain and contact areas can be revealed, and measurements can be made of the magnitude of the lateral extent of a junction under a gate region. Figure 60 shows a polysilicon gate/source–drain

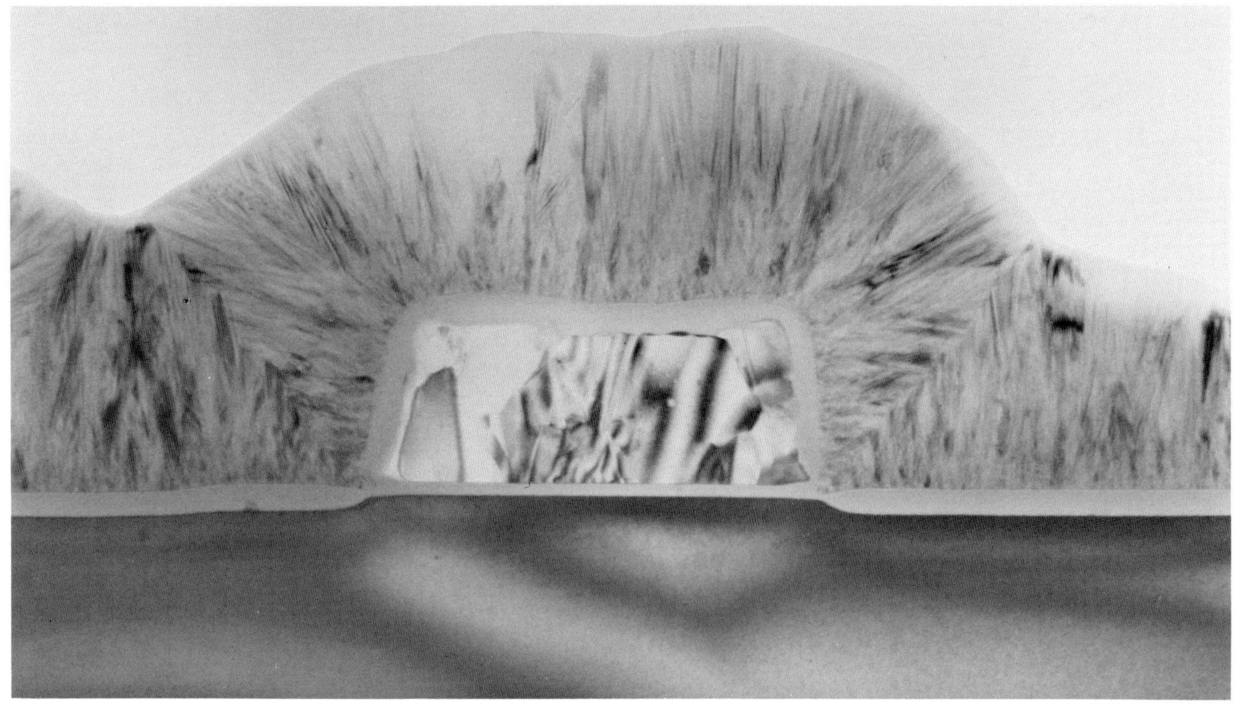

5000Å

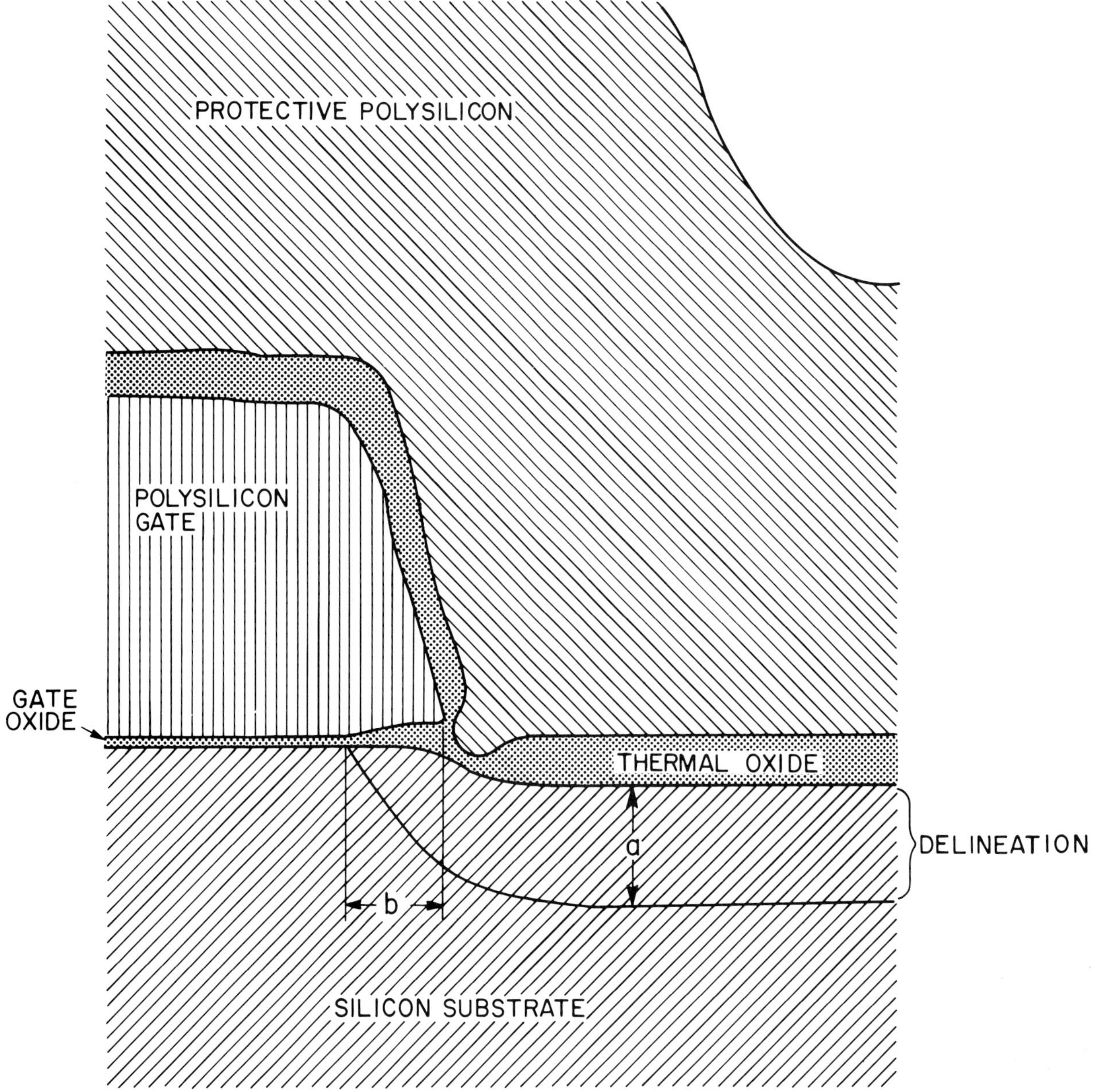

FIGURE 61. Polysilicon gate and source–drain region after n^+p junction delineation. The junction was formed by arsenic implantation and diffusion.

sample before (above) and after (below) treatment with the preferential etchant for 6 sec. Figure 61 shows another delineated sample at higher magnification. The ratio b/a measured from these photographs is 0.8–0.9, showing that the penetration of the junction under the gate is 0.8–0.9 times the junction depth.

The shape of the delineation at the edge of a diffused region depends on the

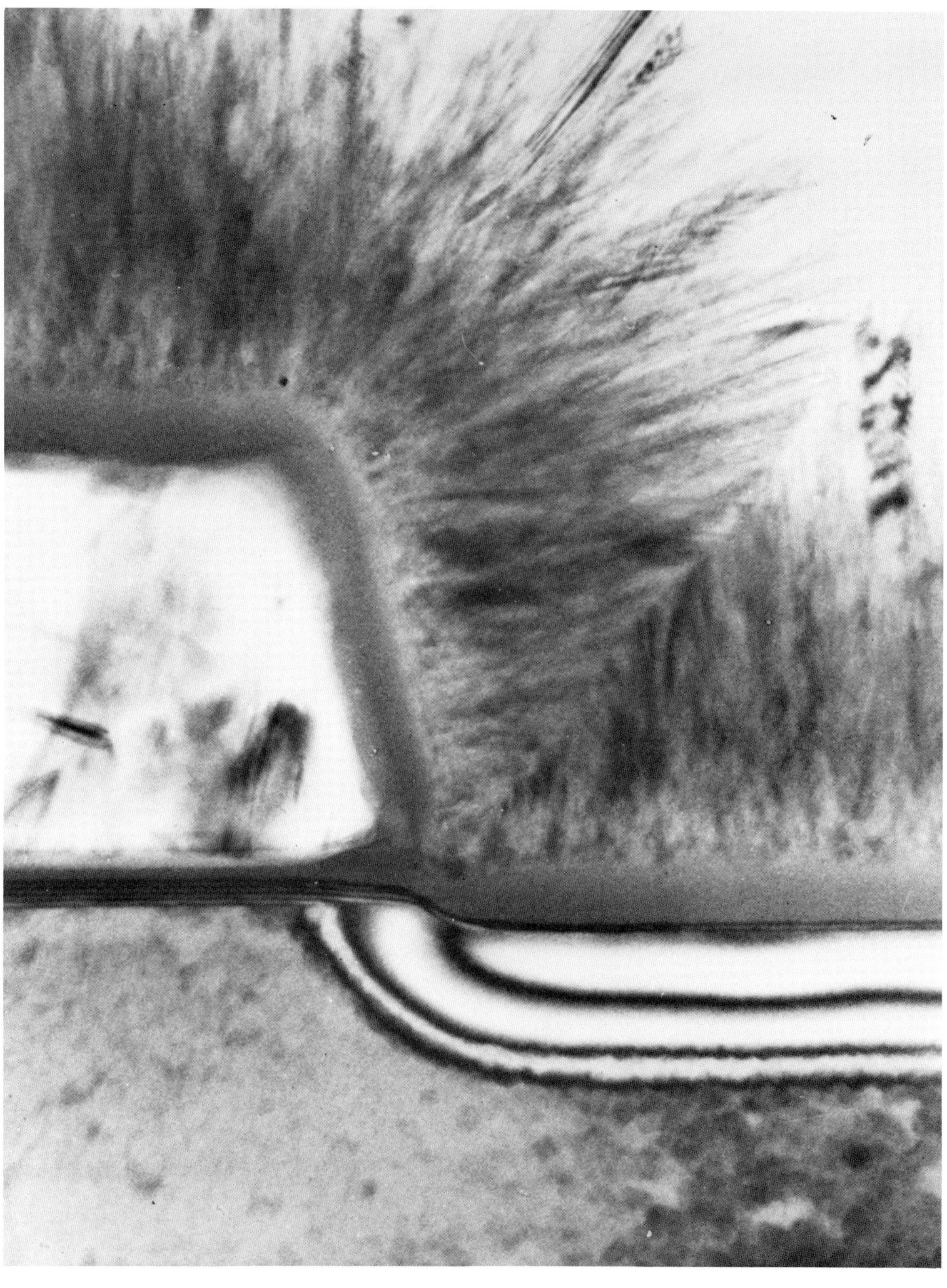

2000Å

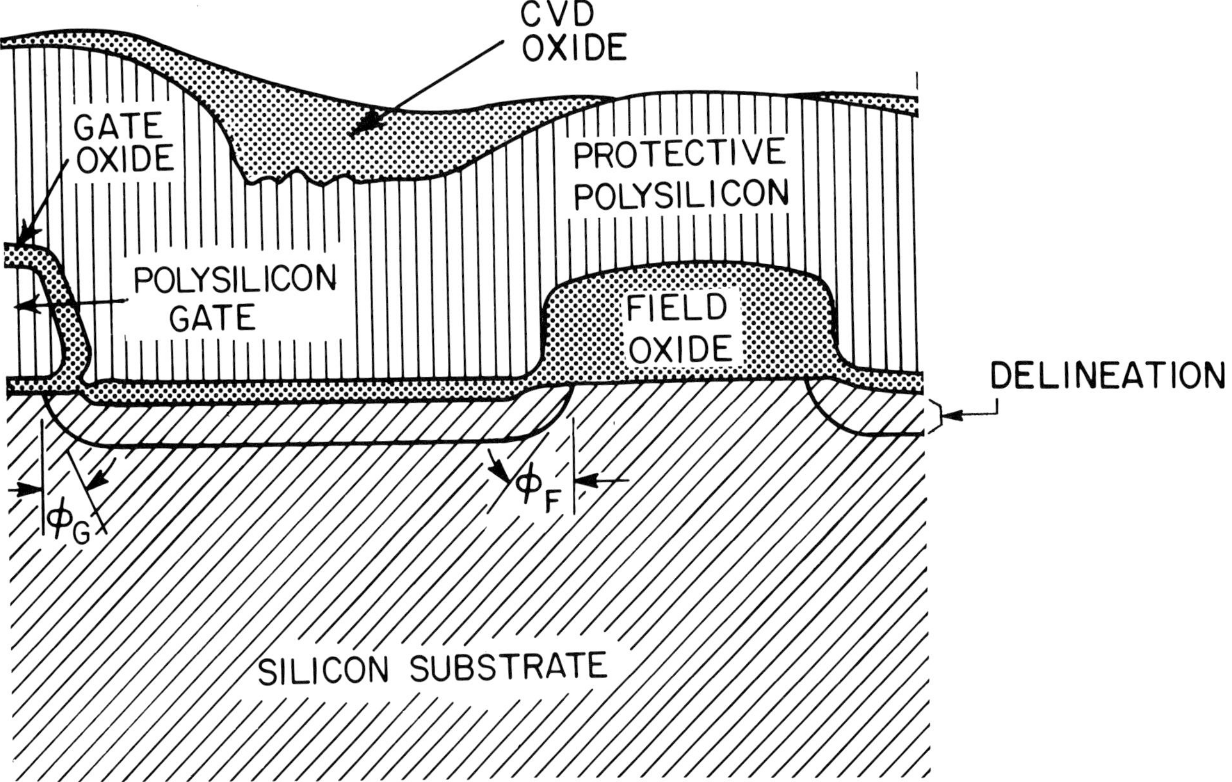

FIGURE 62. Part of a TEM test pattern showing regions at the field oxide and polysilicon gate after junction delineation. The junction curves up more gradually under the field oxide (i.e., $\Phi_F > \Phi_G$). The junctions were formed by arsenic implantation and diffusion.

local structure. In Figs. 62 and 63 the delineation line curves to the surface more gradually under a field oxide edge than under a polysilicon gate edge. This phenomenon probably results from differences in the stress distribution in these morphologically different regions.

The junction depth under phosphorus-doped polysilicon at a substrate contact is expected to be larger compared with a source–drain region because of additional doping due to phosphorus diffusion. In Fig. 64, which shows both junctions after delineation, the difference is apparent.

Junctions of the opposite type (p^+n) are more difficult to delineate. Although a solution of HF, HNO_3, and HOAC (acetic acid) in the volume ratio 1:3:8 preferentially etches p^+ material, all oxides are quickly dissolved and, therefore, the procedure is not as satisfactory as the etching of n^+p junctions.

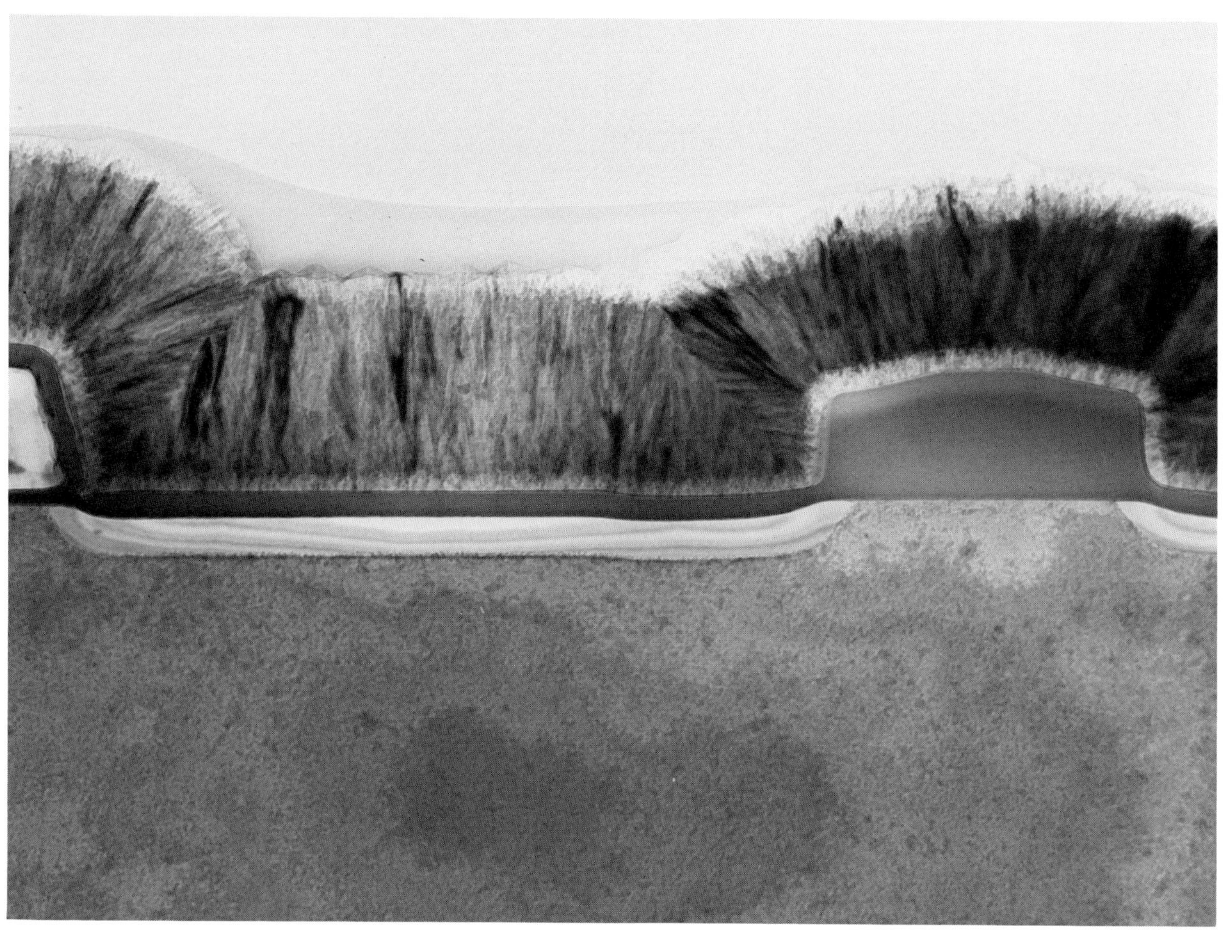
5000Å

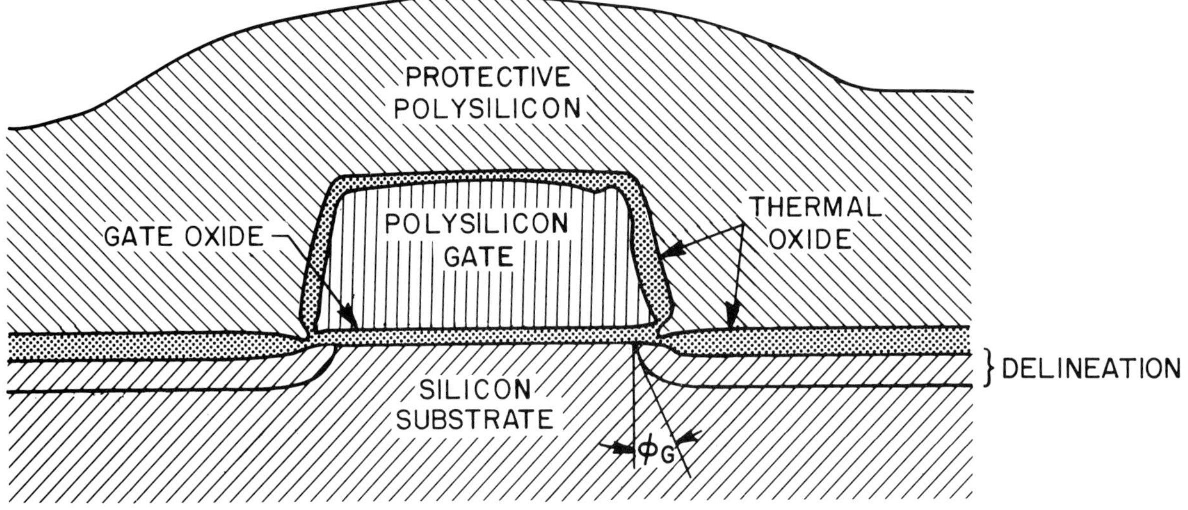

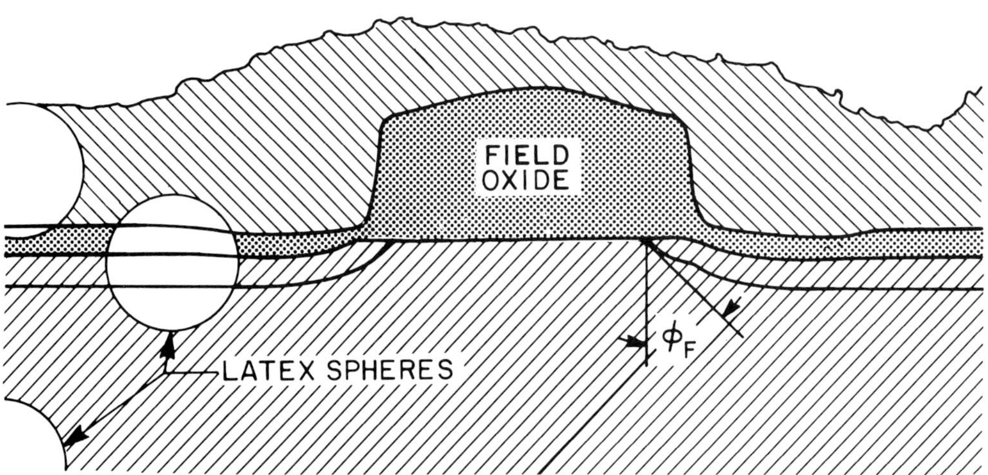

FIGURE 63. Regions near the polysilicon gate (upper) and field oxide (lower) after junction delineation. Junctions were formed by arsenic implantation and diffusion.

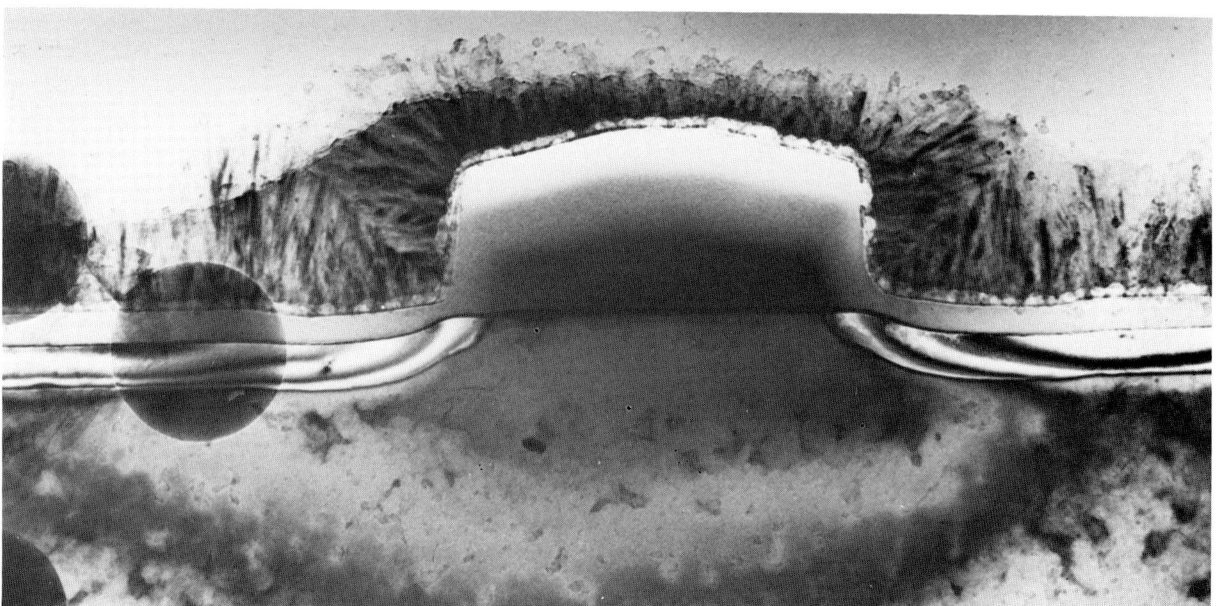

5000Å

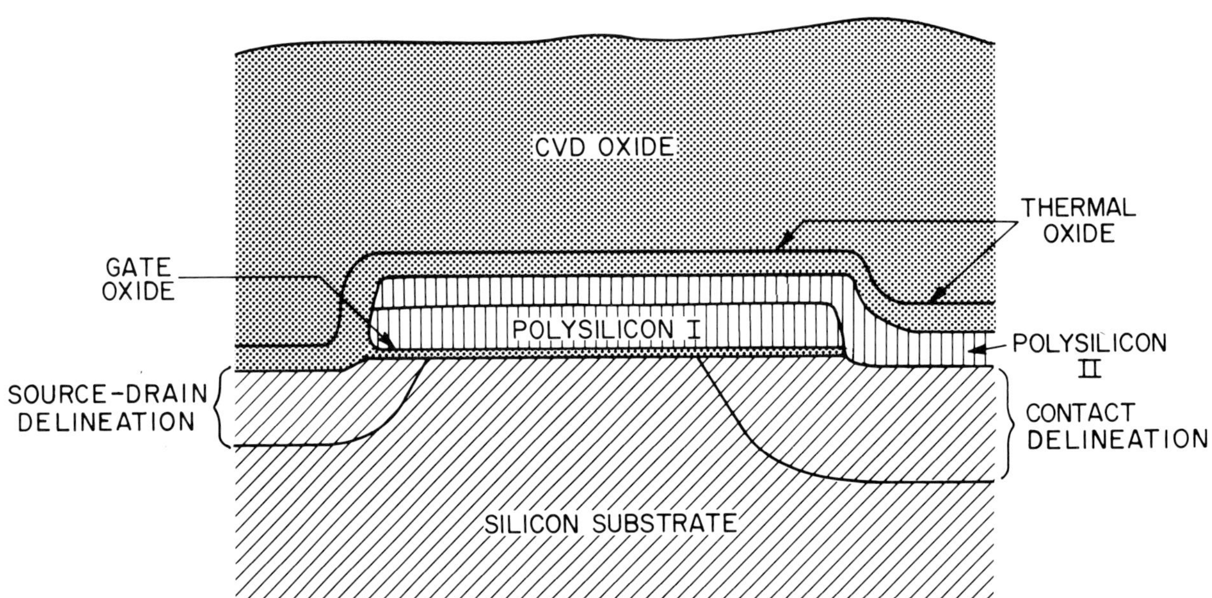

FIGURE 64. The source–drain and first level metal-to-substrate contact regions after junction delineation. The source–drain junction was formed by arsenic implantation and diffusion, and the contact region was additionally doped by phosphorus diffusing from polysilicon II.

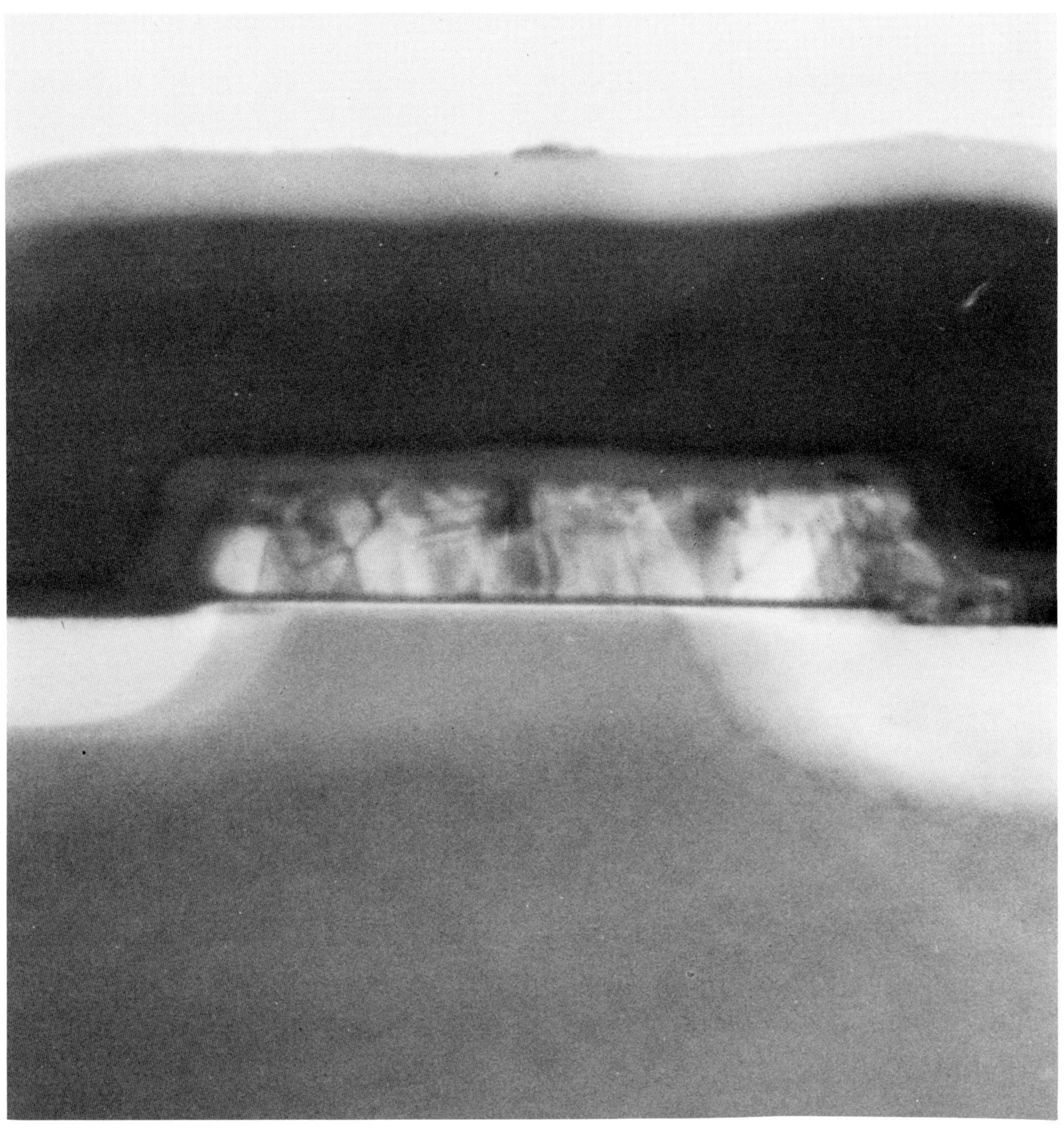

5000Å

CHAPTER SIX

IMPLANTATION DAMAGE

Ion implantation is replacing chemical methods for the creation of doped layers because of the high degree of control it offers, the cleanliness of the procedure, and the large variety of elements that can be implanted. Implantation dose and energy are tailored to device needs and processing considerations, such as the requirement to dope certain regions while masking other areas with resist, oxide, or other layers. Most implanations are performed at 30–150 keV, and dosages vary from a low of 10^{11} ions cm^{-2} to a high of 10^{16} ions cm^{-2}.

In 1954 Shockley[52] proposed implantation as a means of doping semiconductors. Shockley also predicted that radiation damage would result from implanation and that devices would have to be annealed to remove the damage. Radiation damage results in scattering centers that reduce carrier mobility and act as generation–recombination sites. As it passes into the silicon substrate and loses energy the implanted ion creates atomic displacements of the host lattice. With increasing beam current, an amorphous region is eventually created within the silicon. An early calculation of the depth distribution of the radiation damage[53] shows a Gaussian distribution, and, depending on the implanted ion, implant dose, energy, and substrate temperature, the resulting amorphous region can be completely buried within the silicon or intersect the upper surface. The lower edge of the amorphous region is expected to be about twice the projected range of the implanted ion.[54]

The depth and nature of the damage is increased, however, when implanting through an oxide or nitride layer, which is often the case in VLSI processing.

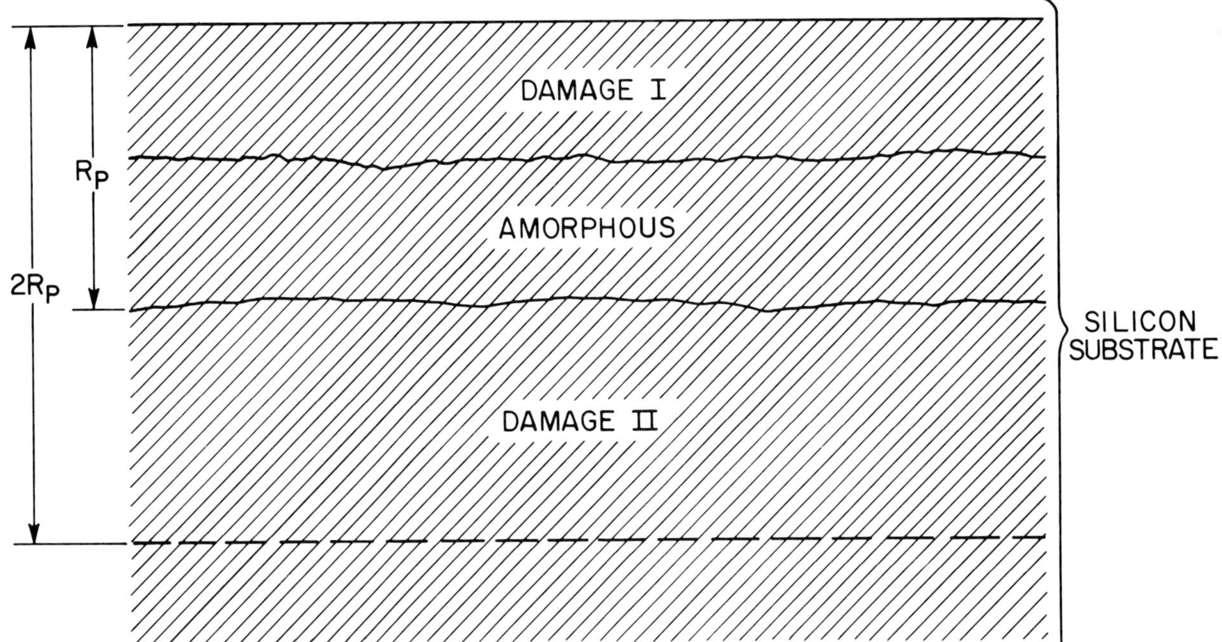

FIGURE 65. Damage caused by 200-keV argon implantation into silicon. An amorphous layer 1000 Å thick separates the upper damaged layer (damage I) from the lower damage layer (damage II). R_p and $2R_p$ indicate the projected range and twice the projected range of argon, respectively.

Oxygen or nitrogen is displaced into the silicon substrate by momentum transfer and influences the characteristics, including annealability, of the defect structure that is produced.[55,56]

Figure 65 shows the defect character for unannealed silicon implanted with 200-keV argon at a dose of 1×10^{16} ions cm^{-2}.[57] The projected range R_p and standard deviation ΔR_p are 2025 and 642 Å, respectively.[58] The upper damage region (damage I) consists of heavily microtwinned material; the lower damaged region (damage II) contains dislocation tangles, particularly dislocation loops in the lower region. The central amorphous region ends at a depth corresponding to the projected range.

1000Å

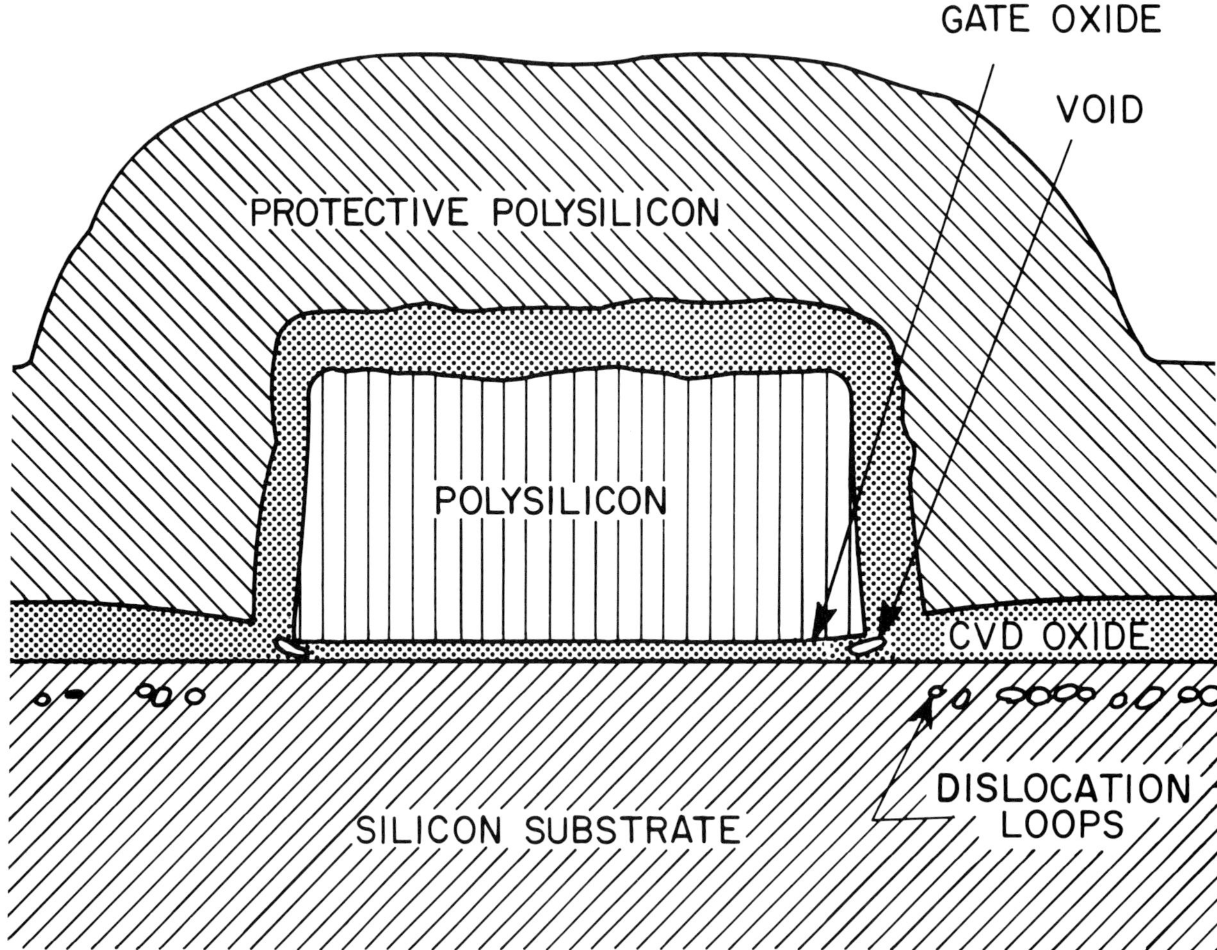

FIGURE 66. Gate region after arsenic implantation and annealing at 850°C. Dislocation loops are visible in the source–drain regions, and oxide voids are present due to undercutting of the gate and incomplete CVD oxide coverage.

Amorphous silicon can be converted into single-crystal material by high-temperature annealing, using the single-crystal substrate as a template for epitaxial regrowth. The amount of residual damage decreases with increasing annealing temperature. Figure 66 shows a source–drain region after arsenic implant at 30 keV and a dose of 7×10^{15} ions cm^{-2}, followed by 17 hr of annealing at 850°C in nitrogen. Dislocation loops are present at a depth of 500 Å, with an area density of 5×10^{10} cm^{-2} and a loop diameter of 250 Å. The depth (measured to the loop centers) is slightly greater than twice the projected range of 30 keV for arsenic in silicon (230 Å).[58] Comparing this result with the

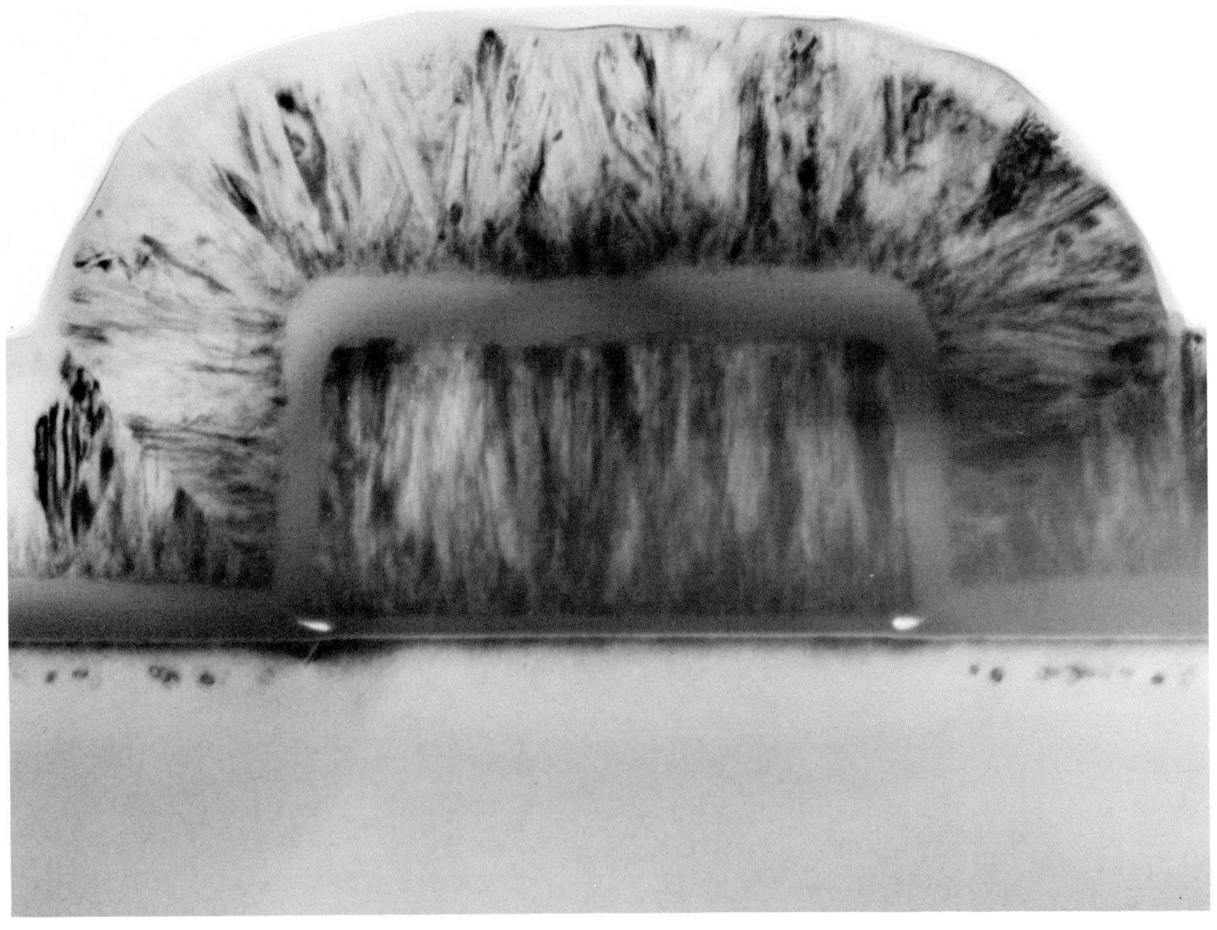
5000Å

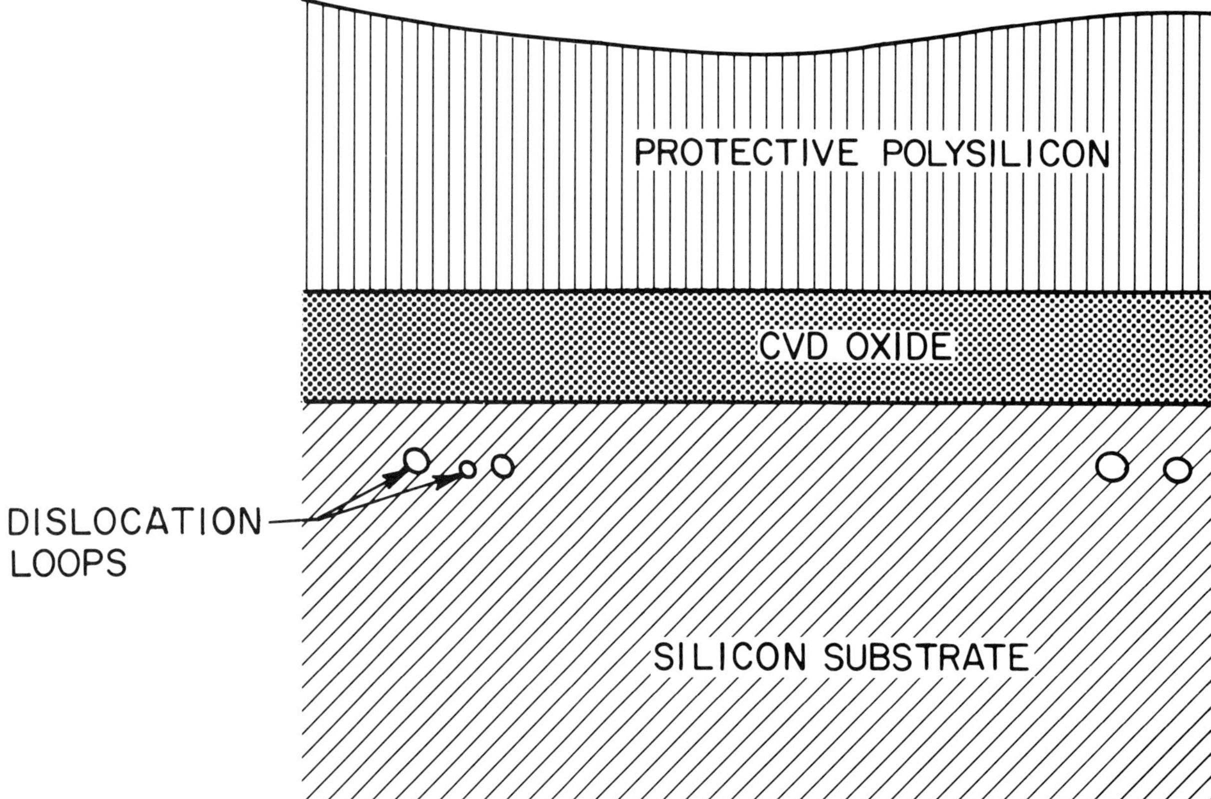

FIGURE 67. Source–drain region after arsenic implantation and annealing. The higher annealing temperature of 900°C caused more annealing of the dislocation loops than for the sample shown in Fig. 66.

defect structure left after a 900°C anneal (Fig. 67) shows that the dislocation loops are slightly deeper (~700 Å) and are present at a lower density ($\sim 4 \times 10^9$ cm^{-2}) after the higher temperature anneal.[59]

Higher temperature annealing can remove all visible damage in the silicon left by arsenic ion implantation, although other artifacts remain. Figure 68 shows a device structure made with 30 keV and 7×10^{15} ions cm^{-2} source–drain arsenic implantation. The upper surface of the field oxide, ion milled at a slower rate (during TEM sample preparation), creates an image of a dark band.

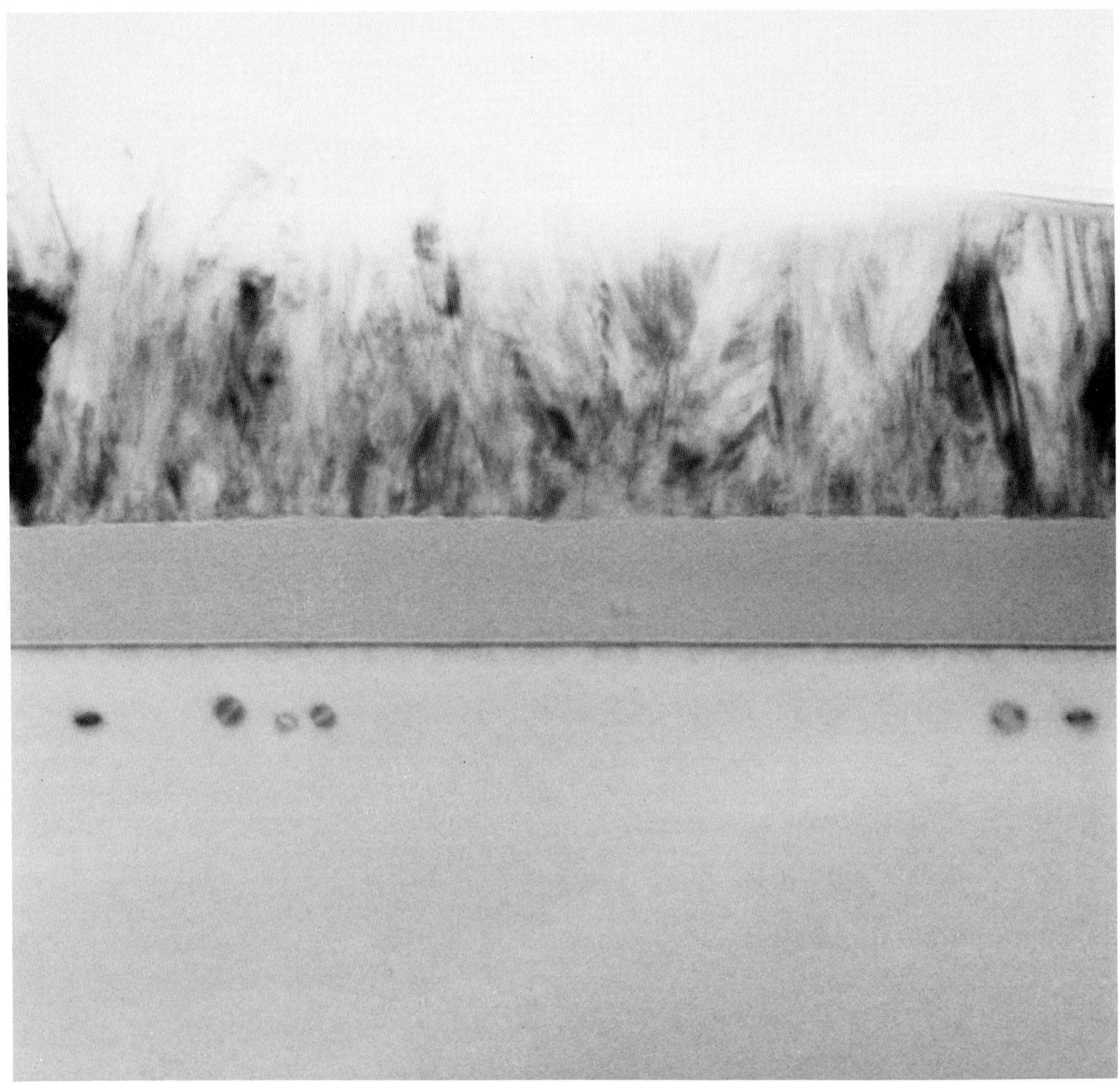
2000Å

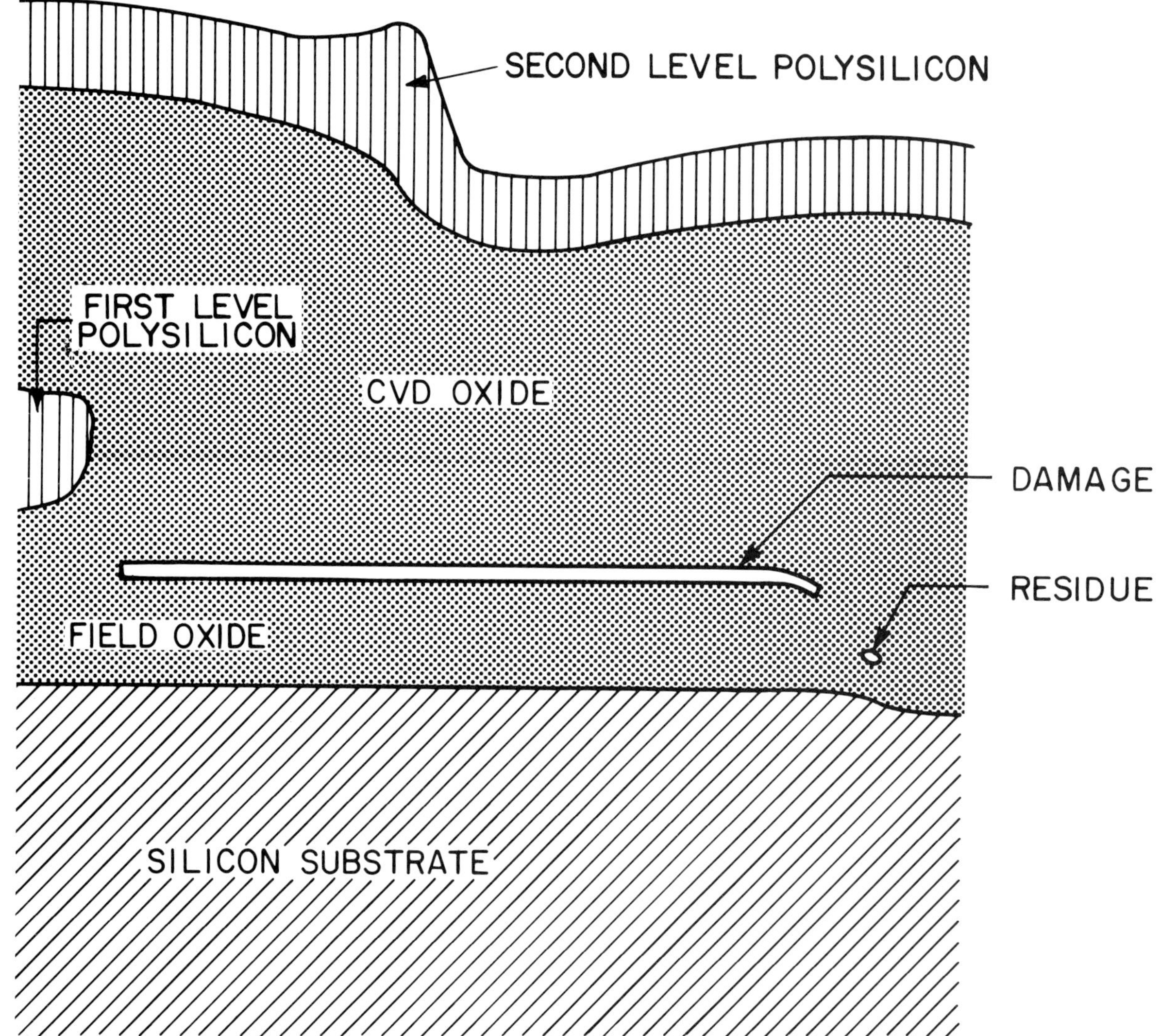

FIGURE 68. An effect of arsenic implantation on oxide properties. A precipitate containing arsenic remains at the edge of the field oxide wall, and implantation produces visible damage on the upper surface of the field oxide (also see Fig. 44).

Microanalysis by X-ray emission spectroscopy shows that this dark band, formed by implantation, contains arsenic. Figure 68 also shows the presence of a small residue of incompletely etched polysilicon caused by the extremely thick polysilicon at the edge of a step (also see Fig. 44). Microanalysis of the residue shows an arsenic content of 3.5 at.%. The arsenic was concentrated in the remaining silicon as the latter was partially consumed during subsequent oxygen anneals (20 min at 900°C in O_2, 10 min at 950°C in wet O_2, and 60 min PBr_3 getter at 900°C), producing the high concentration level that was detected.

2000Å

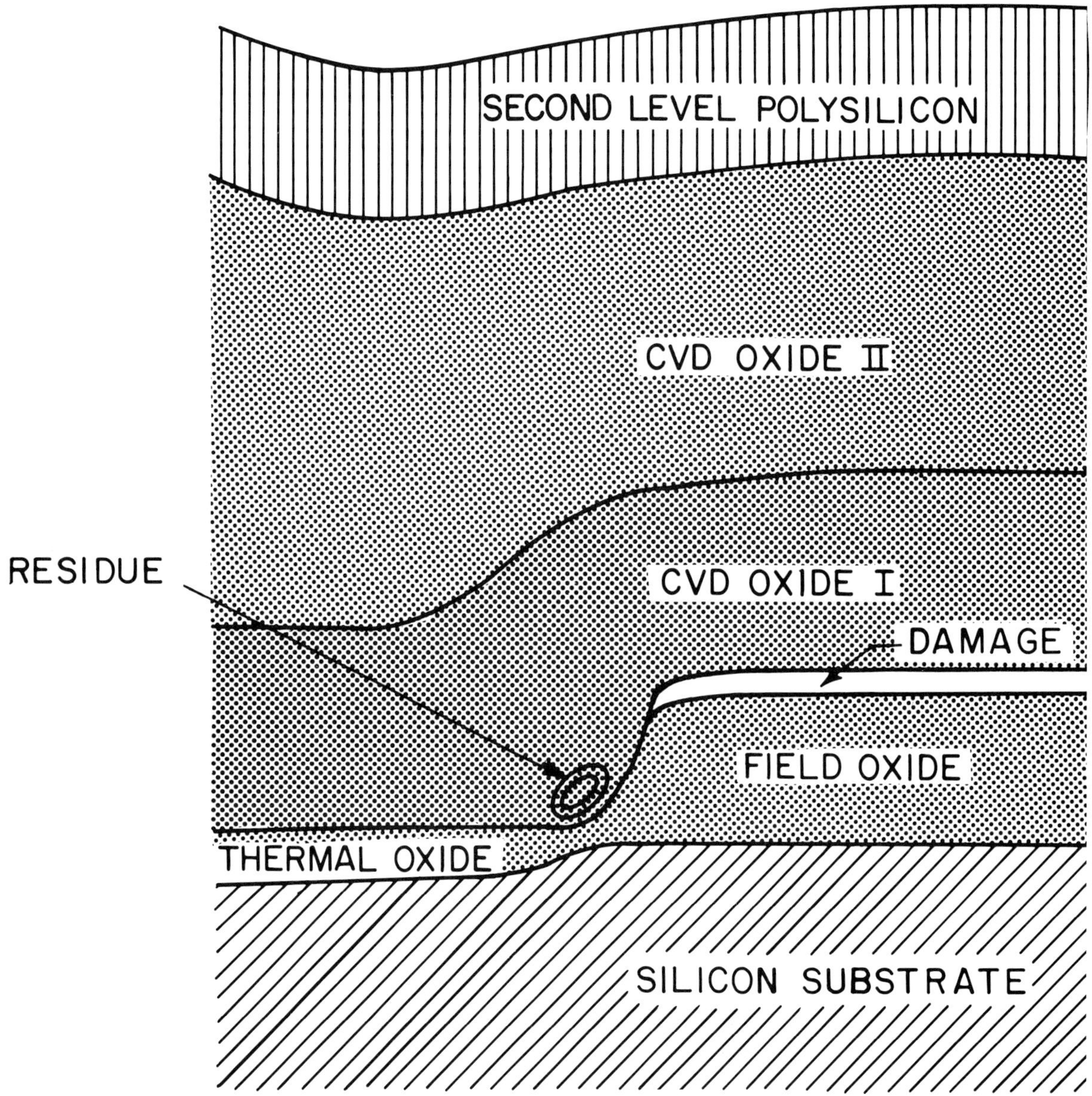

FIGURE 69. Region at the edge of the field oxide after arsenic implantation and anneal. Features are similar to those shown in Fig. 68, except that the residue is a single-crystal precipitate.

The mass of precipitate varies, depending on the slope of the edge of the field oxide wall and on the extent of reoxidation. Figure 69 shows a similar configuration for another sample similarly implanted and annealed, with 10 min at 900°C in wet O_2 replacing the 950°C treatment used for the sample in Fig. 68. The precipitate, a single crystal of heavily arsenic-doped silicon, is larger. The dark band of arsenic-rich oxide is 315 Å thick, and the lower edge is at a depth corresponding to twice the projected range of 30-keV arsenic. Mi-

3000Å

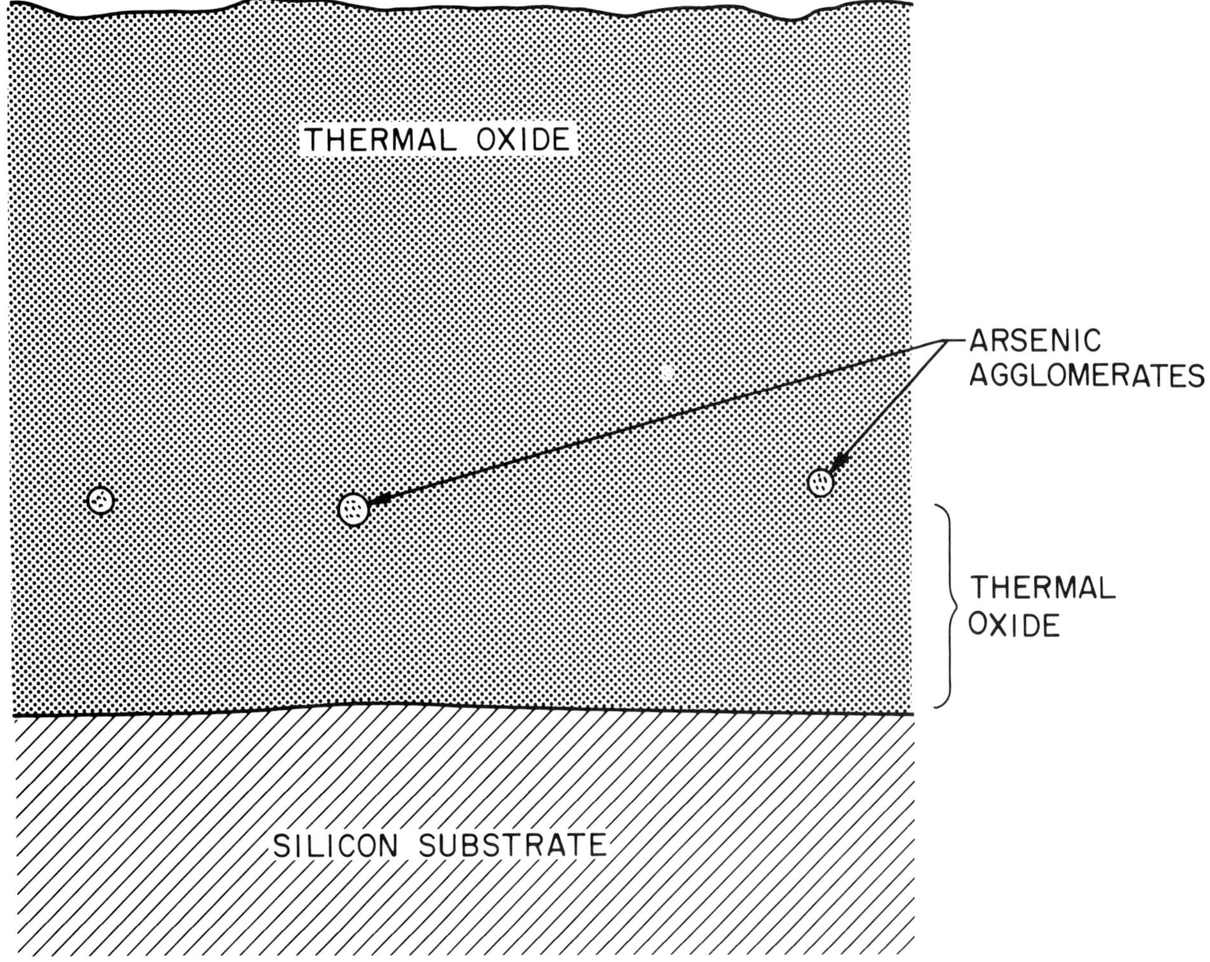

FIGURE 70. Arsenic-implanted polysilicon after complete oxidation of polysilicon. Arsenic has accumulated into agglomerates that are left buried within the oxide.

croanalysis shows that the band region contains 1.9 at.% arsenic. Since the concentration of arsenic would be 10 at.% if all implanted arsenic appeared within a 315-Å deep region, most of the arsenic implantation must have diffused deeper or volatilized during anneal.

Arsenic preferentially segregates in silicon during oxidation,[60] and completely oxidized, arsenic-doped silicon shows evidence of this segregation. Figure 70 shows a sample after complete oxidation of 2500-Å arsenic-implanted polysilicon in wet O_2 at 900°C. Implantation was performed at 30 keV at a dose of 1×10^{15} ions cm^{-2}. The polysilicon is on 1000-Å SiO_2 (additional oxide was grown from the substrate silicon following complete oxidation of the polysilicon), and the arsenic precipitate in Fig. 70 corresponds to the position of the original lower polysilicon/SiO_2 interface.[61]

2000Å

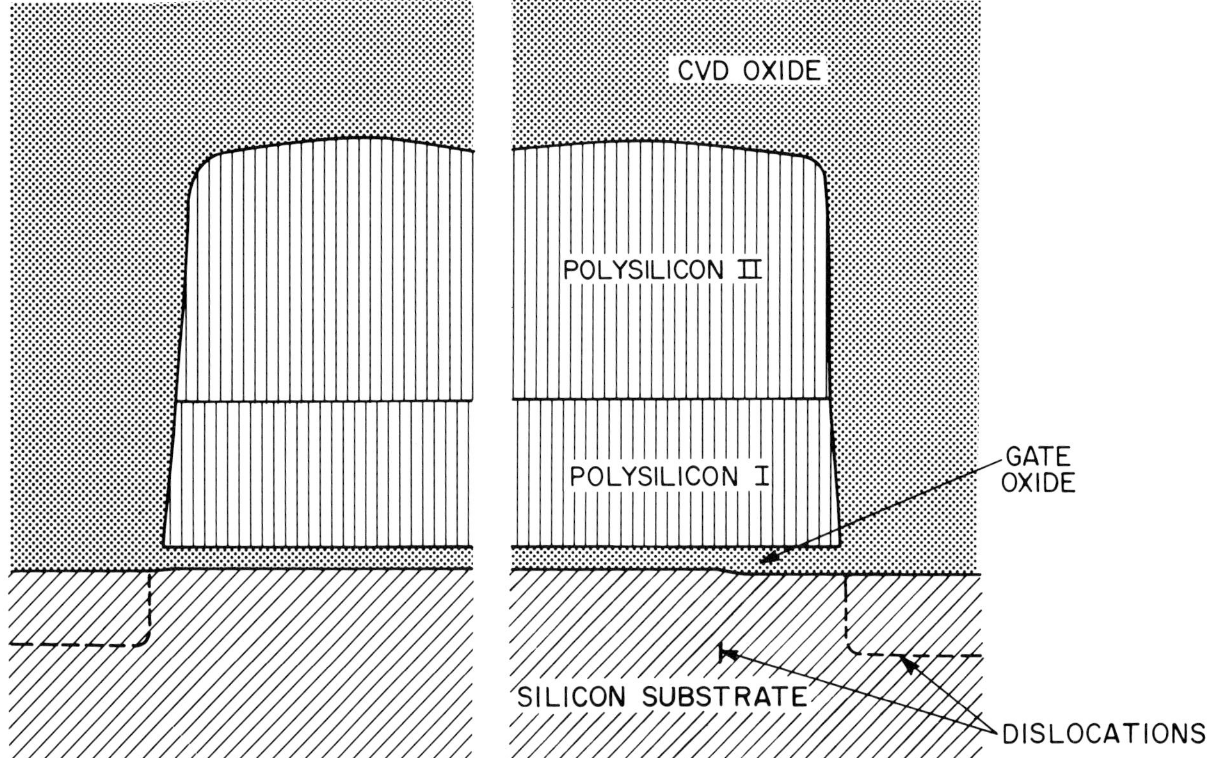

FIGURE 71. Region at the edge of a gate after boron implantation and annealing. Dislocation damage is visible in the source–drain area, and a few dislocations are present a short distance under the gate.

p-Channel source–drain regions in CMOS structures can be made by boron implantation. Since light elements, such as boron, are implanted much deeper than heavier elements of the same energy, ionic compounds of the dopant can be implanted to decrease the implantation depth without having to use an inconveniently low implant energy. Figure 71 shows the gate edge of two samples implanted[59] with 50-keV BF_2 at a dose of 2×10^{15} ions cm^{-2}. The effective implant energy varies with the mass, and the 11.2-keV effective boron energy corresponds to a projected range in silicon of 365 ± 182 Å. After a 10-min anneal in nitrogen at 900°C, a dislocation network appears at a depth of 780 Å, which is slightly more than twice the projected range. It is uncertain whether the few dislocations found at a short distance from the gate edge are related to the implantation. Using a higher energy for BF_2 implant places the amorphous band deeper within the substrate and, after annealing, two bands of defects form (Fig. 72). The sample in Fig. 72 was implanted at 135 keV and a dose of 5×10^{15} ions cm^{-2}, and was annealed at 800°C for 15 min in nitrogen.[62]

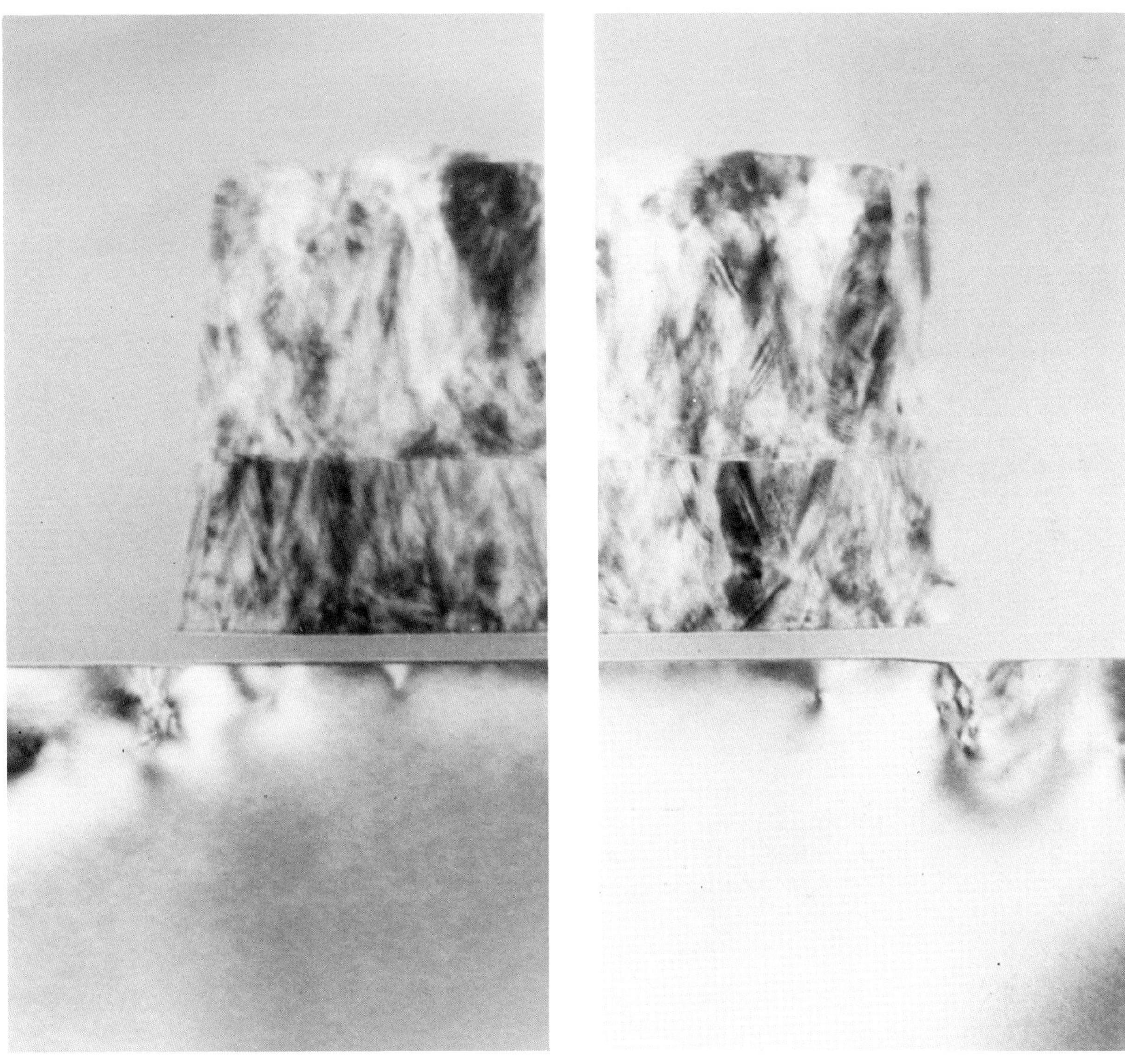

2000Å

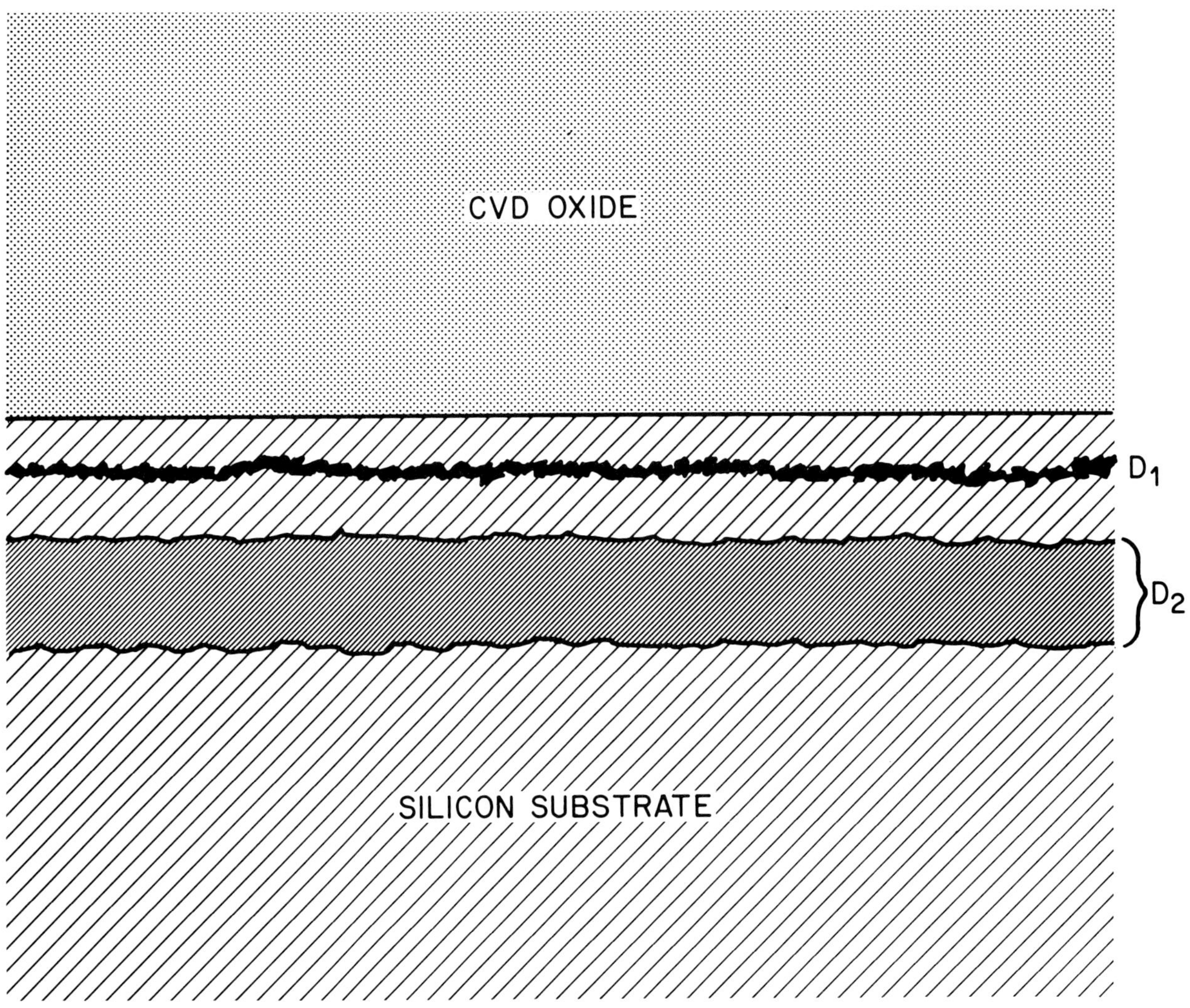

FIGURE 72. Damage left after BF_2 implantation and annealing. Two distinct bands of defects are present: a narrow band 670 Å below the surface D_1 and a broad band that begins 1200 Å below the surface D_2.

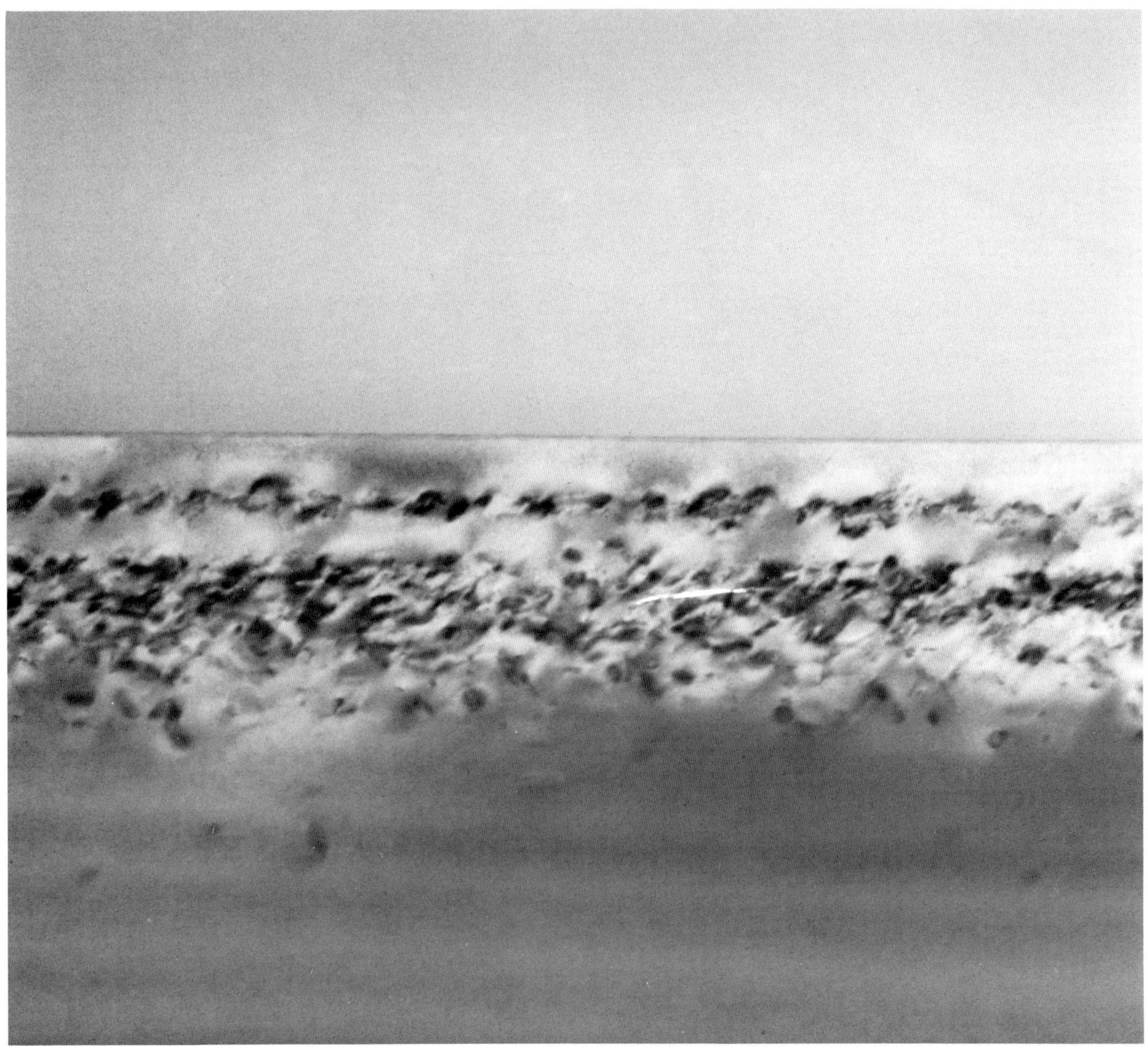

2000Å

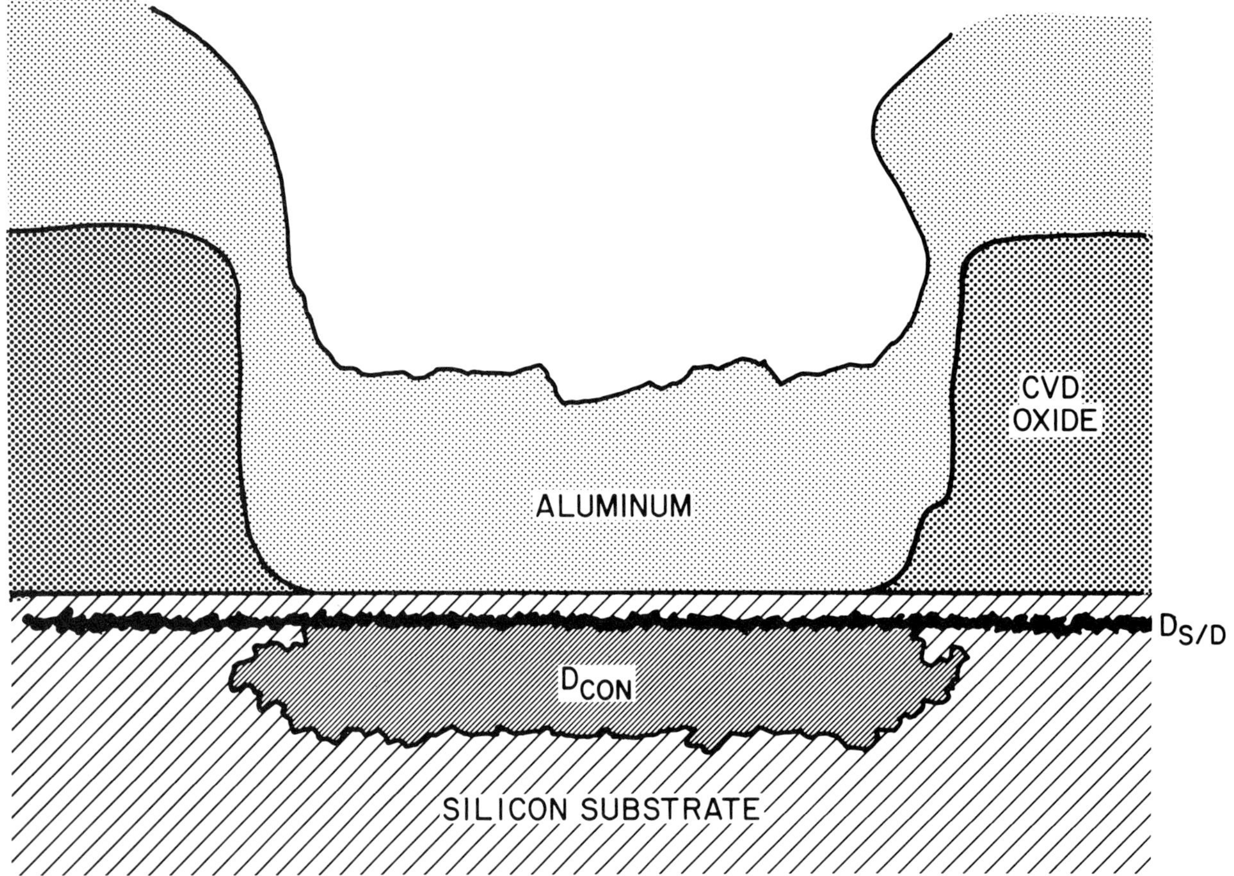

FIGURE 73. Region at an aluminum–substrate contact showing the damage left by boron implantation. A dislocation array $D_{S/D}$ is present in the source–drain region, and a deeper dislocation array D_{CON} formed by three successive implants is present in the contact area.

Aluminum contacts with the silicon substrate are subject to problems of electrotransport (Section 4.3). It was recognized that locally degrading the crystal perfection of silicon in contact window sites during p^+ doping of the contact area may eliminate the creation of silicon pits during alloy formation. Figure 73 shows the effect[62] of a triple implant of boron at 50 keV at 1×10^{15} ions cm^{-2}, 100 keV at 2×10^{15} ions cm^{-2}, and 160 keV at 2×10^{15} ions cm^{-2}, followed by 10 min in nitrogen at 900°C. No evidence is found for aluminum-silicon alloy formation at contact regions. The dislocation network in the contact window extends to a mean depth of 5670 Å with some defects reaching a depth of 8160 Å. The bulk of the damage extends to a depth considerably less than twice the projected range of 4432 Å, which differs from the other results reported in this chapter.

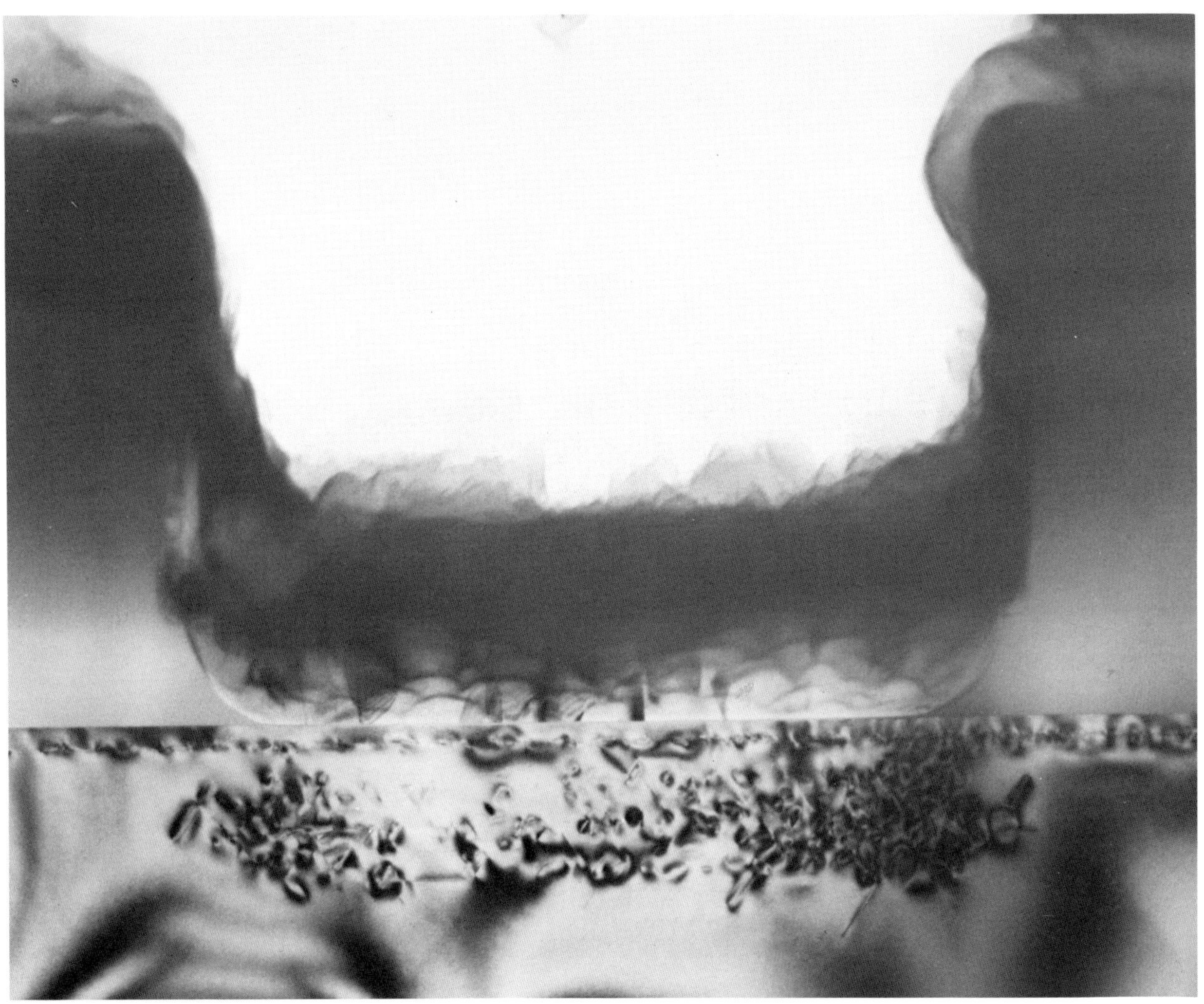
1.0 μm

CHAPTER SEVEN

COMPLETE DEVICES

7.1. INTRODUCTION

The micrographs presented in the previous chapters show specific features of various processing steps (oxidation, metallization, junction formation, ion implantation), emphasizing those features that demonstrate new information, problems, or device failure modes relevant to a particular processing step. The micrographs in this chapter are of complete devices. All morphological features previously described are presented here in context of the complete device structure. The devices presented are *n*-channel MOS and CCD structures and include one- and two-metallization level structures consisting of various combinations of aluminum, polysilicon, and silicide.

7.2 GATE REGIONS

Figures 74 through 76 show micrographs of gate regions of completed devices. The aluminum/polysilicon layers for second level metallization (Fig. 74) maintain high conductivity through narrow lines. First level metallization uses two polysilicon films, according to a processing sequence that requires polysilicon I after gate oxide growth for protection of the active region against contamination during the first level, metal–substrate contact lithography step (Section 1.4). After the windows are opened, polysilicon II is deposited, making contact to the substrate at these window sites.

Planarization methods can prevent the type of severe step edges shown in Fig. 75. Successful planarization requires that the CVD intermediate oxide has

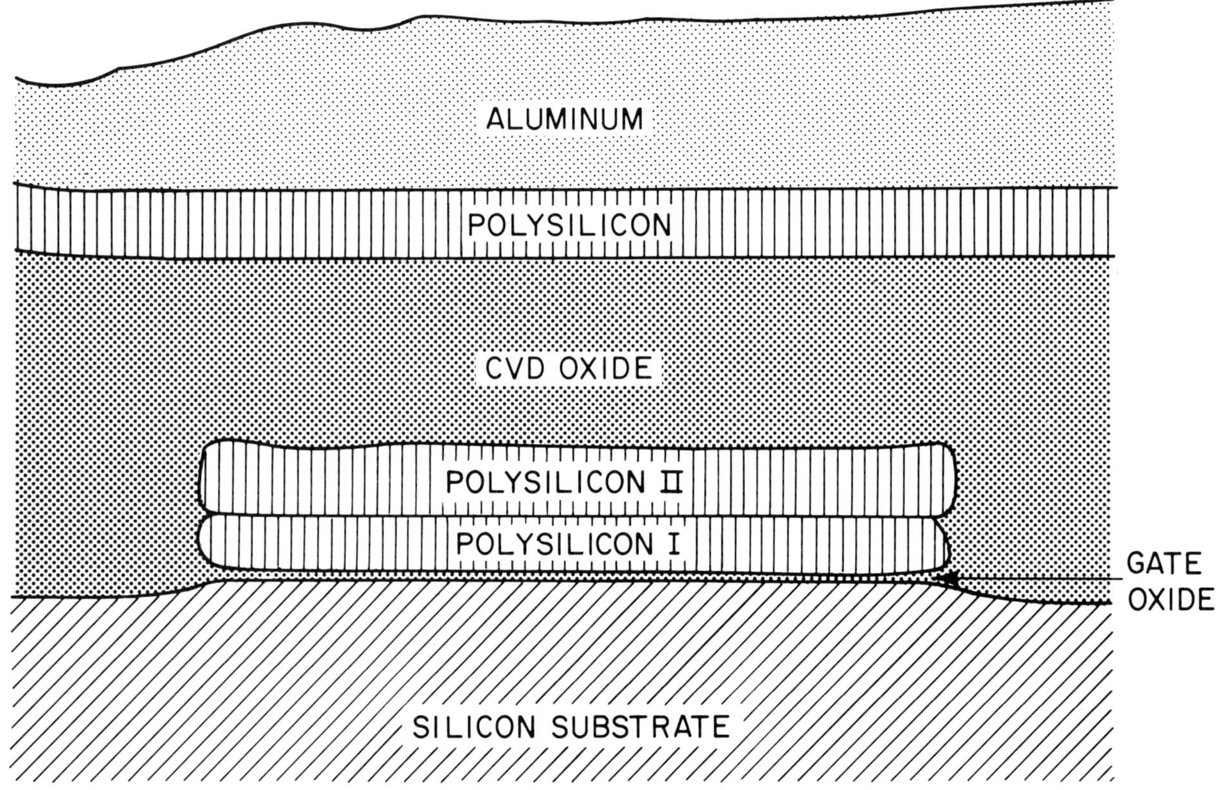

FIGURE 74. Gate region of a complete device.

an etch rate that is isotropic with depth, and that the etch rate be comparable to the etch rate of the polymer used for planarization. Variations in phosphorus content, caused by interruptions during CVD oxide deposition, produced corresponding changes in etch rate (Fig. 76). Differences in ion milling rates of thermal and CVD oxide produce a delineation of these oxides, also shown in Fig. 76.

Because aluminum ion mills at a significantly slower rate than silicon or oxide, in some cases aluminum was chemically thinned (Fig. 75) and in other cases aluminum was completely etched off (Fig. 76) prior to ion milling.

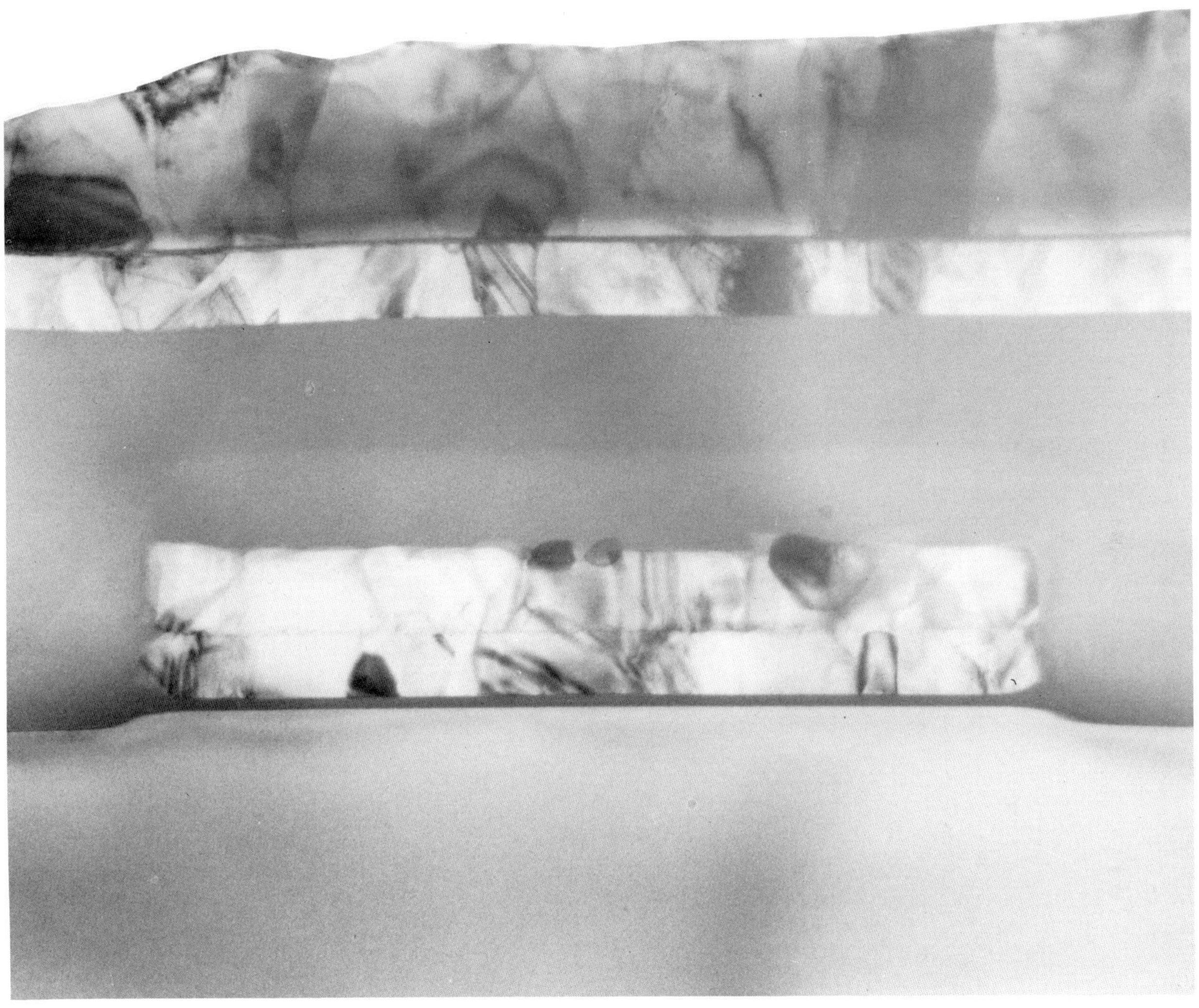

5000Å

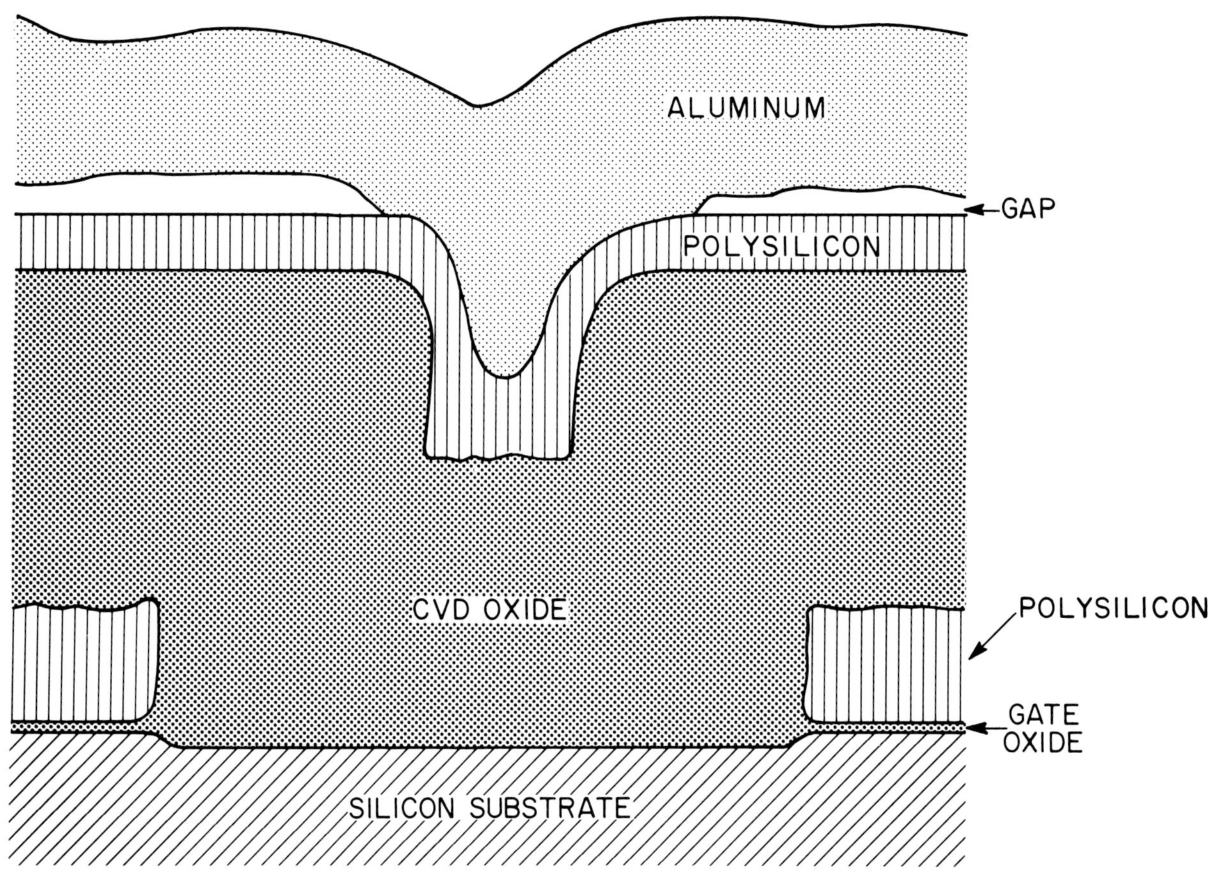

FIGURE 75. Gate region of a complete device containing partially etched aluminum. Planarization of the CVD oxide was not used as indicated by the severe dip in second level polysilicon.

5000Å

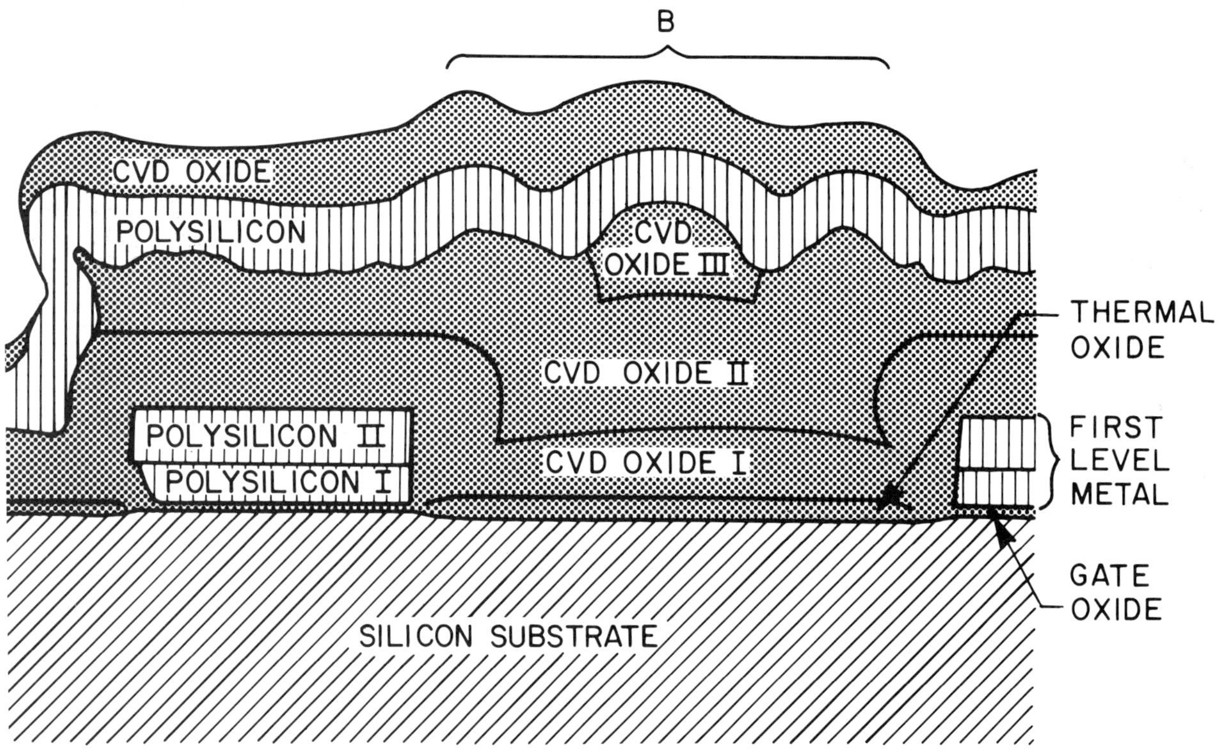

FIGURE 76. Gate region of a complete device; aluminum has been removed. Variations in etch rate of the three oxide layers during planarization cause the bumpy slope of the upper surface. These layers and the regrown oxide layer are clearly delineated.

7.3 FIRST LEVEL METAL–SUBSTRATE CONTACTS

Figures 77 through 79 are micrographs of first level metallization contacts to the substrate. Very thin (20 Å) oxide remaining on the silicon surface at the window prior to polysilicon II deposition has negligible effect on contact resistance, and contact resistance is usually not a problem. When the interface is completely free of oxide, though, intimate metallurgical contact can occur, and subsequent thermal treatment (950°C PBr_3 doping of polysilicon or thermal drive following source–drain implant) can cause epitaxial recrystallization of the polysilicon (Figs. 77–79). Figures 77 and 78 show the grain boundaries and the epitaxial regrowth. In many cases, regrowth extends laterally over the gate oxide for a short distance converting part of a polysilicon II/polysilicon I layer to a single grain.

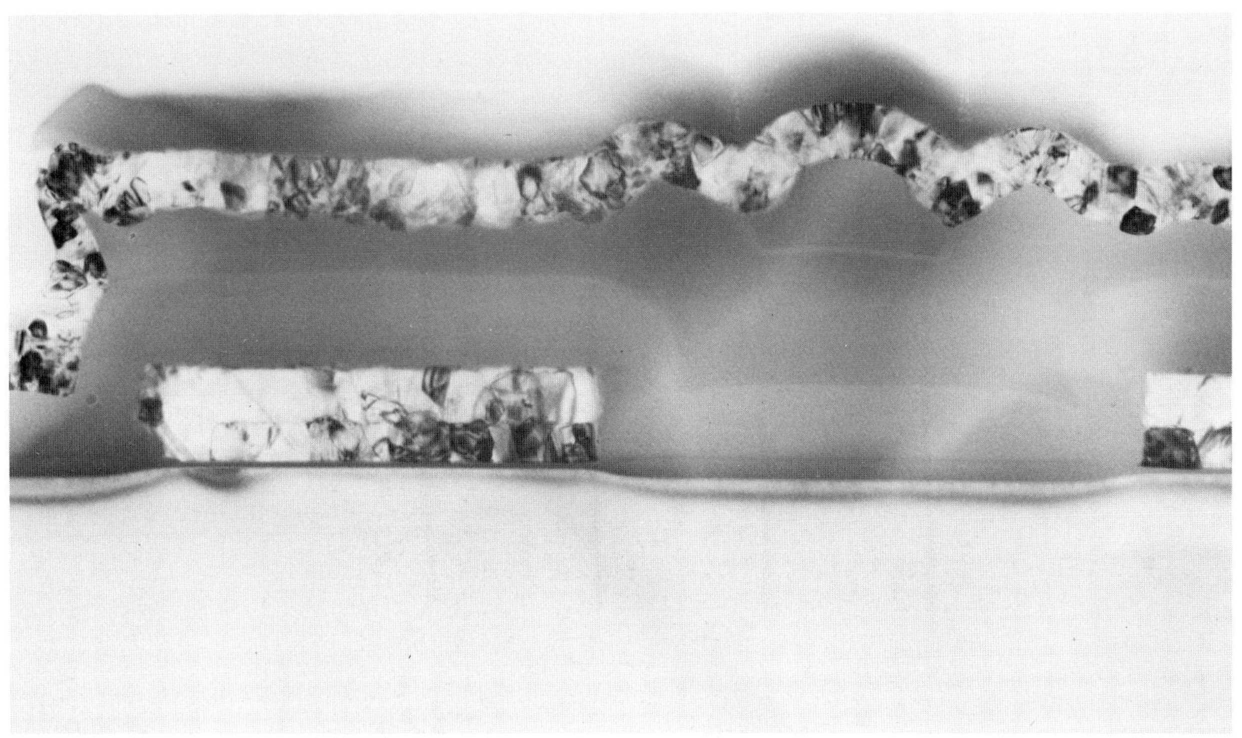
1.0 μm

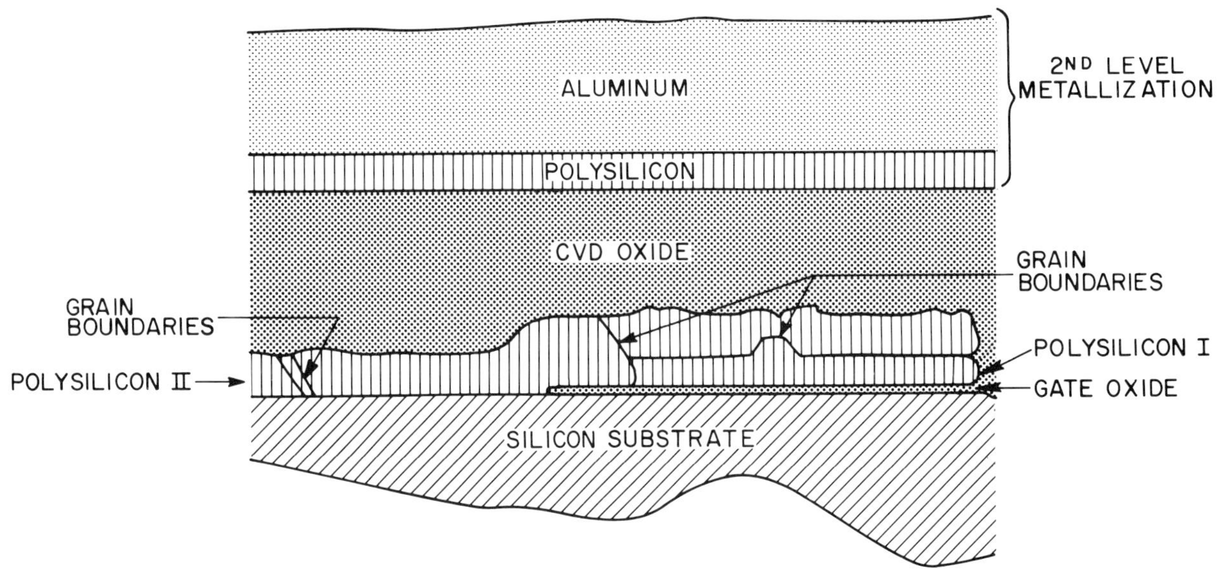

FIGURE 77. A complete device showing polysilicon II contact with the substrate. Epitaxial regrowth forms large grains; grain boundaries in the regrown region are indicated. Epitaxial regrowth also propagates laterally over the gate oxide.

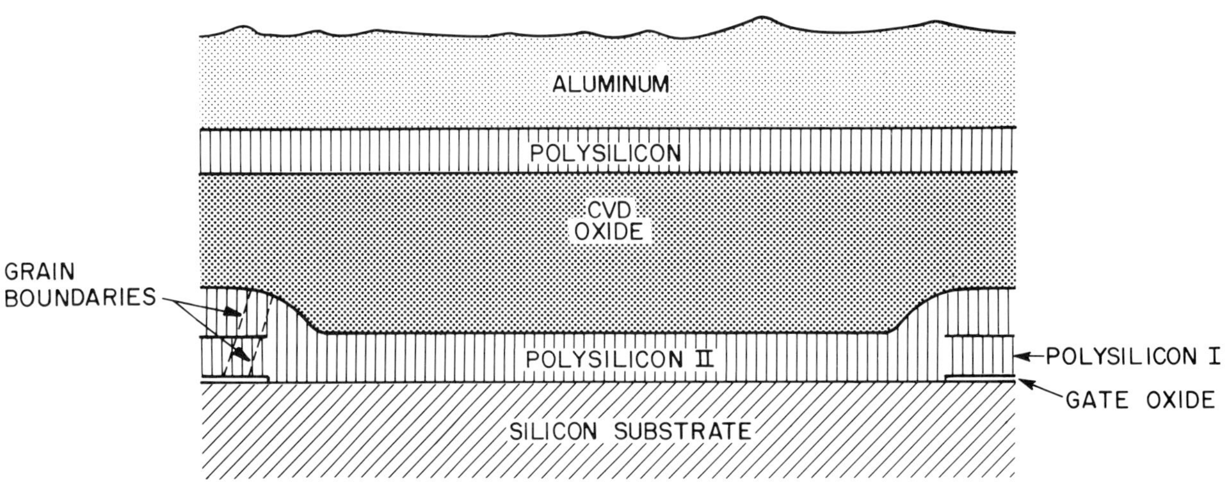

FIGURE 78. A complete device showing epitaxial regrowth of polysilicon II over the substrate, which extends laterally over the gate oxide where grain boundaries intersect both polysilicon layers.

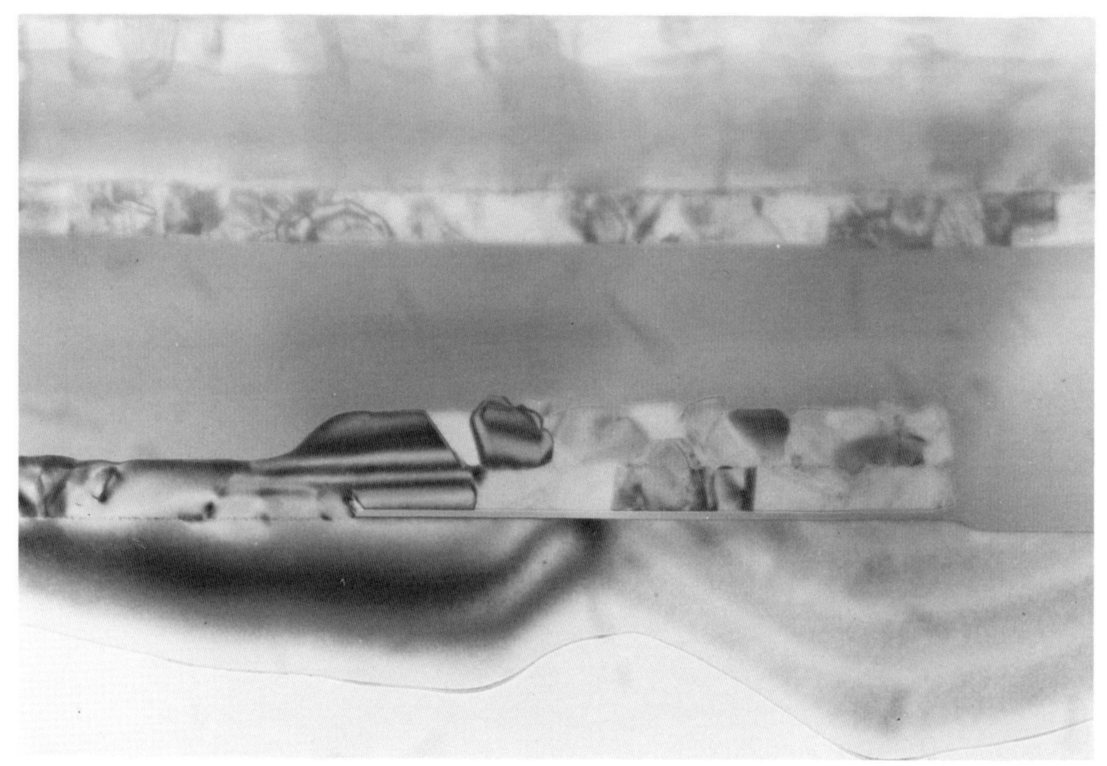
1.0μm

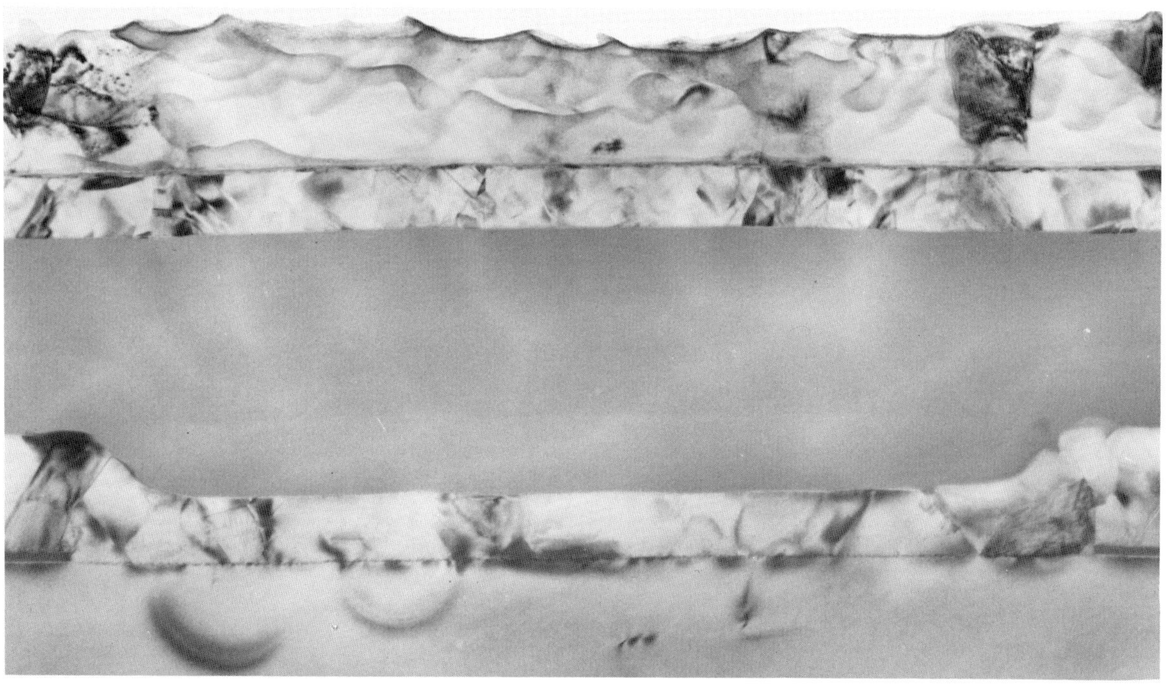
1.0μm

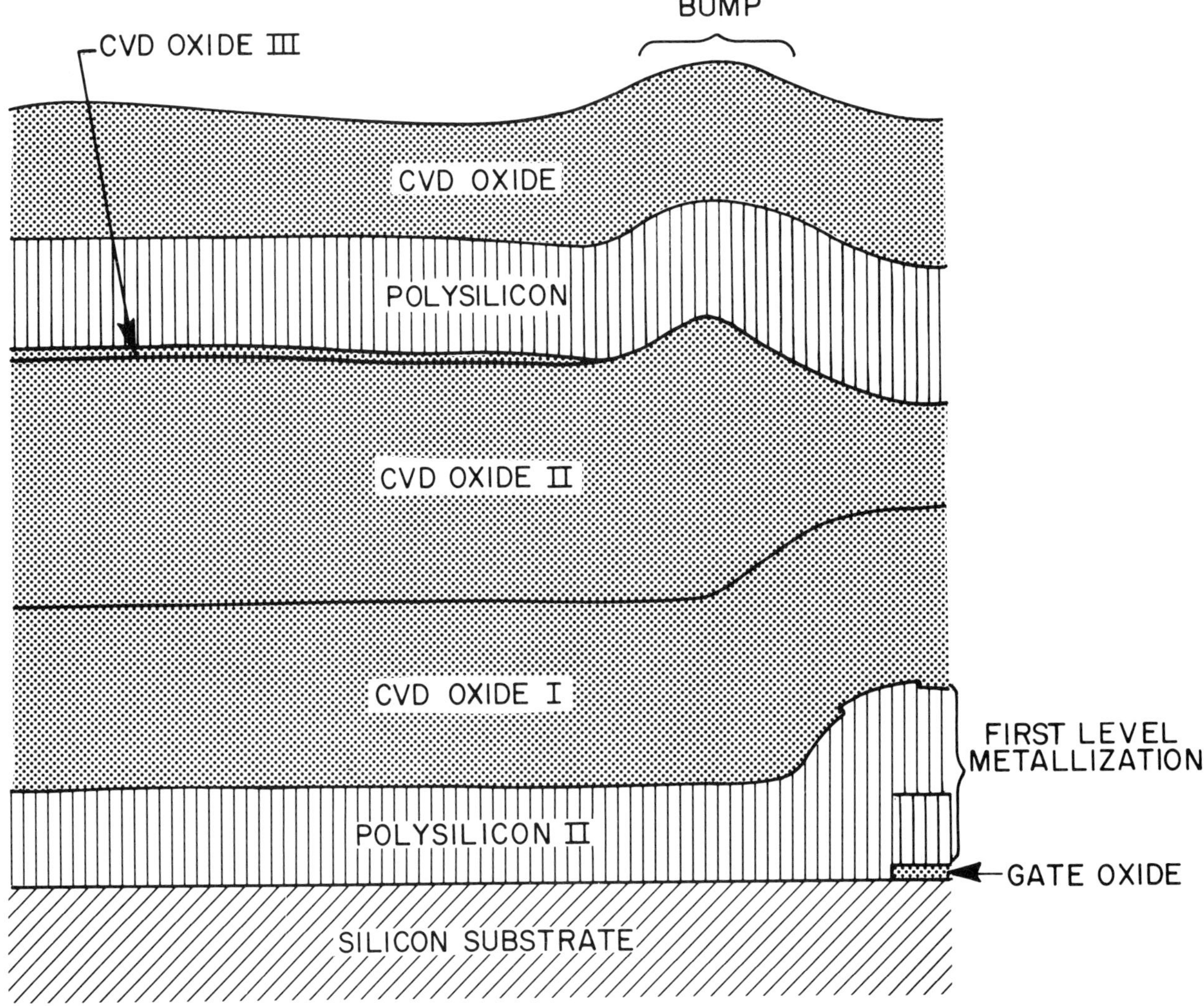

FIGURE 79. A complete device showing epitaxial regrowth of polysilicon II over the substrate. The bump is caused by variation in the etch rates of the CVD oxide layers.

Figure 79 shows a region with a polysilicon contact to the substrate for the same device shown in Fig. 76. Variations in the phosphorus content of the CVD oxide caused the peculiar shape of the top-level metallization (polysilicon only, the aluminum was removed) at the edge of the first level metallization. CVD oxide II, which has a higher phosphorus content, etches faster than CVD oxide III, and a step is created at the surface when CVD oxide II is reached during the planarization etch.

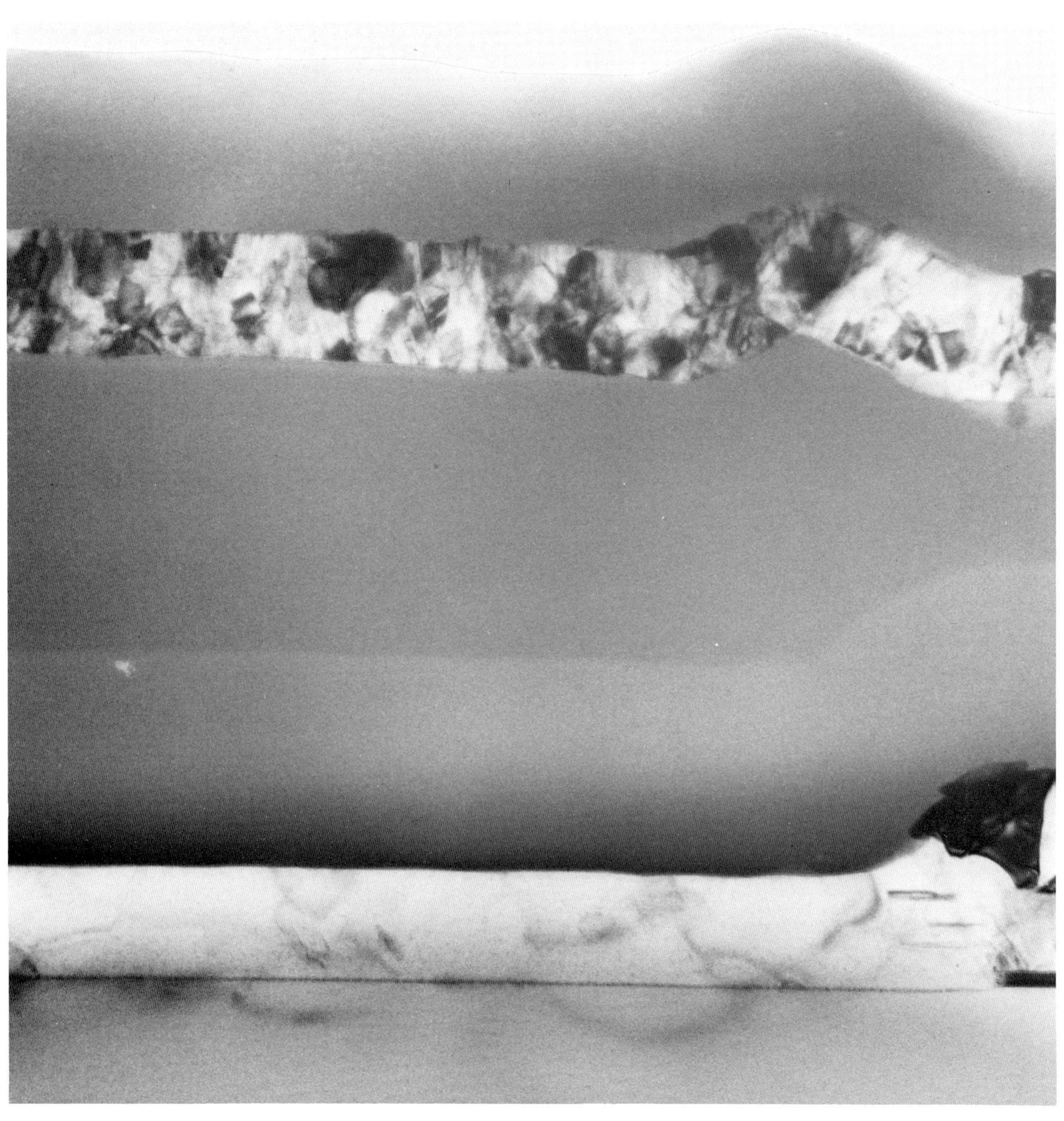
5000Å

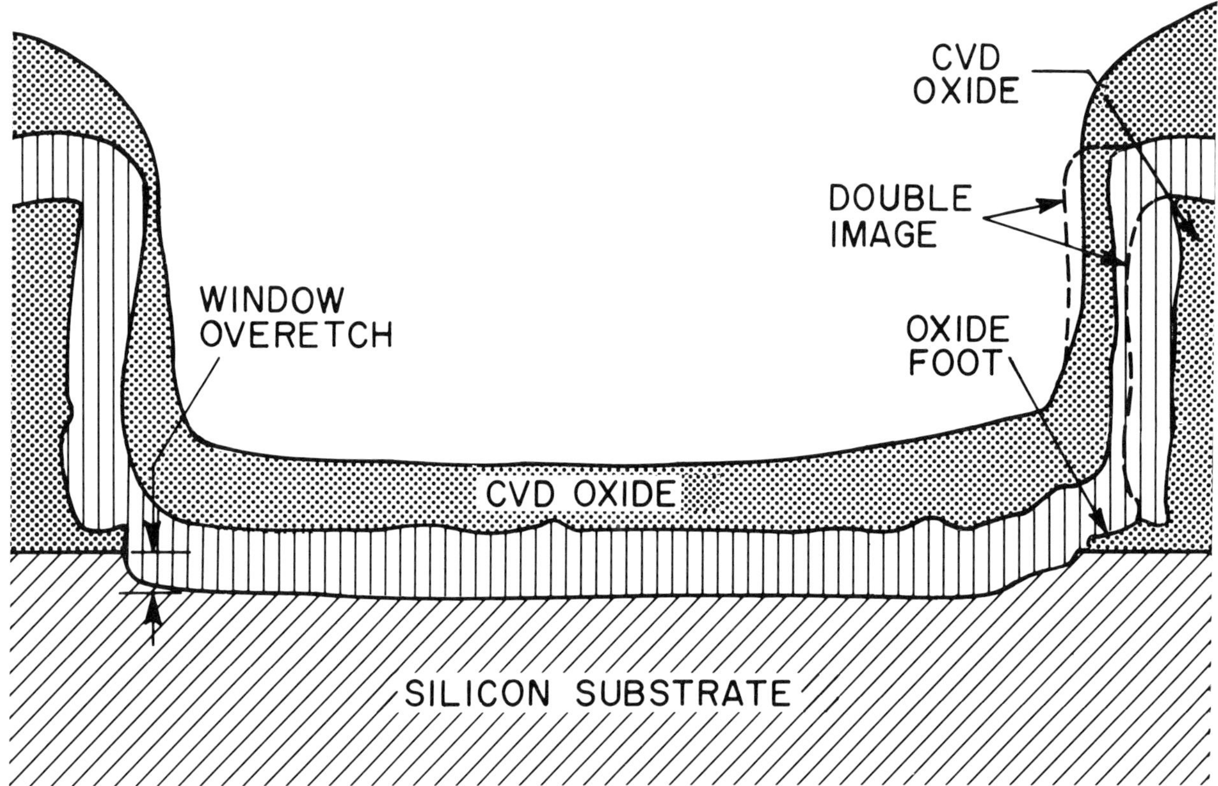

FIGURE 80. A complete device showing second level metallization contact with the substrate. The window was overetched by 2000 Å. A slight chemical etch after reactive ion etching of the windows produces a differential etching of the CVD and thermal oxides leaving a "foot" of oxide. The double image of the vertical polysilicon wall was caused by local undulations of the edge of the oxide due to undulations in the resist edge.

7.4 SECOND LEVEL METAL–SUBSTRATE CONTACTS

Figures 80 through 83 show complete devices with second level metallization contact to the substrate. These figures show different degrees of window overetching; Fig. 82 shows the most severe case, and Fig. 81 shows a nearly ideal case. Prior to metallization deposition, reactive ion etching at the window definition step is sometimes followed by a chemical BHF cleaning step. Chemical etch rates vary with the type of oxide, and phosphorus-doped CVD oxide, which etches laterally at a fast rate, leaves an oxide step at the slower etching, thermal oxide layer (Fig. 80). Figure 80 also shows a double image of the right vertical polysilicon wall but not of the left wall. Undulations in the edge of the resist that define the window cause this double image; it is not an artifact of TEM study, such as might be caused by sample curling.

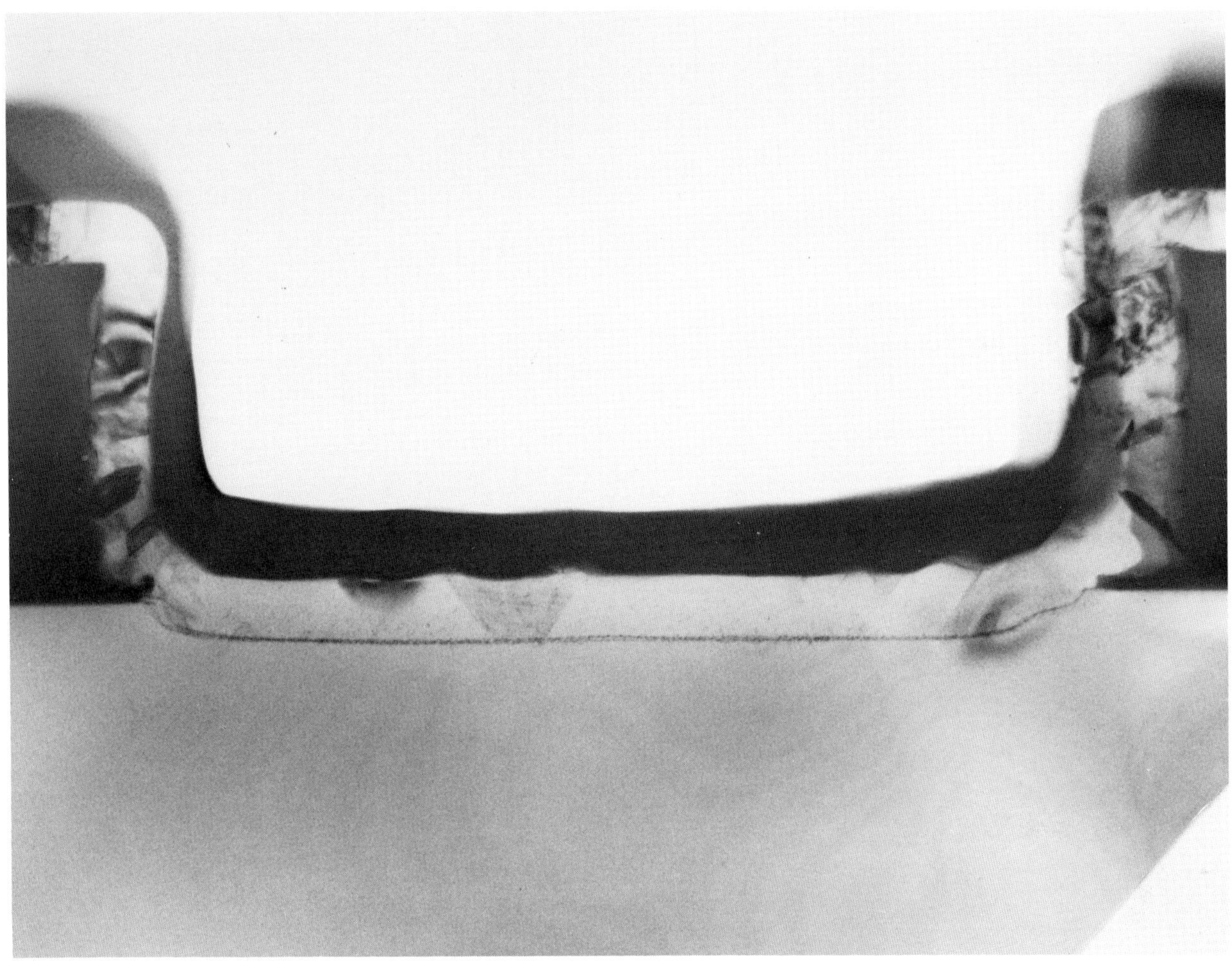
1.0 μm

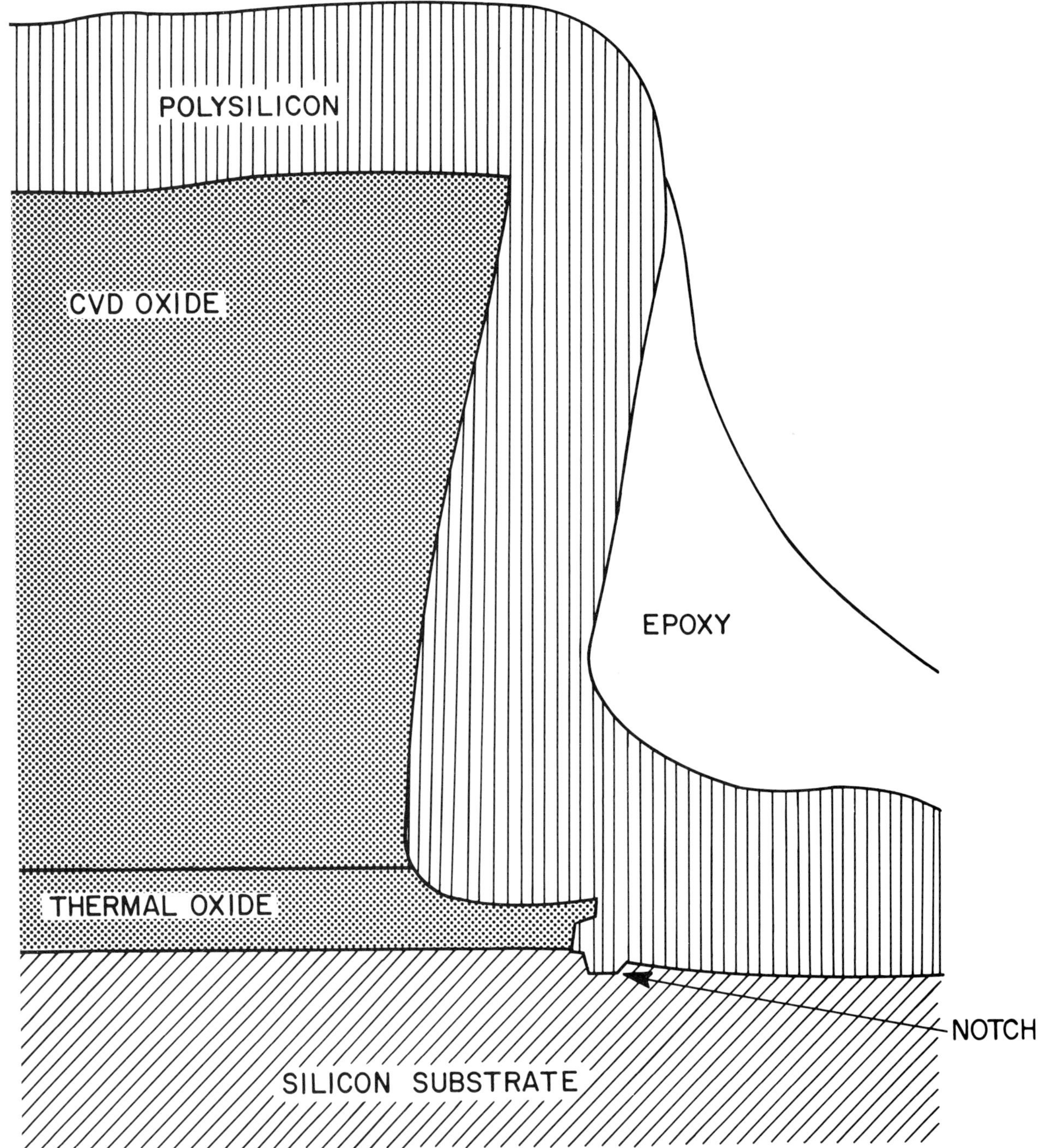

FIGURE 81. A complete device showing the edge of a window at second level metallization contact with the substrate. A notch is sometimes seen in the substrate adjacent to the oxide wall because of local enhancement of reactive ion etching.

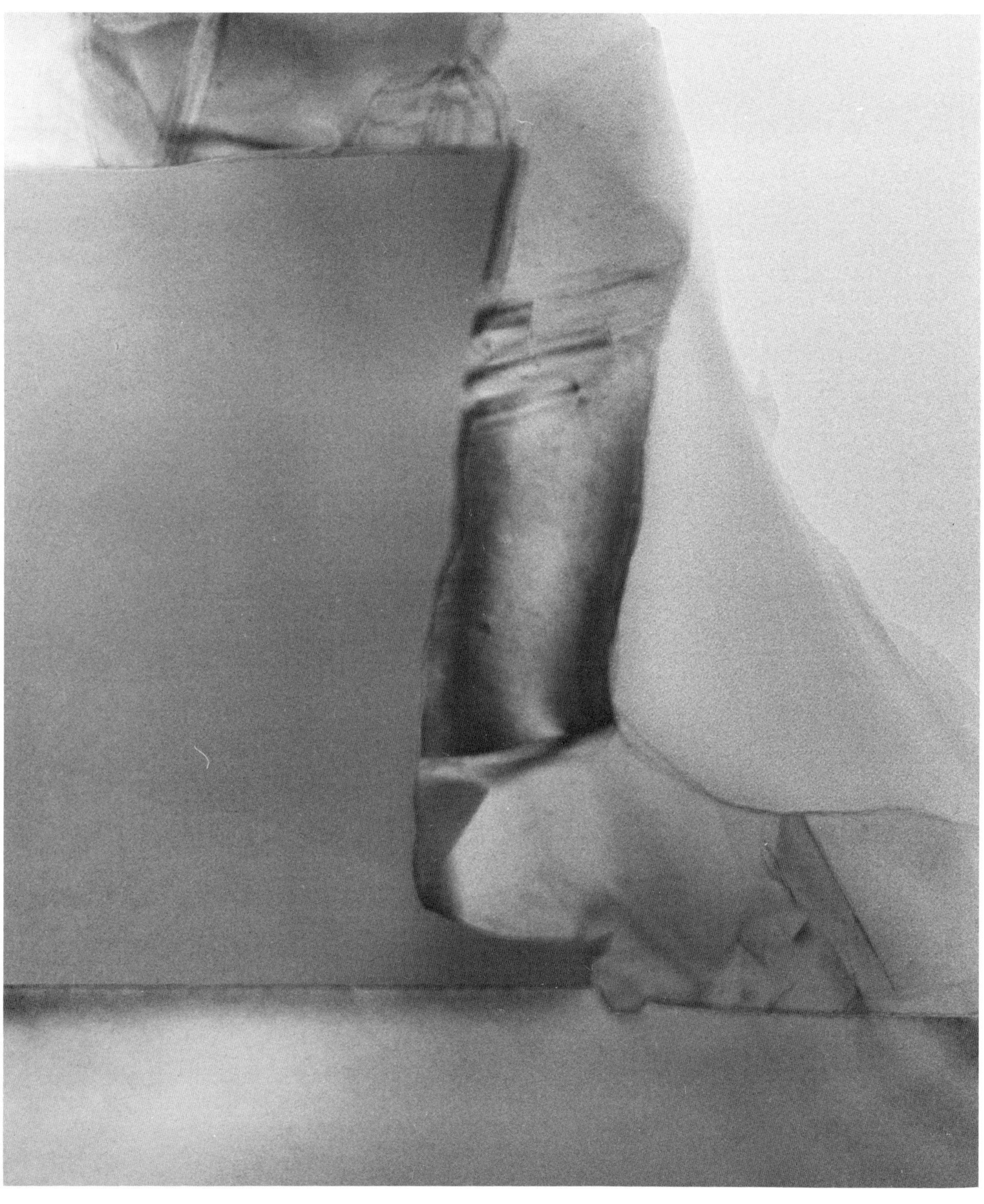
5000Å

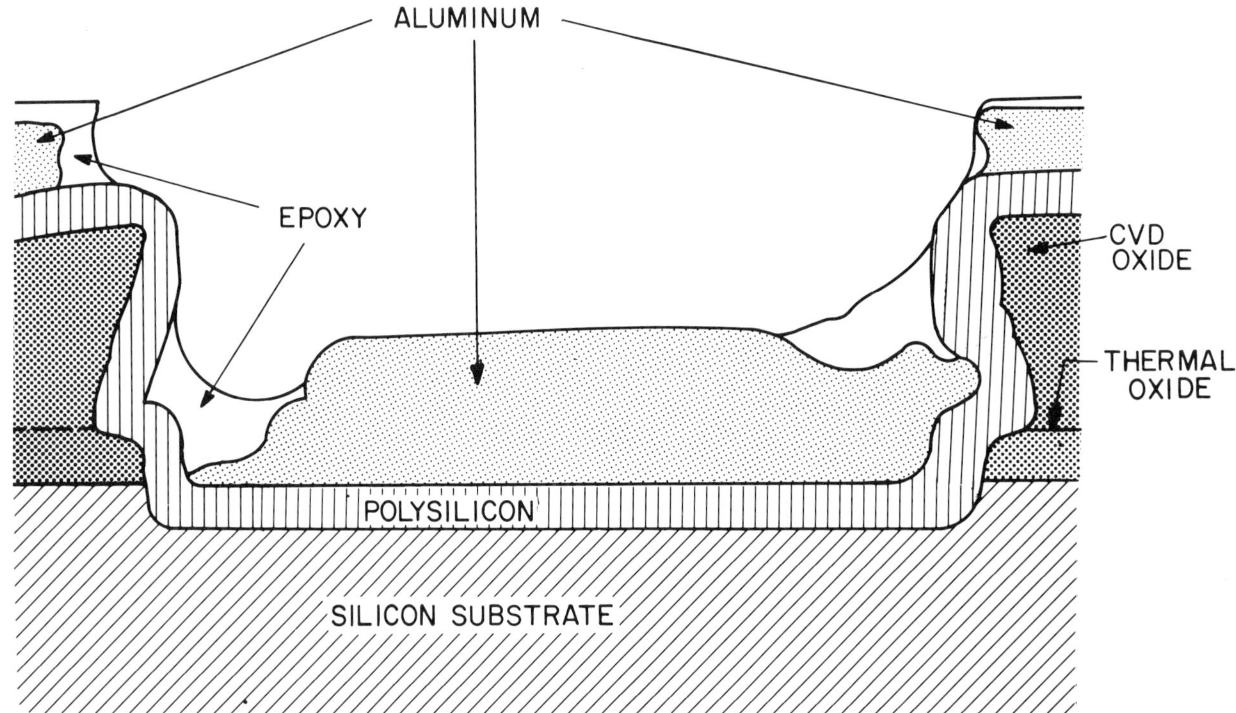

FIGURE 82. A complete device showing second level metallization contact with the substrate. Extensive overetching of the window causes deep (3500 Å) penetration of the contact. Increasing the phosphorus content with CVD oxide depth produces a negative slope at the oxide wall, which in turn results in poor aluminum step coverage.

Reactive ion etching sometimes produces enhanced etching just inside the edge of the defining mask leading to the appearance of a notch (Fig. 81). Also shown in Fig. 81 is a CVD oxide wall with negative slope. The negative slope is caused by the phosphorus content increasing with depth, which produces a corresponding variation in etch rate. Polysilicon conforms well to this wall with no thinning, but aluminum covers such a tapered wall very poorly (Figs. 82 and 83).

1.0μm

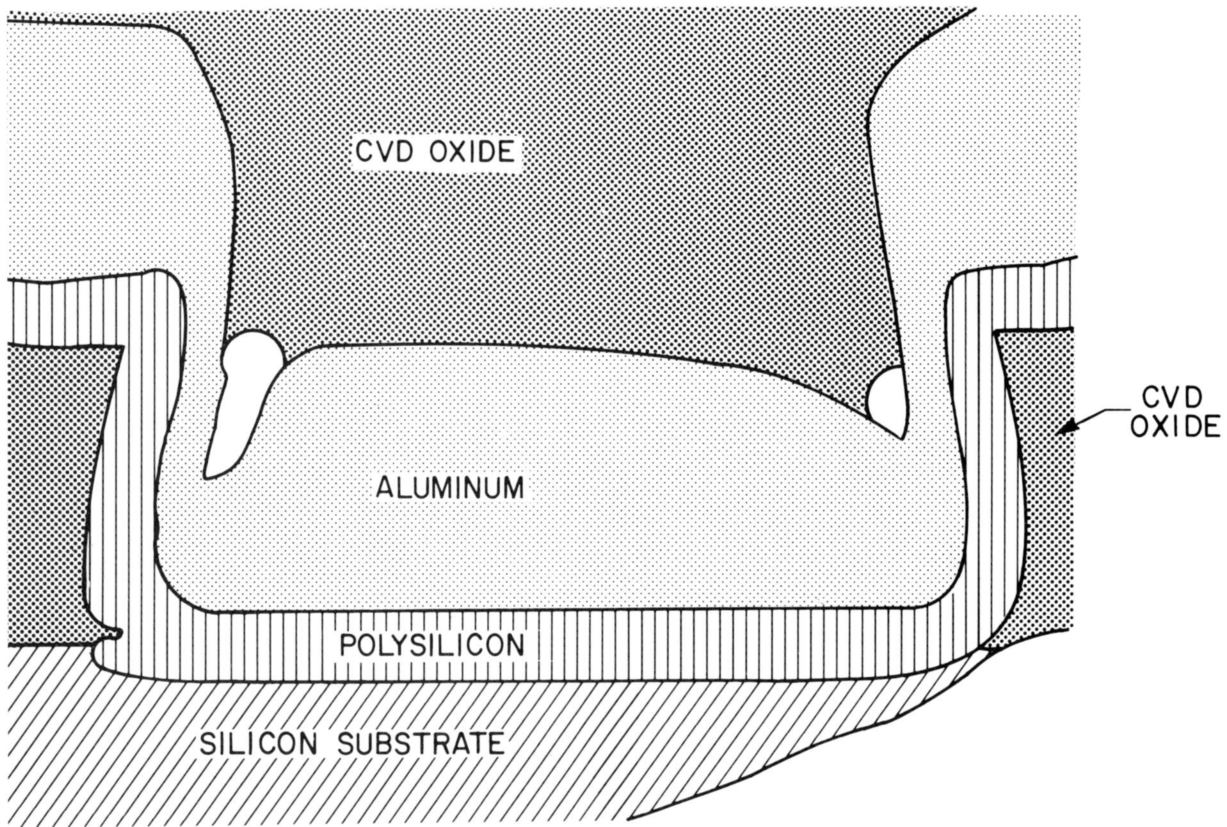

FIGURE 83. A complete device showing second level metallization contact with the substrate. The micrograph indicates poor aluminum coverage at the oxide wall for the same reason as described in Fig. 82.

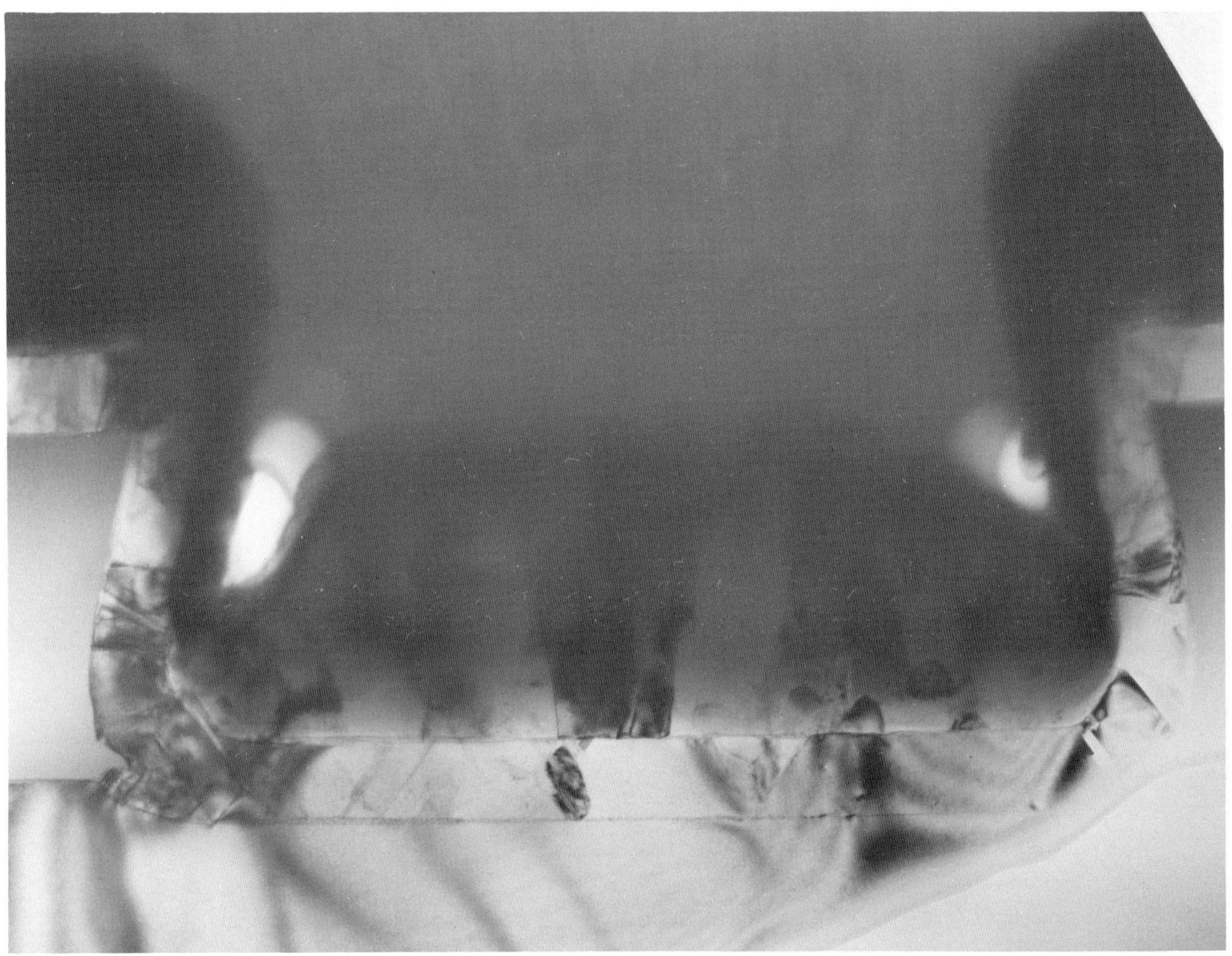

1.0μm

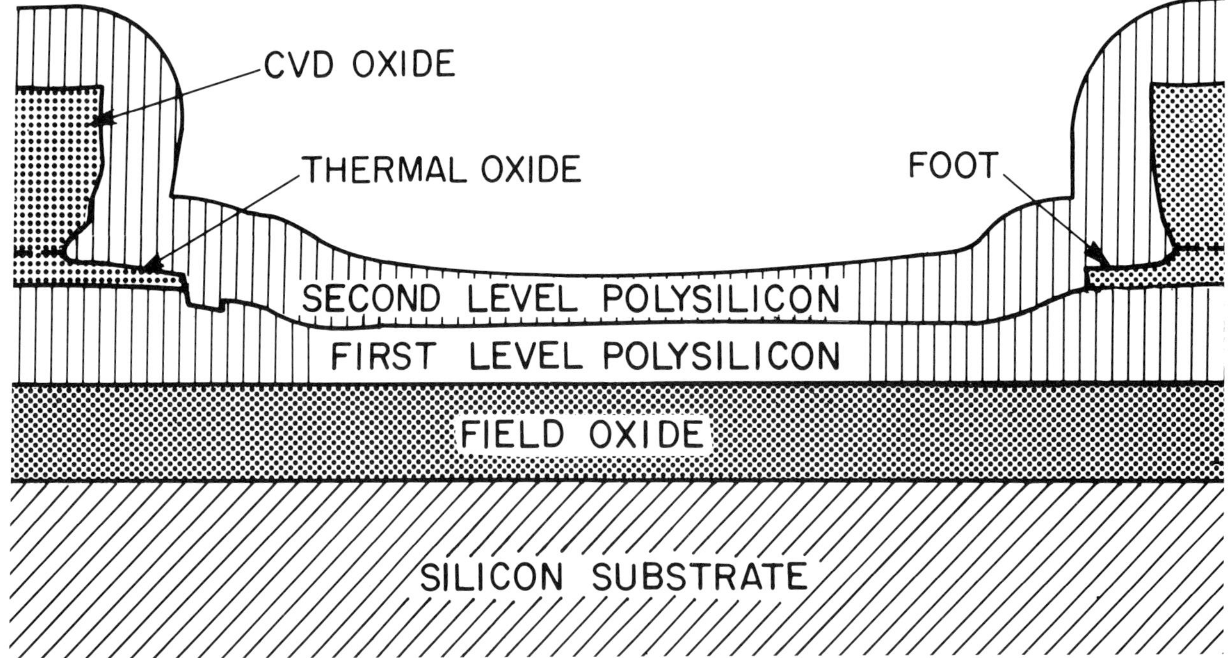

FIGURE 84. A complete device showing second to first level metallization contact. Differential chemical etching of thermal and CVD oxides produces the oxide foot.

7.5 SECOND LEVEL METAL–FIRST LEVEL METAL CONTACTS

Figures 84 through 92 show second level metallization to first level metallization contact areas. Differential etching of CVD and thermal oxides produce the oxide foot (Fig. 84) as previously described. The shape of the notch produced by enhanced reactive ion etching at the edge of the mask can vary considerably on the same sample (Figs. 85 and 86). Variations in the morphology of resist edges, including asymmetries on opposite sides of the same window, most likely cause this variation in shape. Local loss in anisotropic etching during window patterning may result in a notch feature (Fig. 87) showing an undercut (S, Fig. 88). The shape of the local field at the oxide edge may cause the loss in anisotropic etching.

Extreme overetching produced the contact region shown in Figs. 89 and 90. Contact between the two metallizations is restricted to the angular interface between them, and the contact area is reduced by about one-third compared with a normal contact.

Figures 91 and 92 show second to first level metallization contact. The first level consists of two polysilicon layers for reasons described earlier. Figure 91 shows a normal contact; Fig. 92 shows a considerably overetched contact area. Figure 92 also shows poor step coverage of aluminum compared to polysilicon, once again pointing out the advantage of a dual-film, second level metallization scheme.

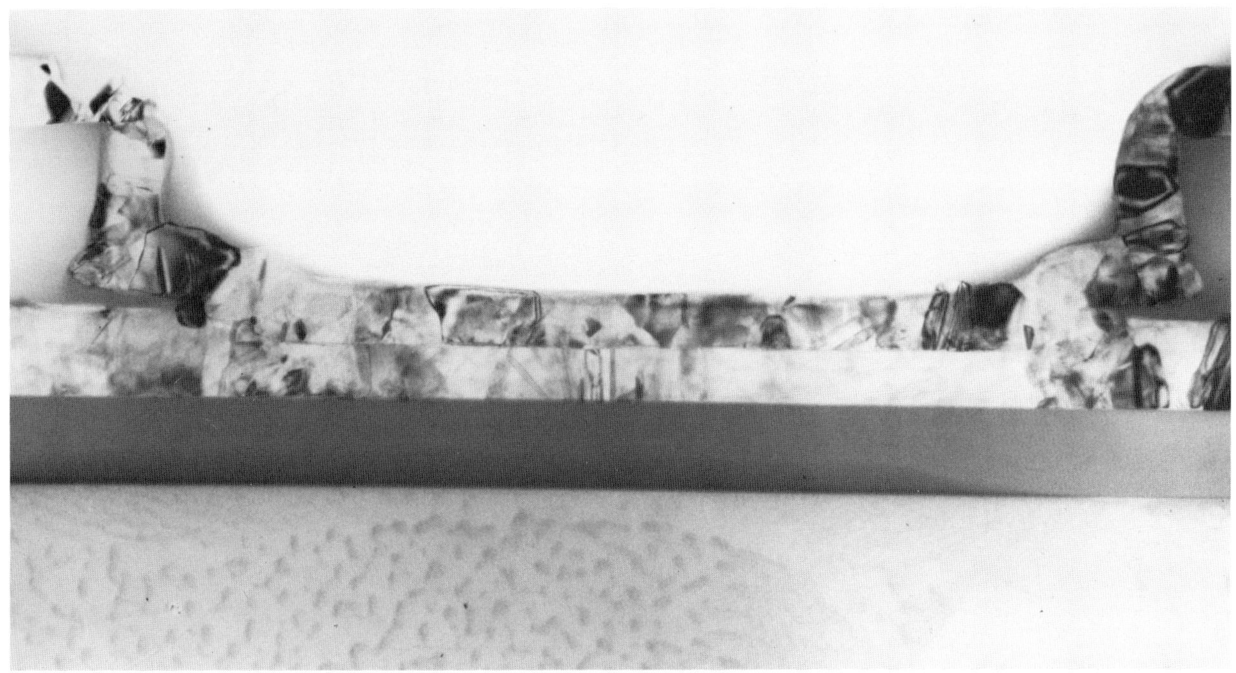
1.0 μm

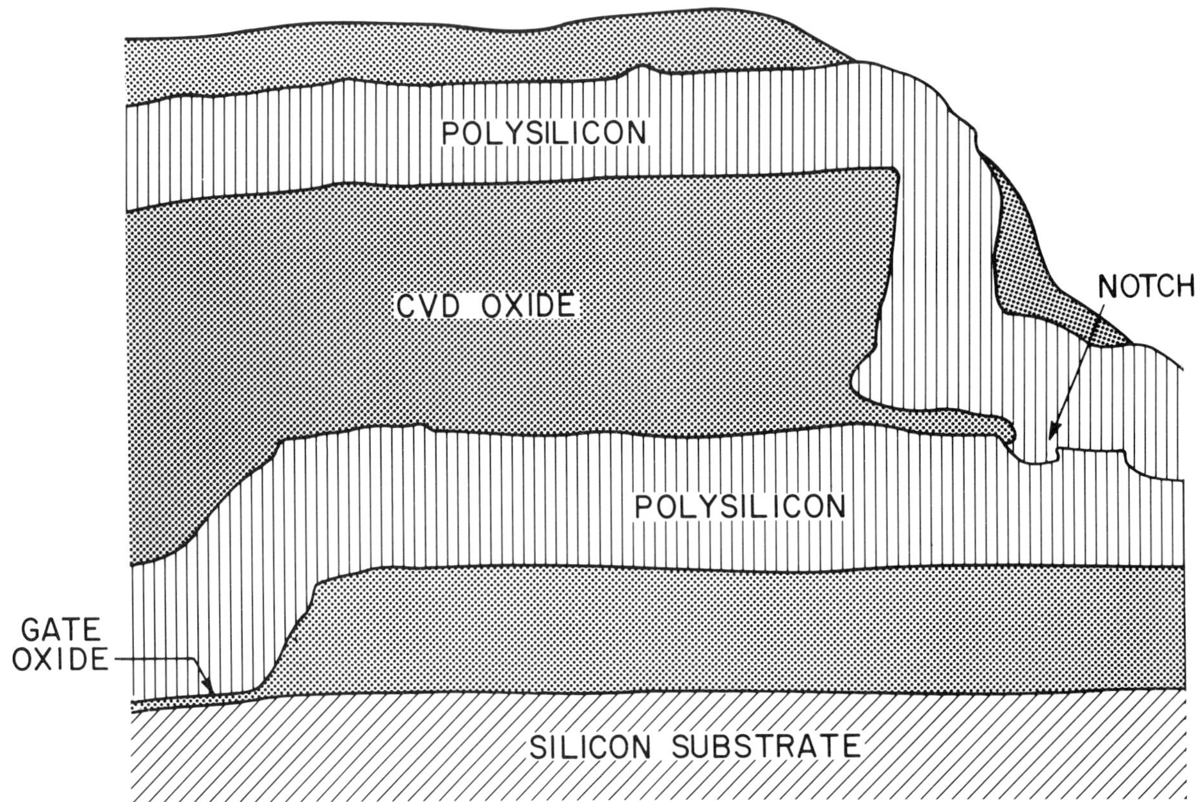

FIGURE 85. An enlargement of the left edge of the contact shown in Fig. 84. Local reactive ion etching enhancement creates a notch. A 250-Å gate oxide is also visible on the left.

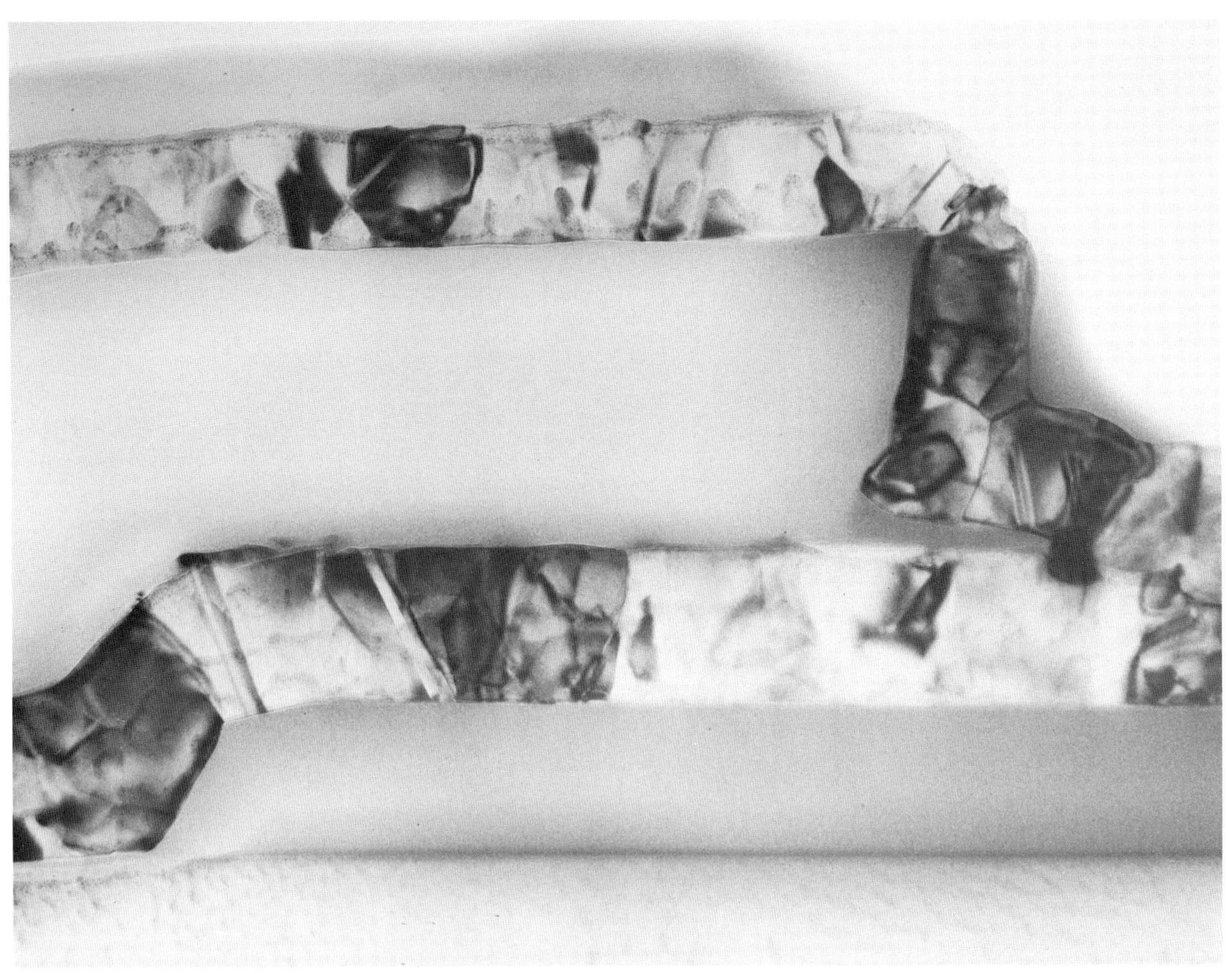
5000Å

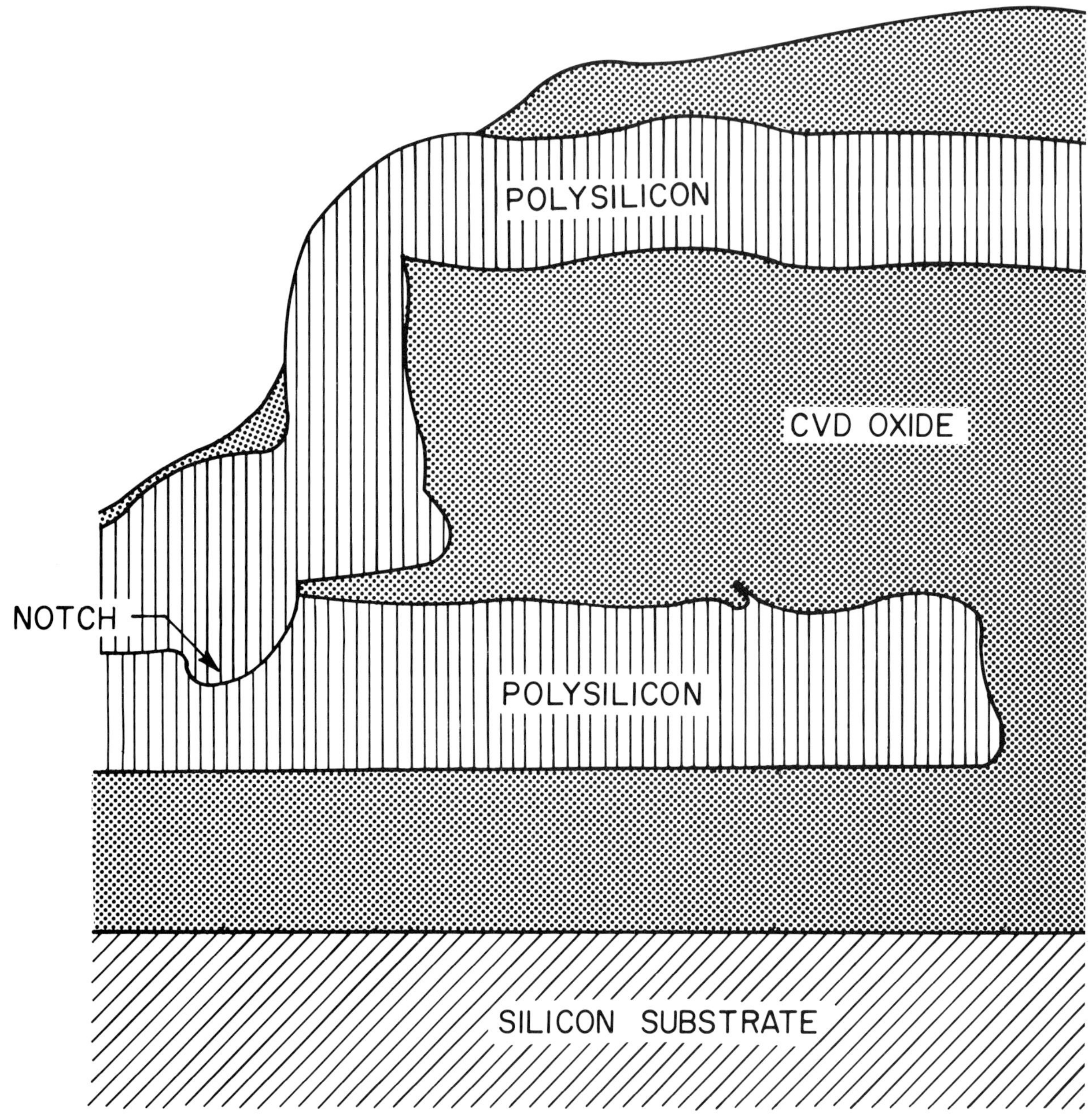

FIGURE 86. An enlargement of the right edge of the contact shown in Fig. 84. Reactive ion etching creates a large, irregularly shaped notch. (Compare with Fig. 85.)

5000Å

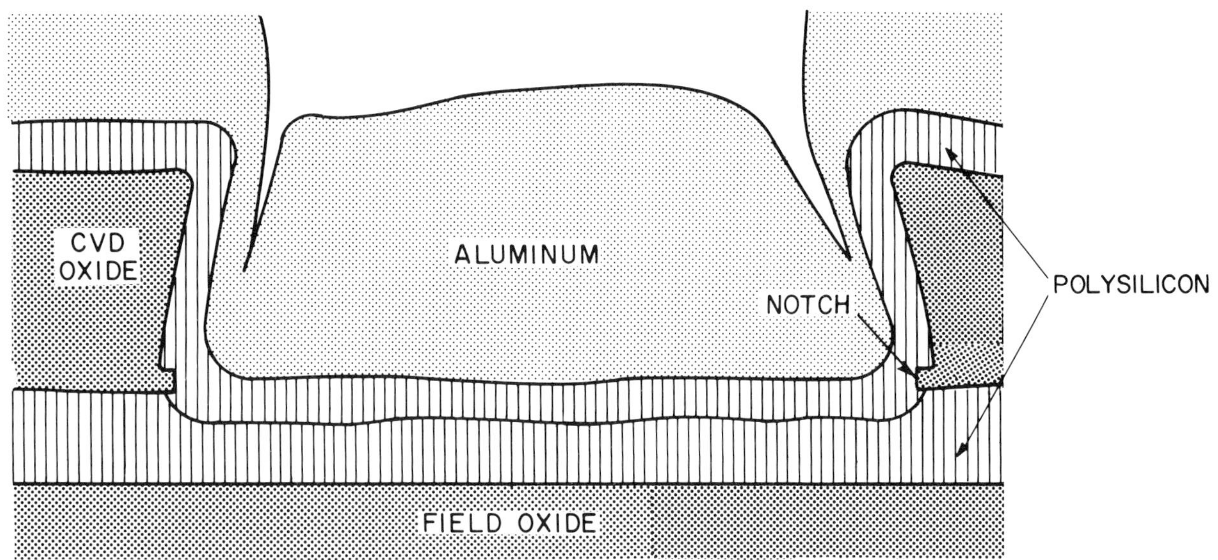

FIGURE 87. A complete device showing second to first level metallization contact and notch.

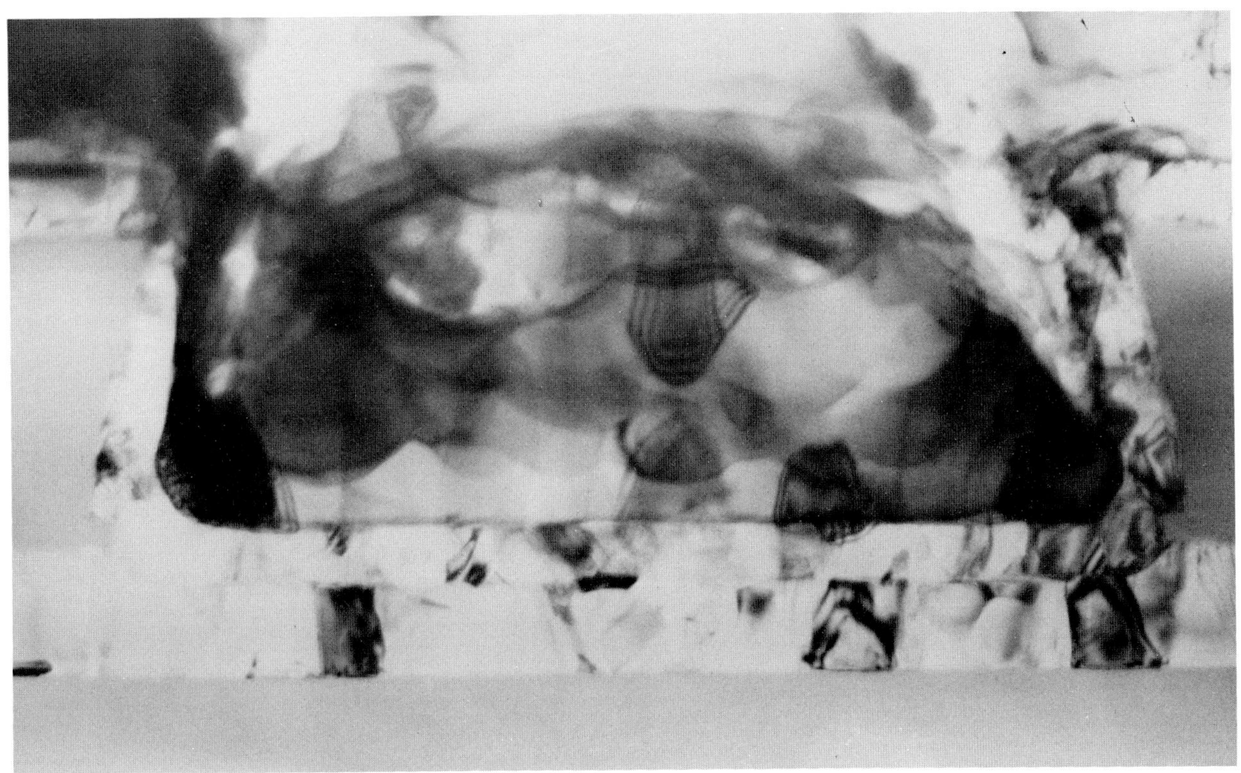
5000Å

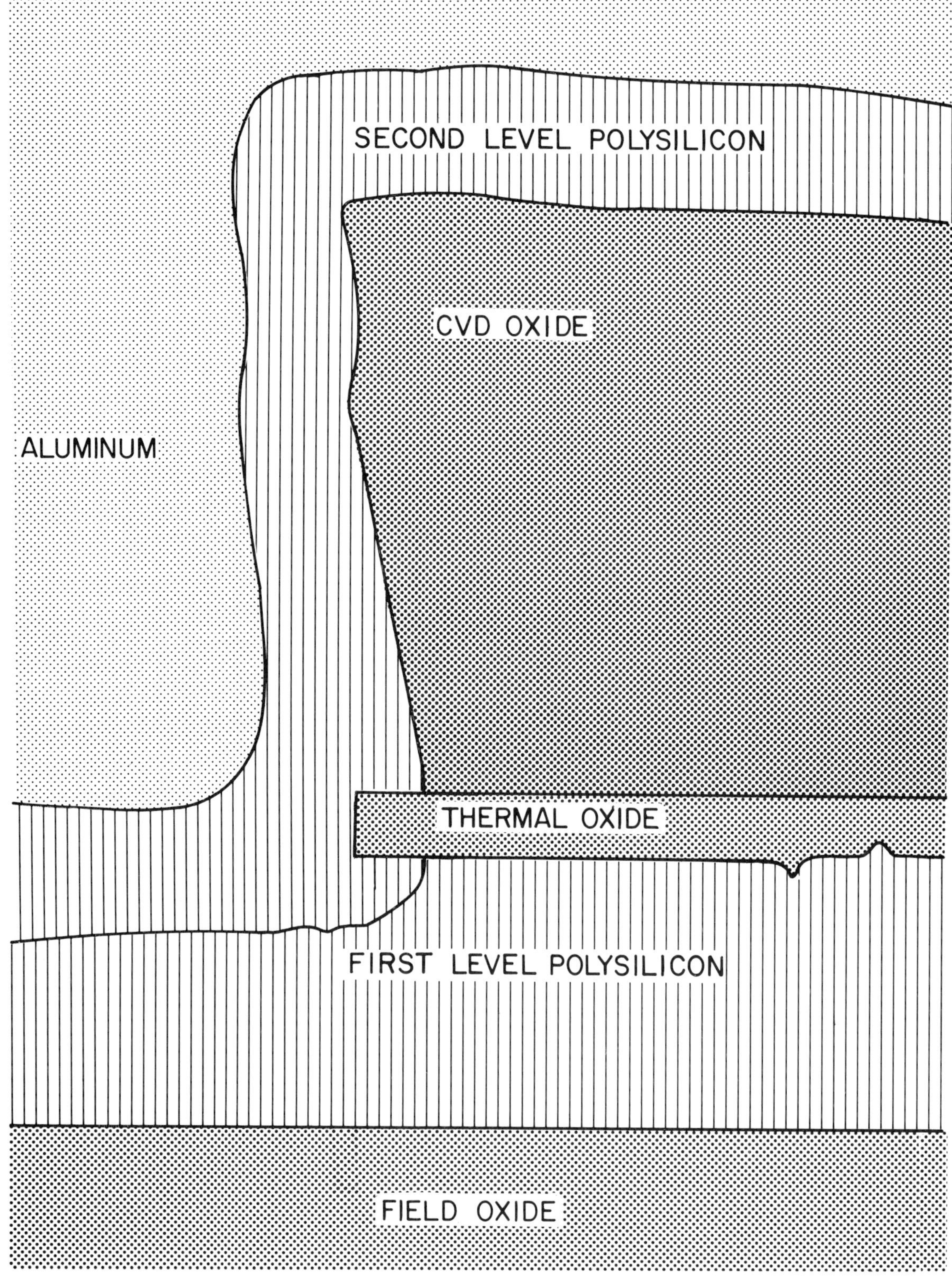

FIGURE 88. An enlargement of the right edge of the contact shown in Fig. 87. The presence of polysilicon in the region below the notch indicates an undercutting by the reactive ion etch step that defines the window.

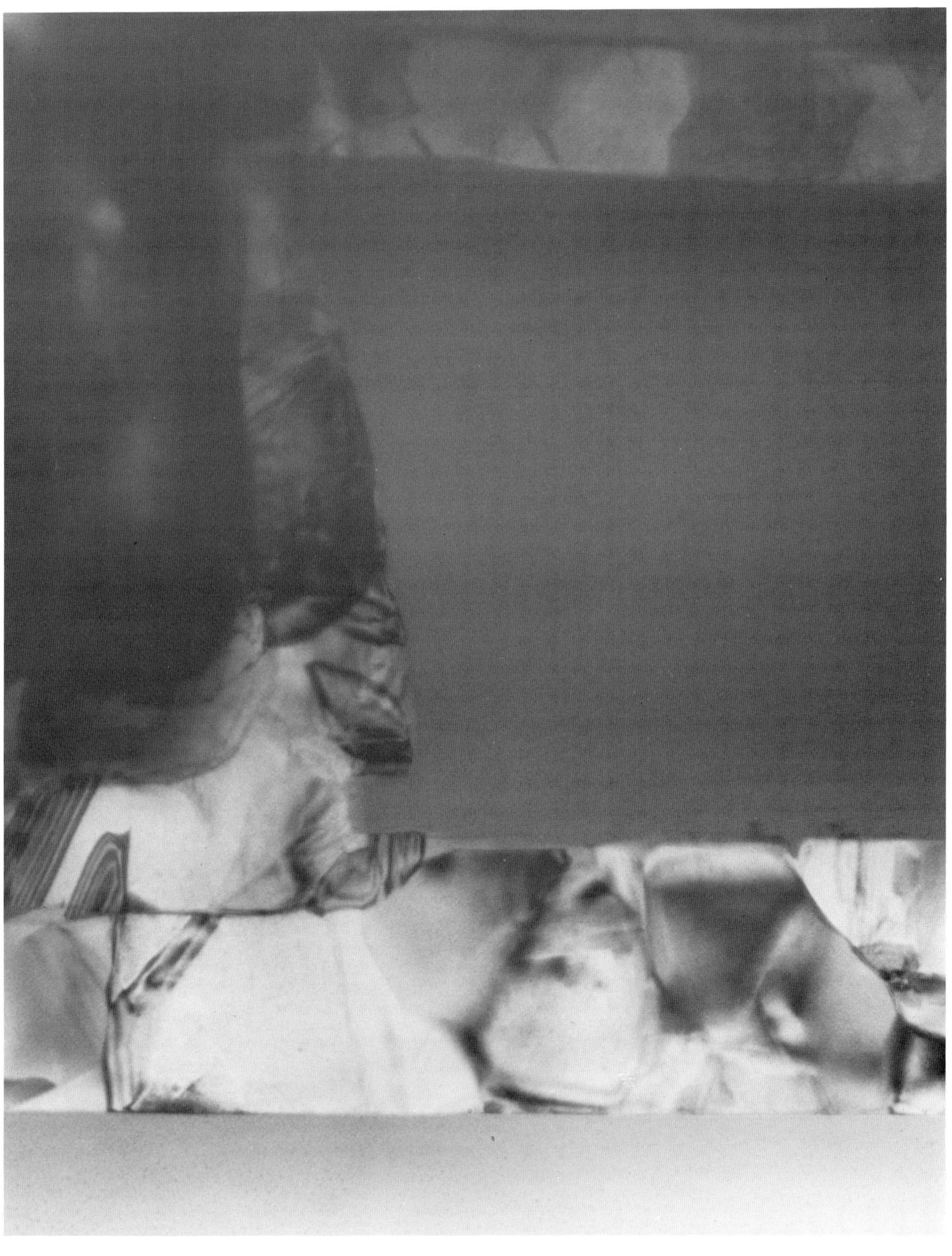
2000Å

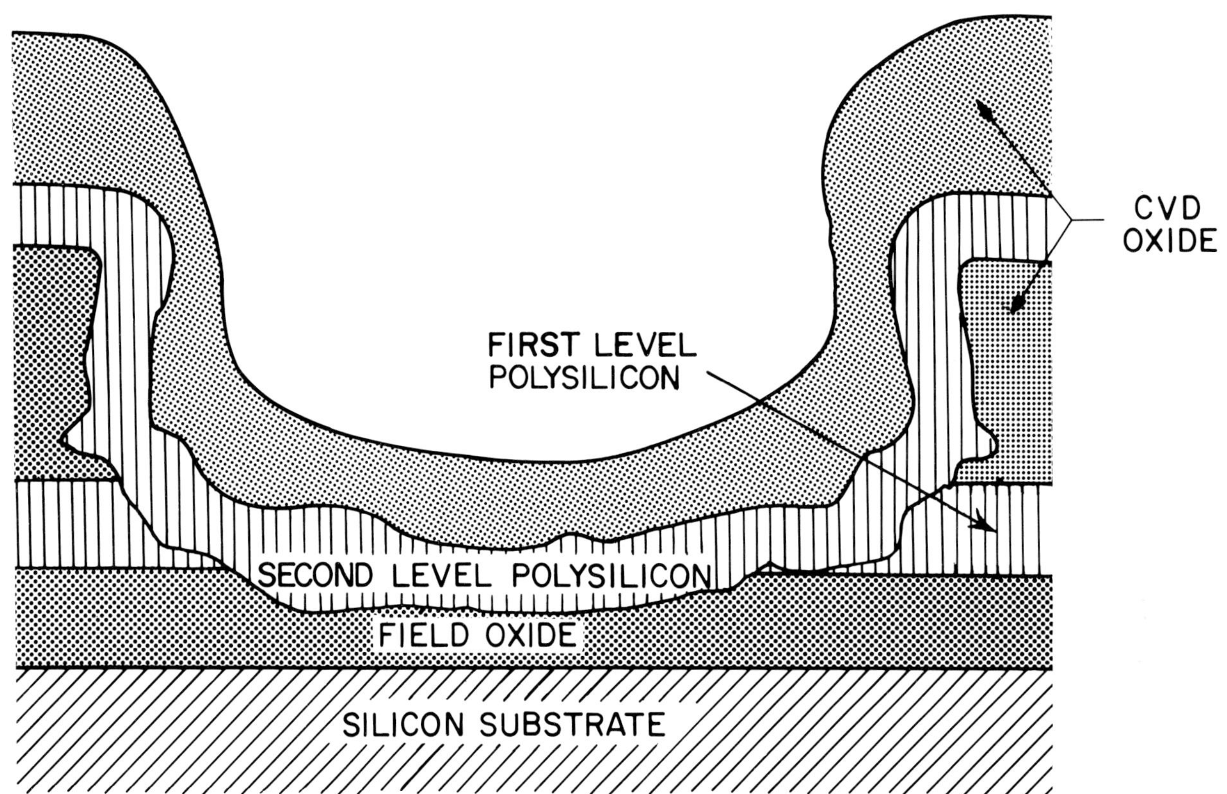

FIGURE 89. A complete device showing second to first level metallization contact. Extreme overetching greatly reduces the total area of contact between the two metallizations.

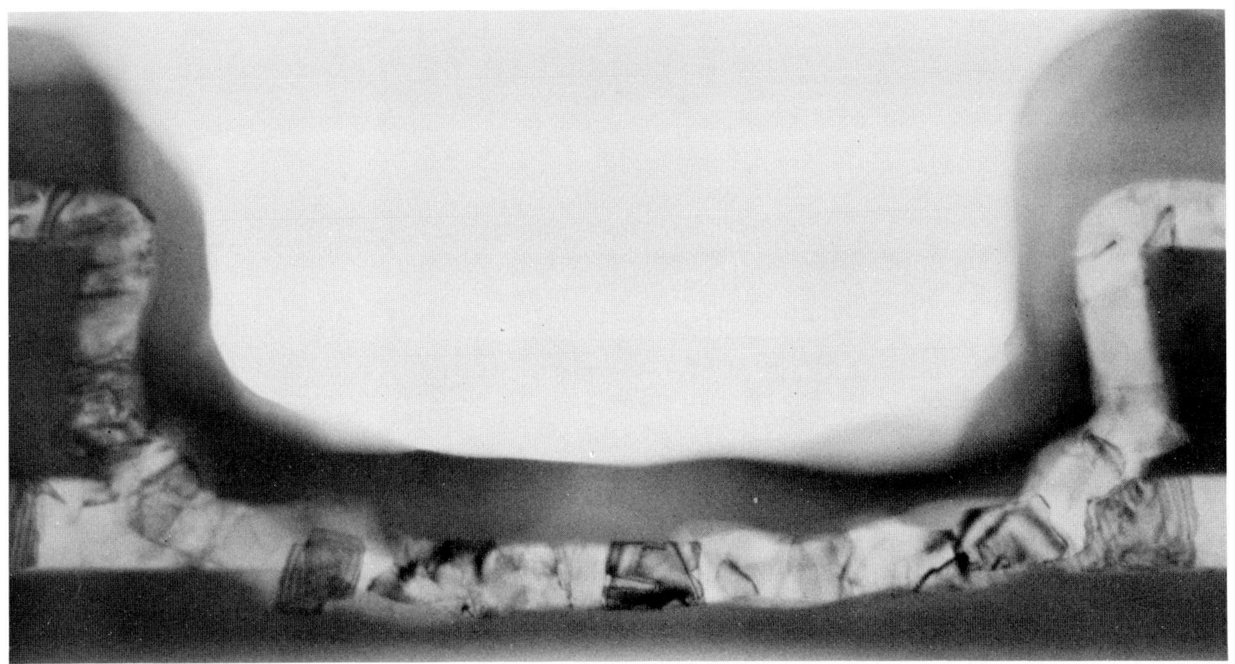
5000Å

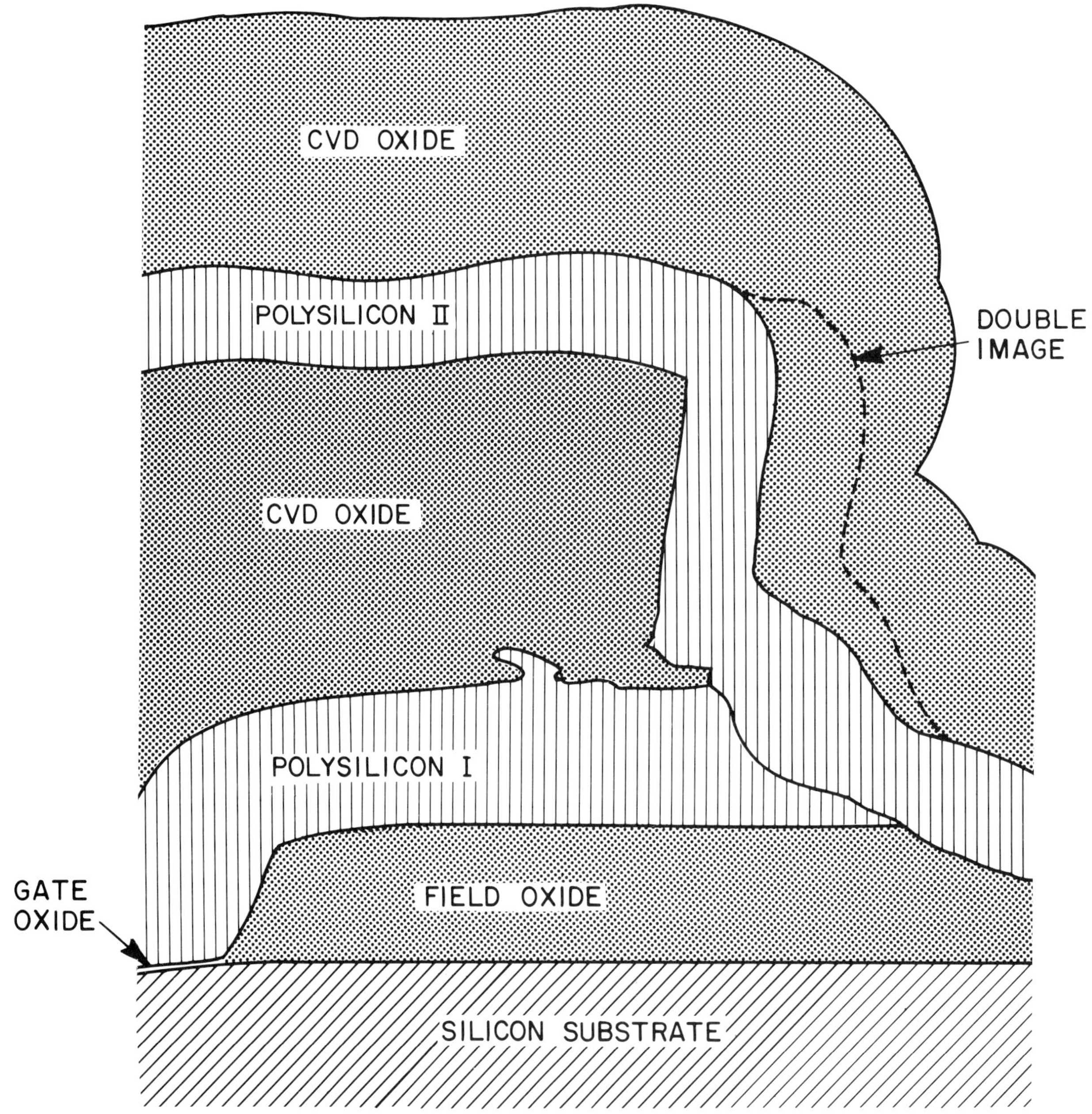

FIGURE 90. An enlargement of the left edge of the contact shown in Fig. 89. Local undulation of the oxide wall causes the appearance of an additional double image in the micrograph.

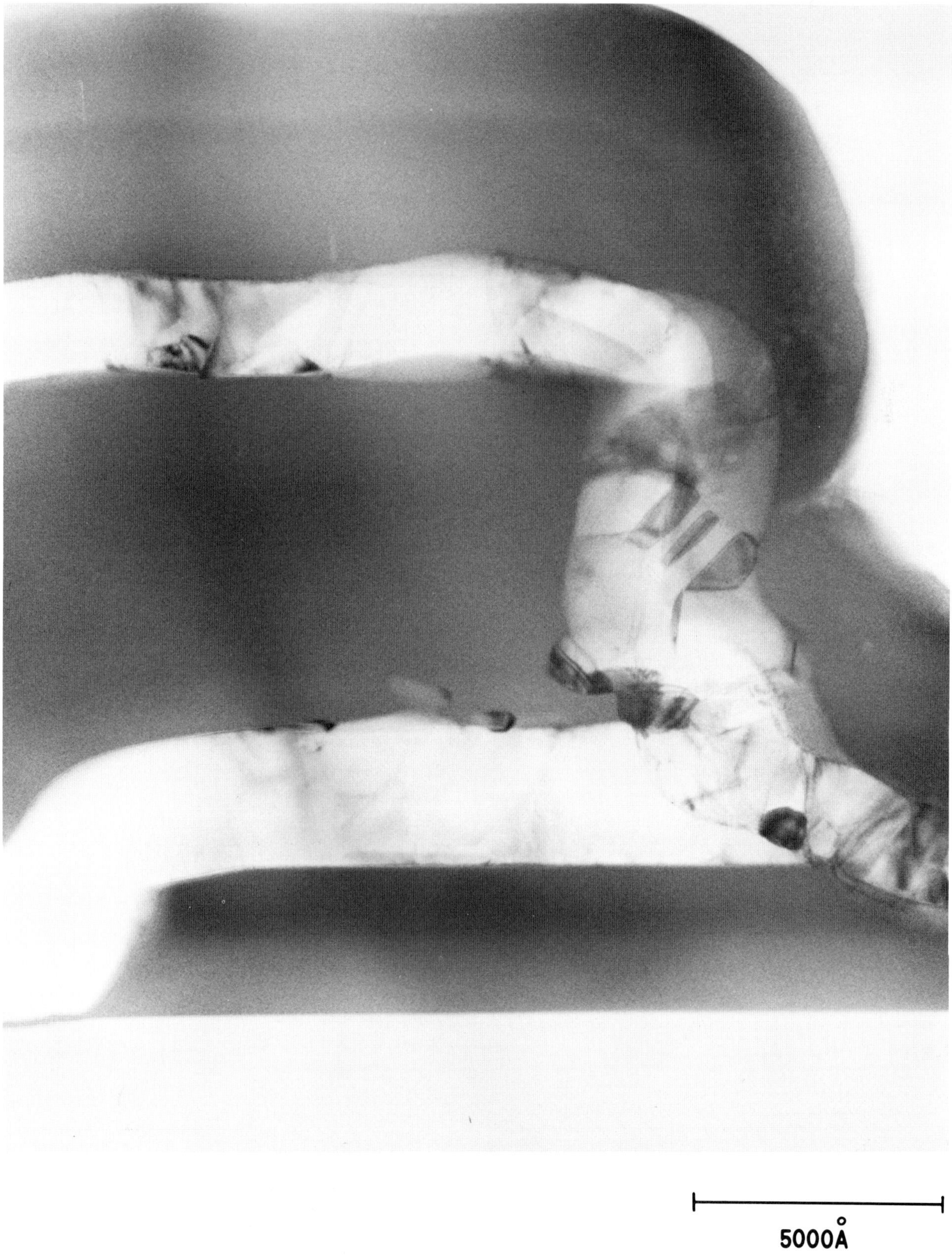

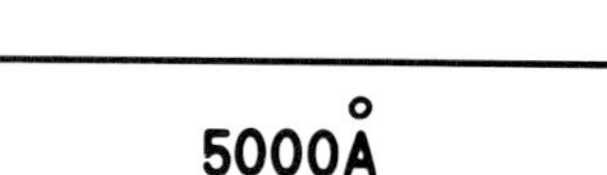
5000Å

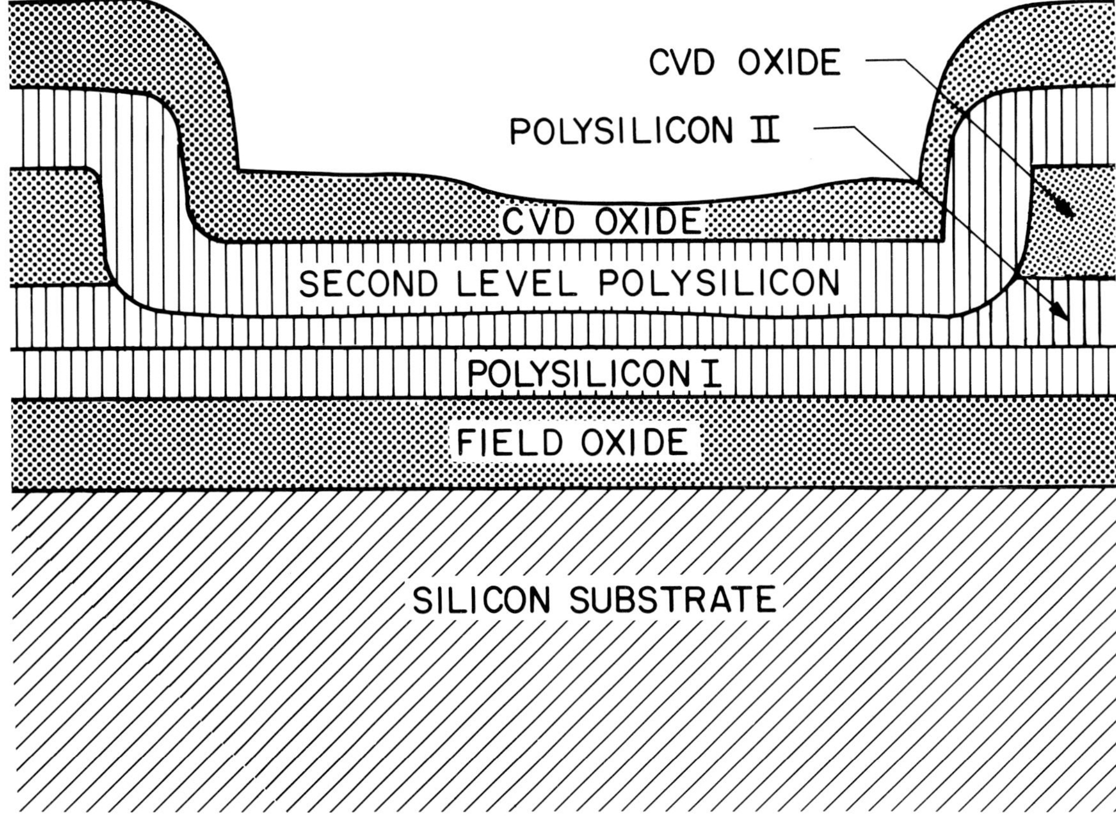

FIGURE 91. A complete device showing second to first level metallization contact. The first level consists of two polysilicon layers.

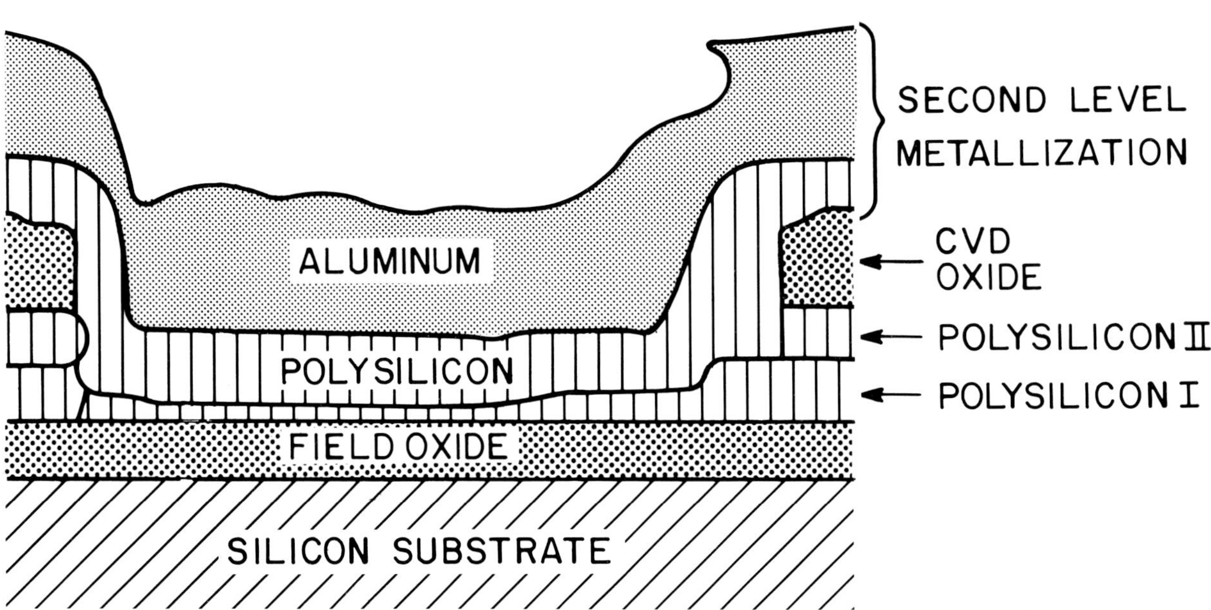

FIGURE 92. A complete device showing second to first level metallization contact. Overetching has produced a contact with the lower polysilicon layer (polysilicon I). Note the poor step coverage by the aluminum.

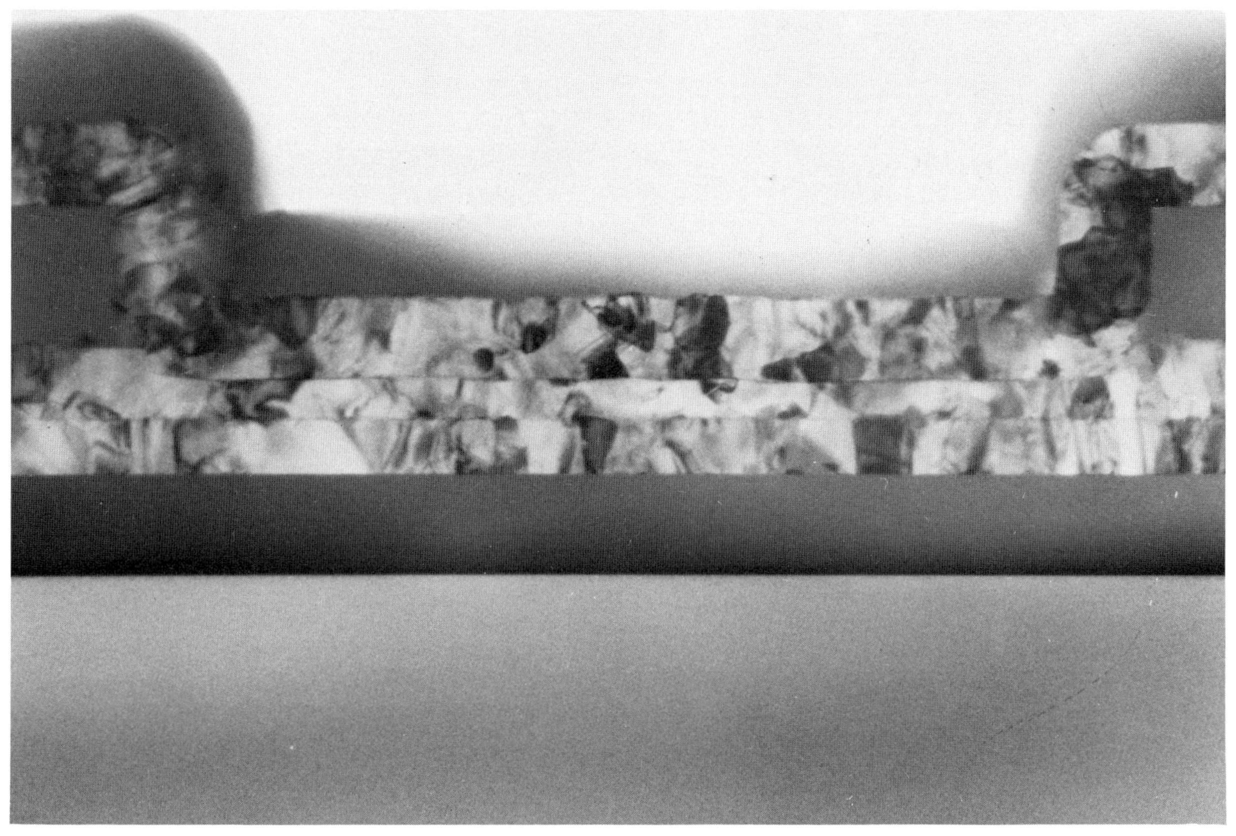
1.0μm

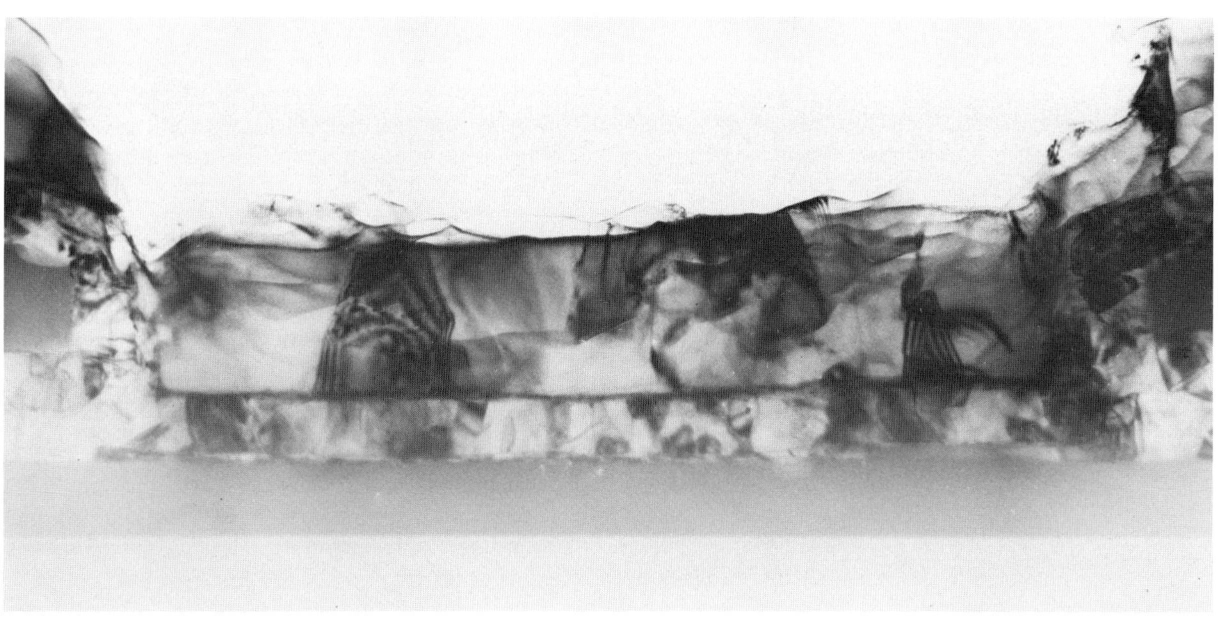
1.0μm

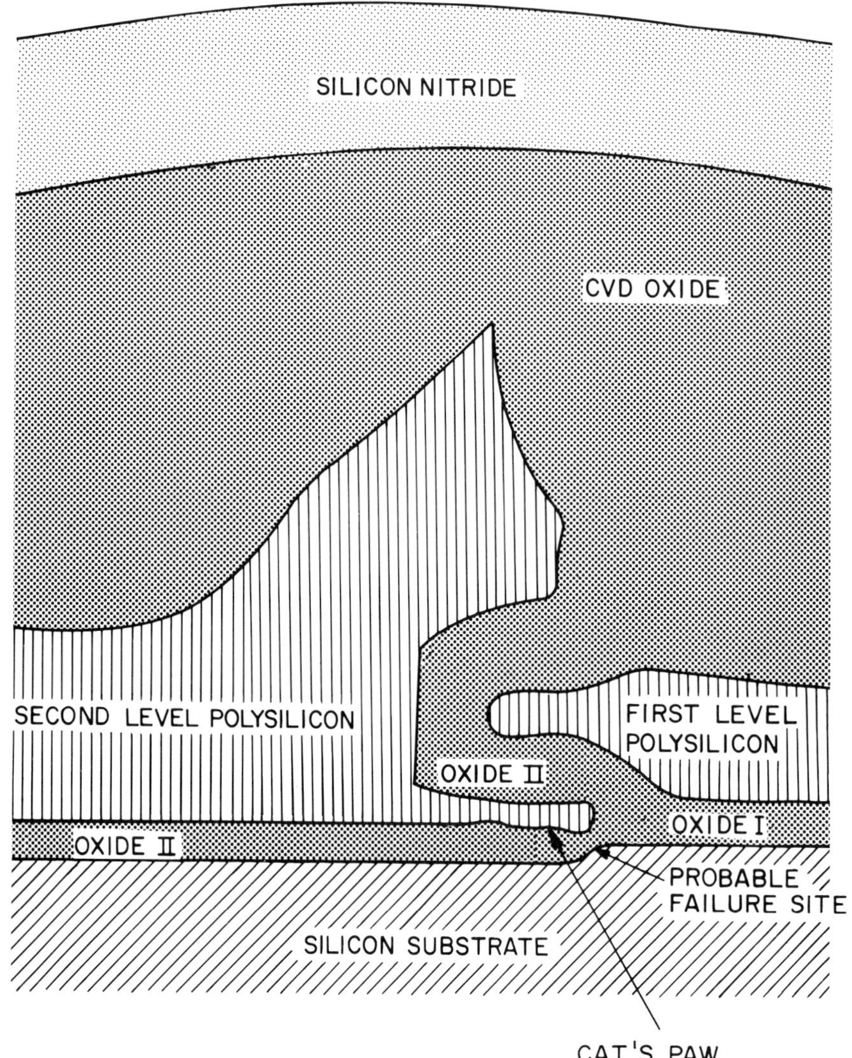

FIGURE 93. A complete device showing a region that has two polysilicon metallization layers in close proximity. Chemical etching to clear the surface areas of oxide growth prior to oxide II growth caused undercutting of the first level polysilicon. The close spacing of the second level polysilicon and the substrate led to the appearance of a cat's paw and a probable failure site.

7.6 GENERAL DEVICE FEATURES

Single-crystal silicon and polysilicon oxidation characteristics that give rise to potential failure modes were described in Section 3.2. Figures 93 through 95 show some of these features appearing in finished devices, and Fig. 96 illustrates a solution to one of these problems. Second level polysilicon sometimes assumes the configuration of a lying cat with an outstretched paw[63] (Figs. 93 and 94). A chemical-etch undercutting of the edge of the first level polysilicon

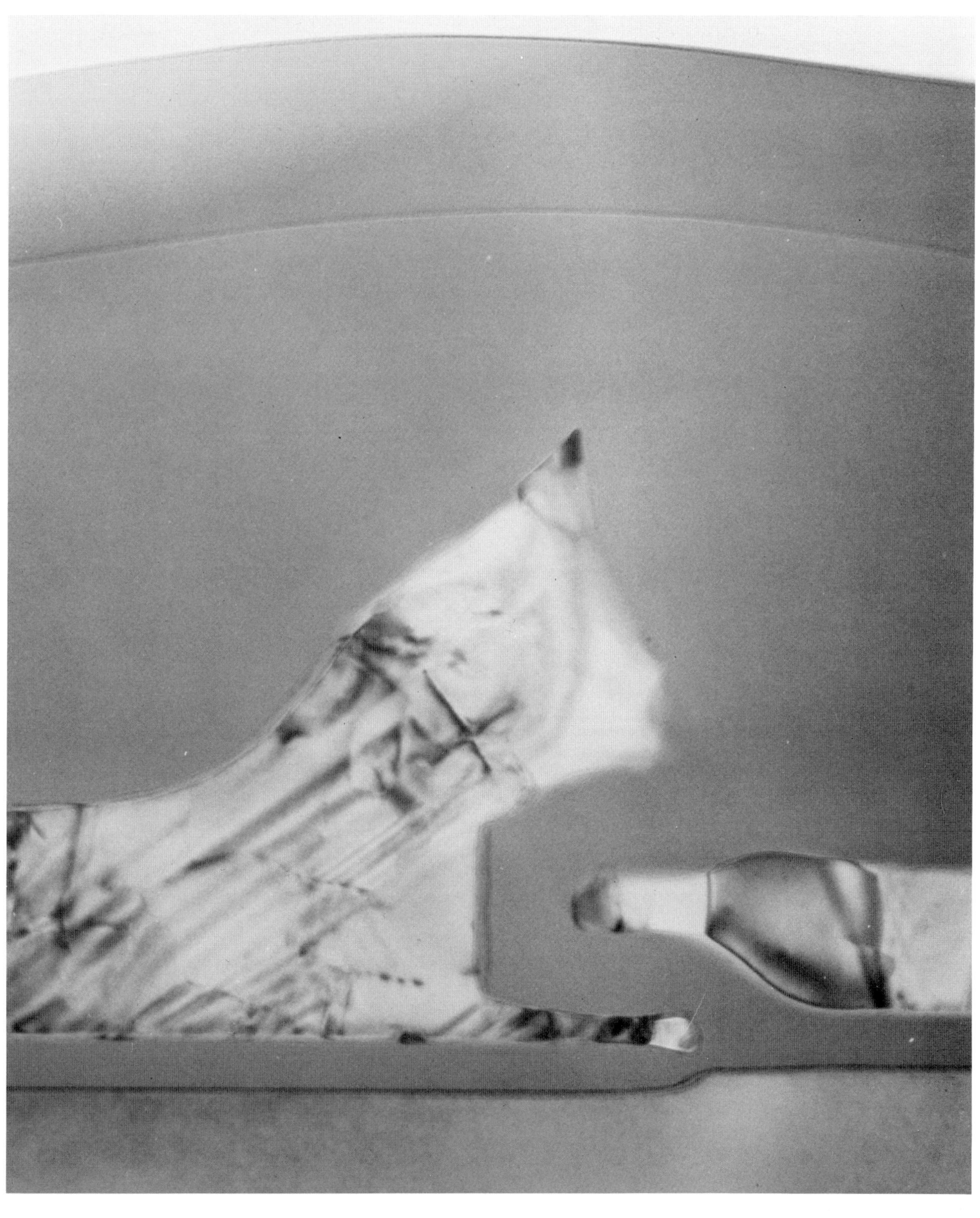

5000Å

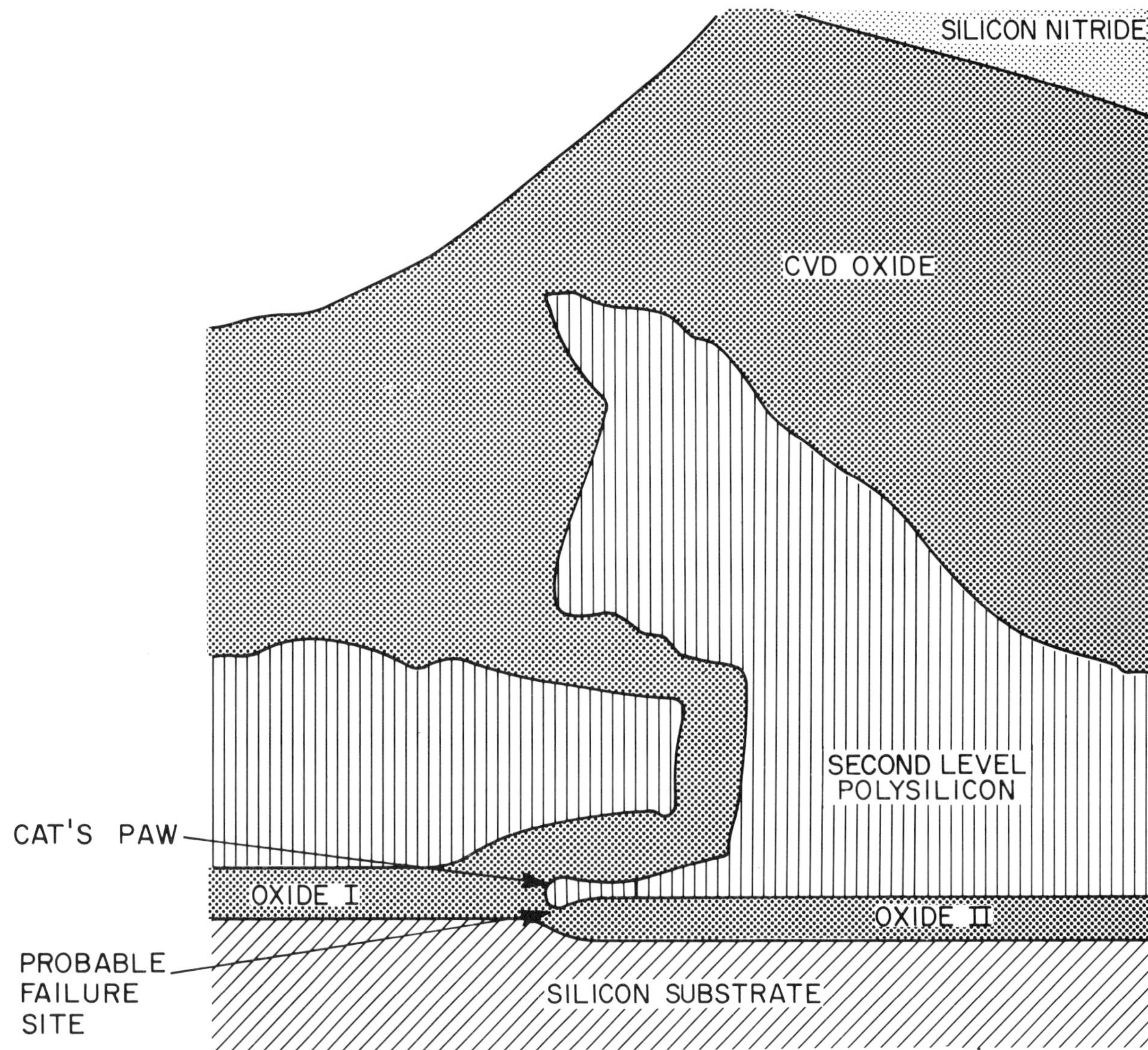

FIGURE 94. A complete device showing a region that has two polysilicon layers in close proximity (compare with Fig. 93). The cat's paw and probable failure site are indicated.

leaves a slot-shaped void following oxide II growth; second level polysilicon deposition fills in the slot forming the "cat's paw." The growth rate of oxide II was reduced at the end of the slot by a geometrical barrier to available oxygen, which leaves the second level polysilicon–substrate spacing narrower. A potential failure mode is thereby created.

Texture introduced into polysilicon by its thermal oxidation is detrimental to its dielectric behavior (Chapters 3 and 4). Figure 95 shows an example of this

5000Å

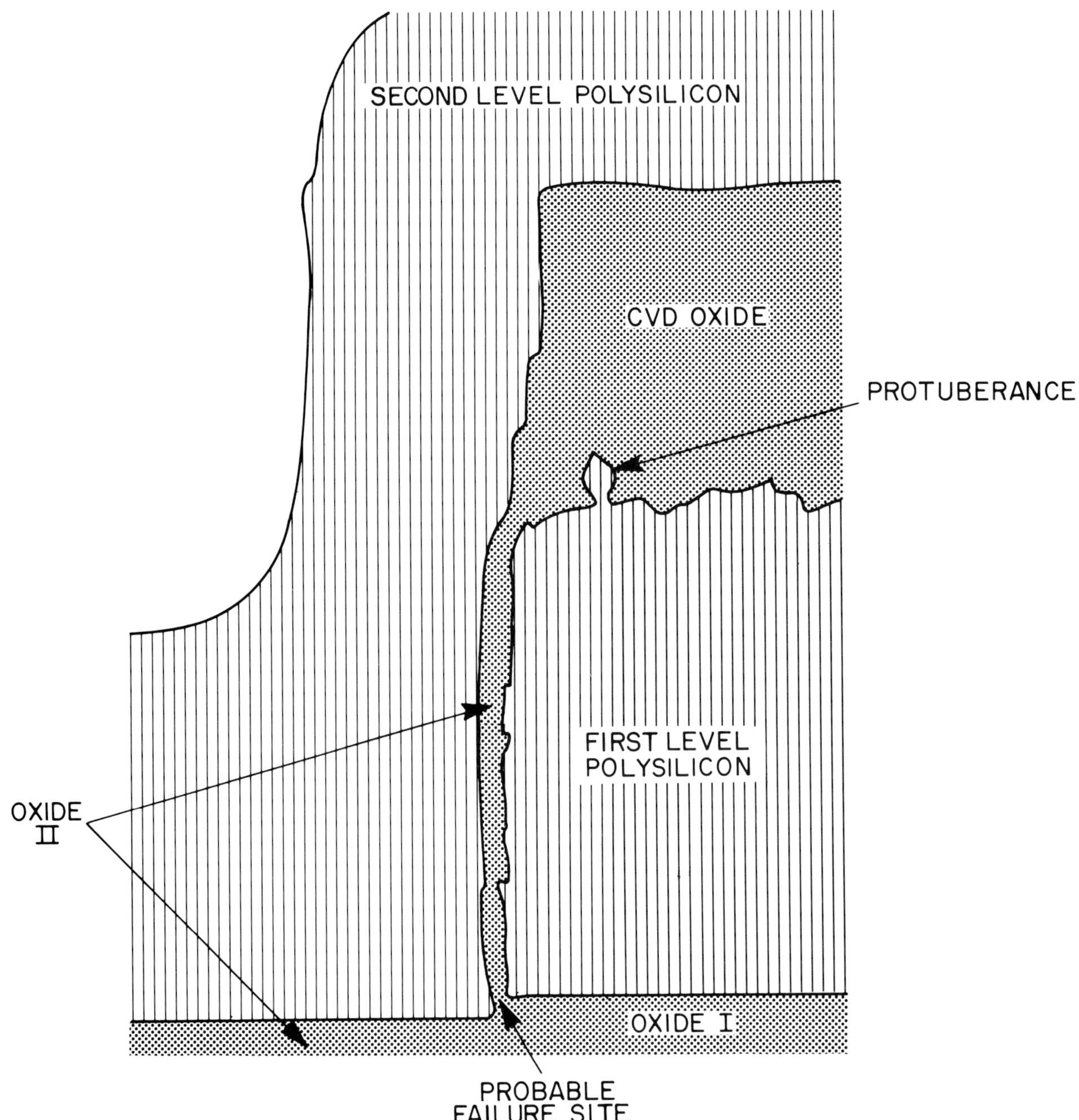

FIGURE 95. A complete device showing a region that has two polysilicon layers in very close proximity. Oxide separating the two polysilicon films (oxide II) was thermally grown from polysilicon on the right, leaving a rough surface containing a protuberance and an emerging horn. The latter feature gives rise to a region of locally thin oxide, which is a probable failure site.

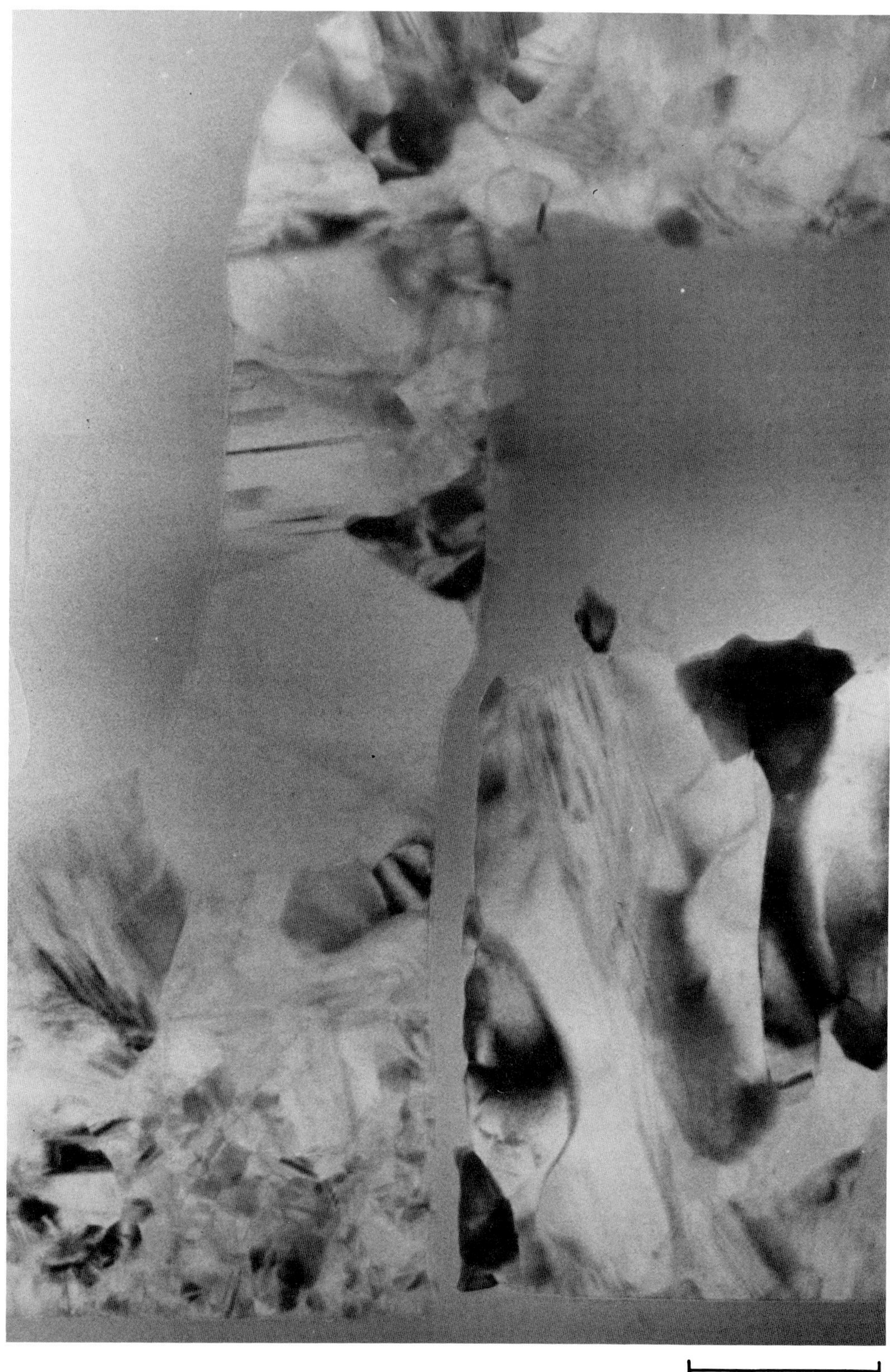
2000Å

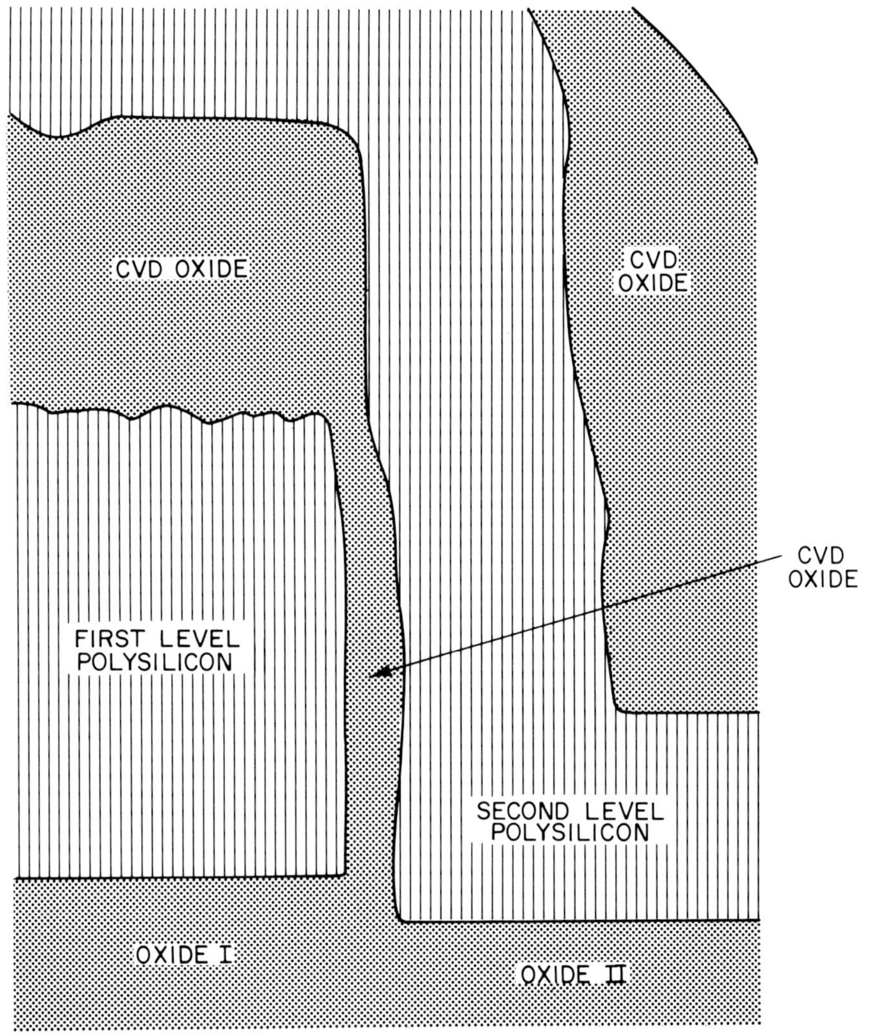

FIGURE 96. A complete device showing a region similar to that in Fig. 95 but processed so that a smooth first level-polysilicon oxide interface remains. Oxide separating the two polysilicon films was deposited (rather than thermally grown) immediately following first level polysilicon patterning.

texture and its impact on device performance. Oxide II was grown in dry oxygen at 1000°C, and capacitor measurements for subsequent second level polysilicon to first level polysilicon show a mean breakdown field of 3.5 MV/cm.[64] An extremely rough texture, including a protuberance and horn formation with a locally thin oxide (Chapter 3), is visible on the side wall and top surface of the oxidized polysilicon. The texture at the side wall is undoubtedly responsible for the low dielectric strength of this oxide, and the thin oxide region is particularly vulnerable as a potential failure site. This problem is avoided by changing the processing procedure. The thermal oxide was re-

2000Å

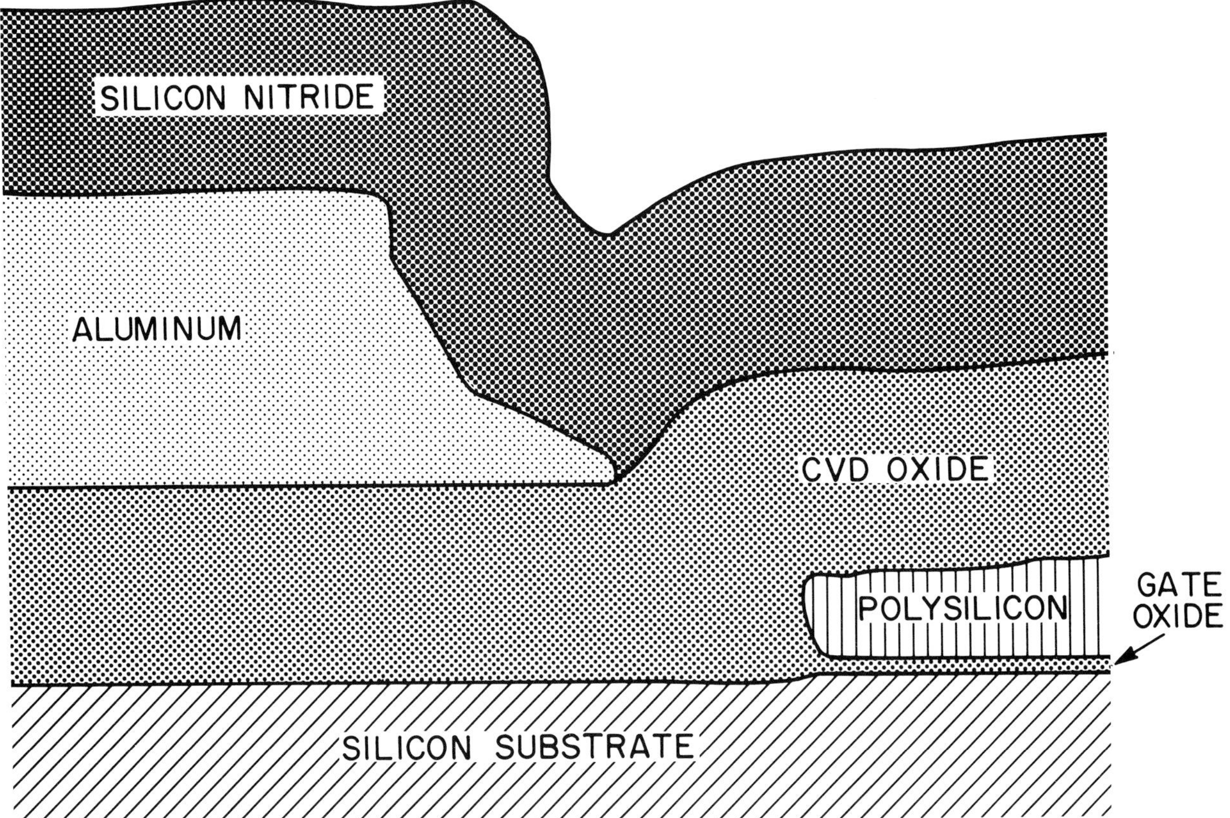

FIGURE 97. A complete device showing the effect of patterning by chemical etching only. Note the taper of the aluminum metallization edge.

placed by CVD oxide at the side wall of first level polysilicon. The polysilicon wall maintains the smooth texture created by reactive-ion etching of the polysilicon (Fig. 96), and the mean breakdown field of second level polysilicon to first level polysilicon capacitors increases to 6.5 MV/cm.[64]

Figures 97 and 98 illustrate an older procedure and a recent development in VLSI technology, respectively. Before the use of plasma and reactive-ion etching methods, wet chemical methods were used exclusively for pattern delineation. Figure 97 shows the effect of patterning by chemical etching with aluminum as the second-metallization layer and silicon nitride as an encapsulating dielectric. The taper of the edge of the aluminum, prohibitive in most of the current VLSI devices, was acceptable in older IC technology. This feature should be compared with the wall tapers formed by reactive-ion etching and shown in most of the other micrographs in this book.

The push for higher conductivity as an offset to decreased metallization linewidths and the need for metallization that is stable under oxidation conditions spurred the development of silicide technology (Section 4.4). Figure 98

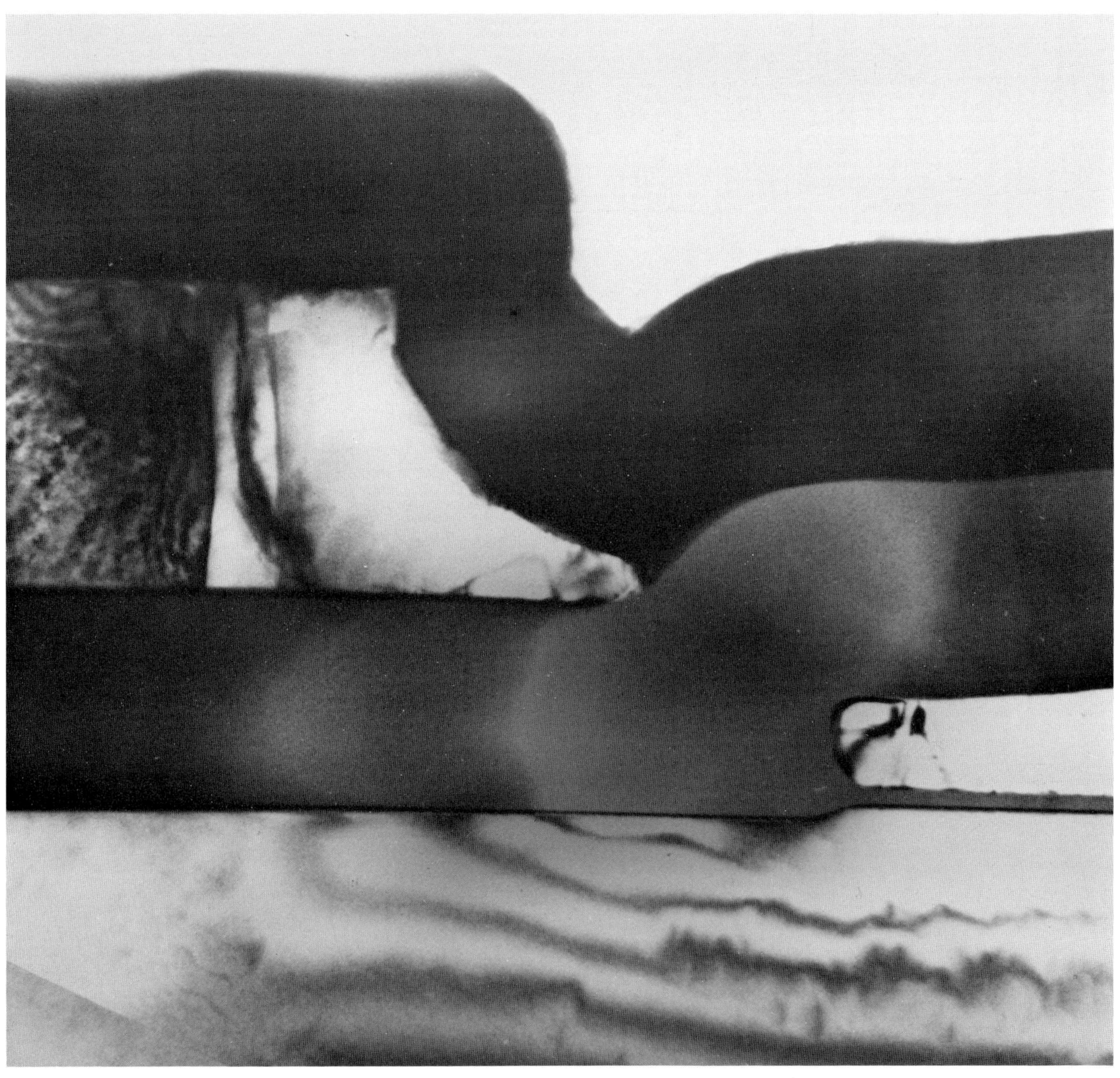

2.0μm

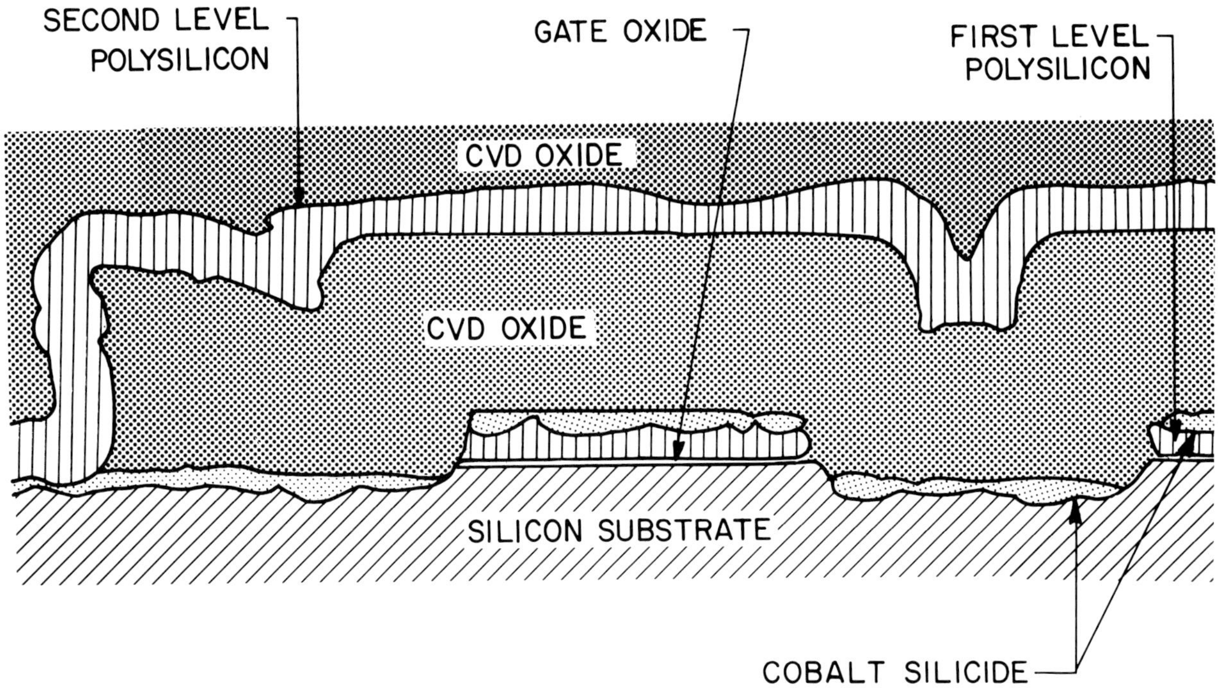

FIGURE 98. A complete device using cobalt silicide as part of the first level metallization.

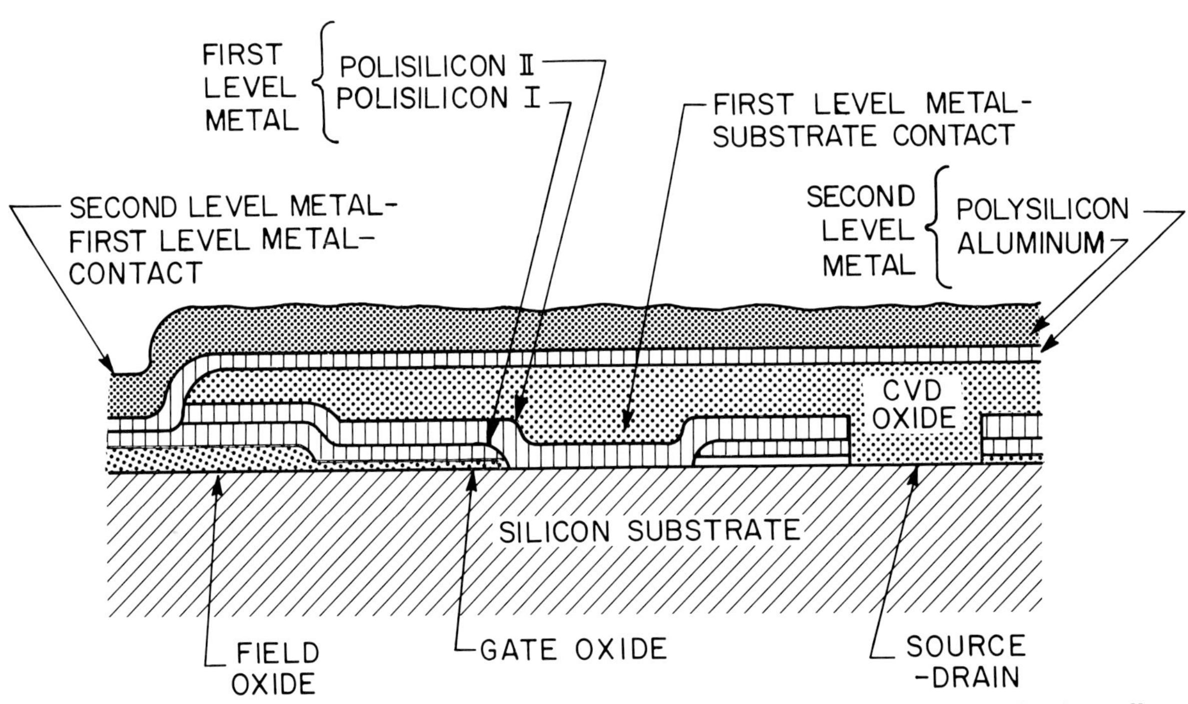

FIGURE 99. A complete device showing two contacts (second to first level metallization and first level metallization to substrate) and source–drain and gate regions. This micrograph demonstrates a successfully planarized device.

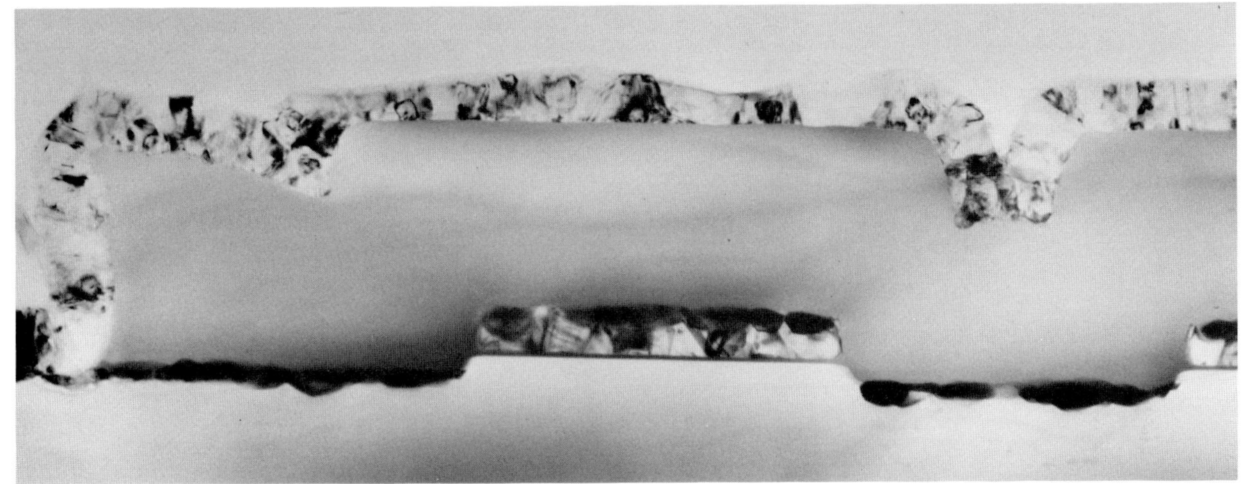

1.0μm

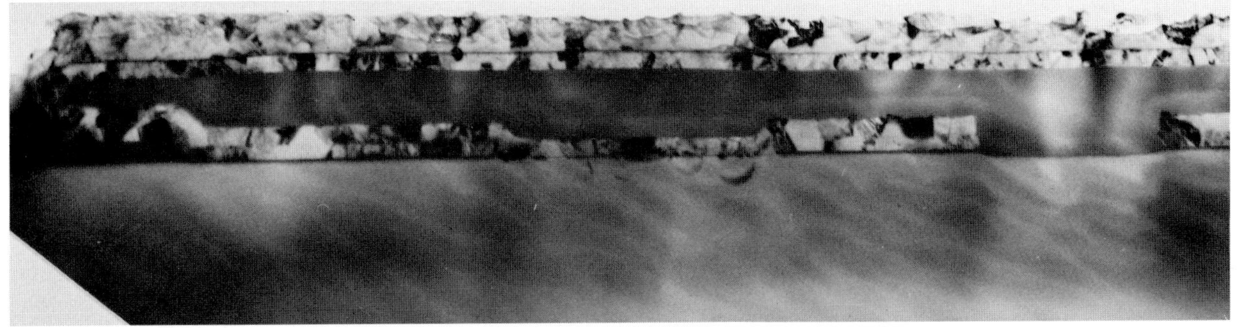

4.0μm

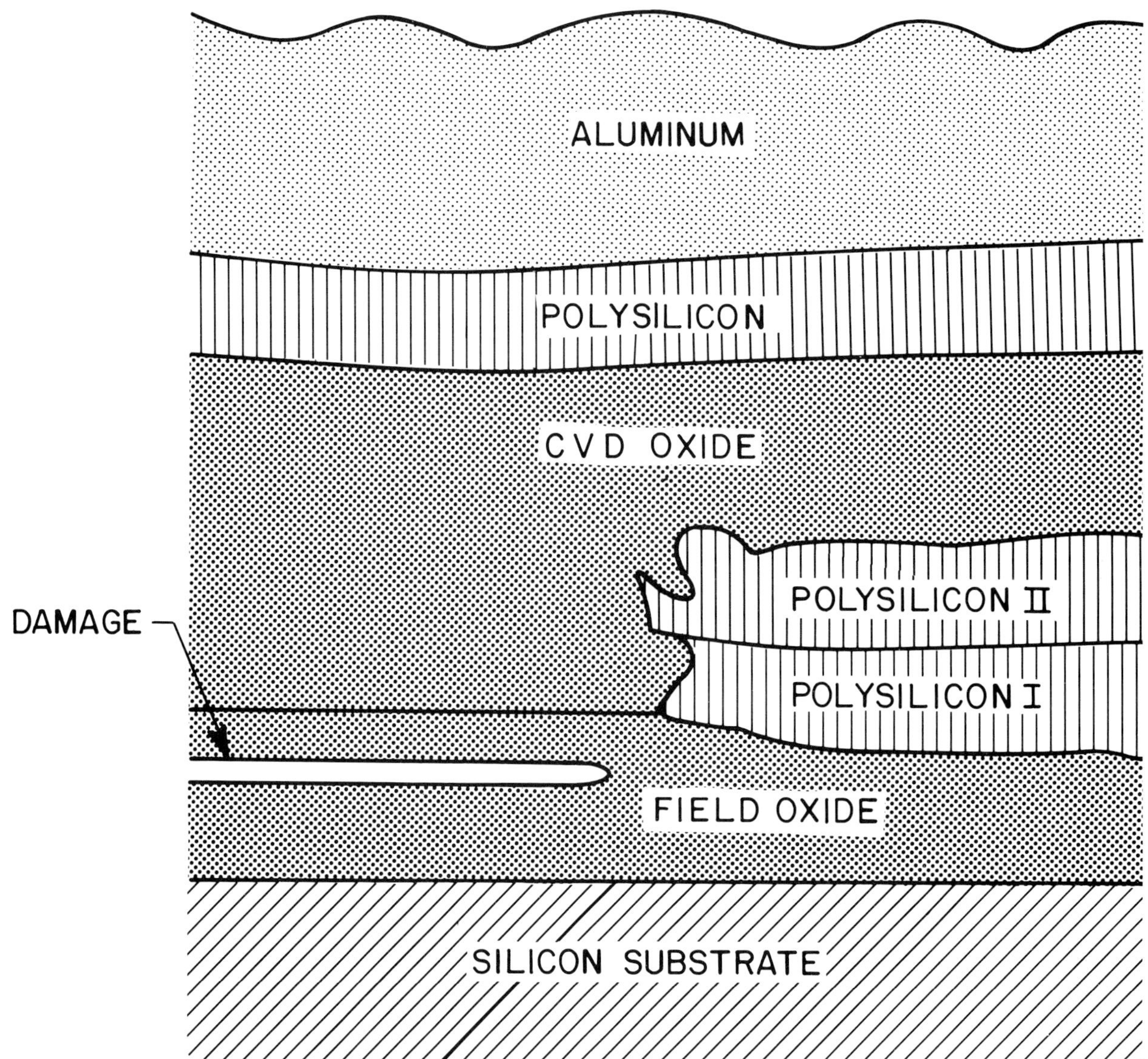

FIGURE 100. A section of the complete device shown in Fig. 99 at higher magnification. Note the absence of a step in the upper metallization layer over the edge of the first level metallization, indicating successful planarization. Implant damage can be seen in the field oxide.

shows a device made with cobalt silicide and polysilicon as the first-metallization layers. The silicide was formed by sintering the metal deposit. Uneven silicide, revealed by micrographs such as this one, is a common problem in the development of new silicides as metallization material.

Finally, Fig. 99 shows a low magnification view of a device that represents successful processing. The figure shows a number of features that were de-

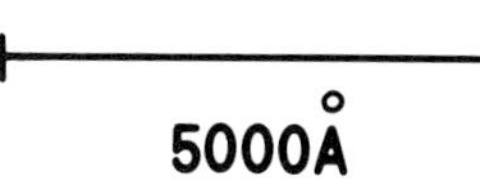
5000Å

scribed earlier, including two types of contacts, the gate and source–drain area, and aluminum–polysilicon, second level metallization. The absence of steps in the top metallization layer shows successful planarization. Higher magnification of a part of this device (Fig. 100) shows residual implant damage in the field oxide, as previously discussed.

CHAPTER EIGHT

DEVICE LEAKAGE AND BREAKDOWN

8.1 INTRODUCTION

Many studies of the relationship between substrate defects and device performance are essentially generic, demonstrating that substrates with a certain crystallographic defect structure form devices that are leaky or in which the minority carrier lifetime is depressed. These studies are valuable, because they have helped to create an understanding of the type of quality needed in substrate material and of the gettering steps required to avoid these problems. A somewhat more difficult problem arises when the substrate defect density is low, nonuniform, or poorly understood, and analysis is needed at the exact site of device failure.

Analysis of the cause of junction and capacitor leakage and breakdown usually involves studying these features with the optical microscope as a first step. Study by other analysis methods follows, if necessary. These problems are often caused by misalignment errors, which are revealed by optical study. However, leakage and breakdown problems, caused by substrate defects introduced during processing, require further investigation, usually with chemical etching or TEM methods.[65–67] For this type of problem the location of the failure in the device needs to be known, its position marked, and additional study made at the substrate site that corresponds to the position of the defect.[68]

Electron-beam induced current (EBIC) studies are useful in locating the leakage site in a *pn* junction,[67,69] Schottky-barrier contact,[70] or MOS capacitor.[71,72] Once located, the site must be marked in a manner that will withstand TEM sample preparation so that the defect site can then be located and studied by TEM.

Two examples of the application of EBIC–TEM techniques to failure analysis problems are described below. The first example involves *pn*-junction leakage,[69] and the other example involves gate oxide leakage and breakdown.[72] In both examples TEM studies show the nature of the defect giving rise to the corresponding leakage problem.

8.2 *pn*-JUNCTION LEAKAGE

In this first example, leakage studies using van der Pauw testers on wafers of different thickness showed a consistent difference in the measured leakage current (Fig. 101). These junctions were formed by diffusing p^+ regions in *n*-type substrates, and the thicker (508 μm) wafers showed a significant lowering of the breakdown field compared with the thinner (355 μm) wafers. SEM studies of a number of these junctions showed EBIC signals corresponding to increased reverse junction current at sites along the p^+n-junction perimeter (Fig. 102*b*), probably because of local enhancement of the tunneling current caused

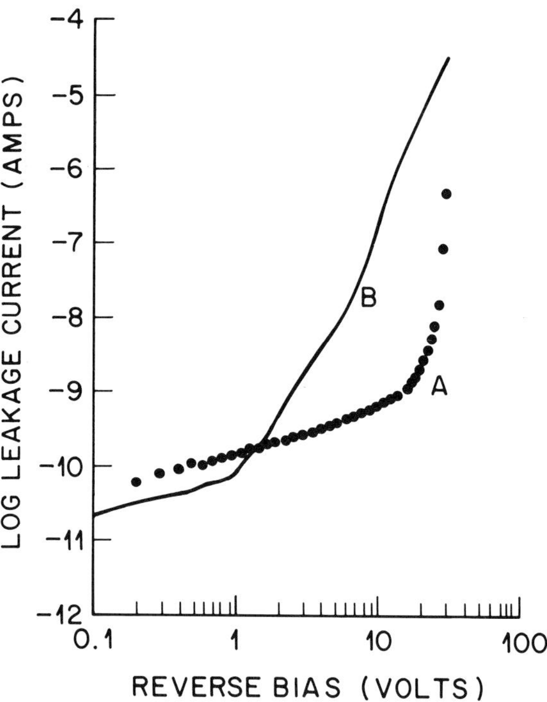

FIGURE 101. Typical leakage currents from p^+n junctions on wafers of two different thicknesses showing a higher leakage on the thicker material. Wafer A is 355 μm thick and wafer B is 508 μm thick.

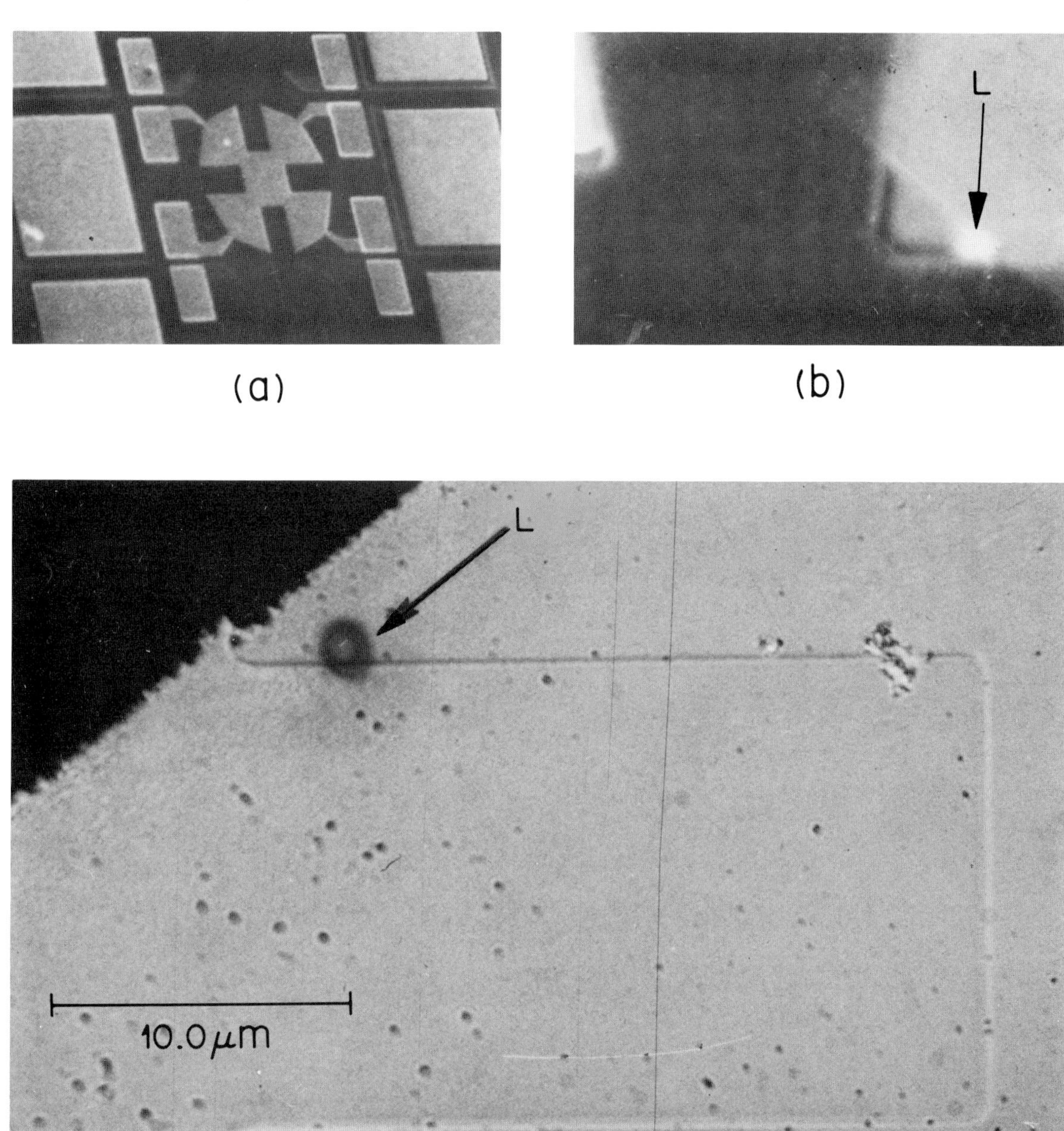

FIGURE 102. Studies of van der Pauw p^+n junctions. (*a*) SEM secondary-electron image. (*b*) Combination secondary-electron and EBIC image from upper-left area of tester showing leakage site L. (*c*) Optical micrograph of defect site after ion milling showing leakage site marked by polymerized hydrocarbon deposit L.

by local defect-induced field perturbation.[73] The presence of EBIC signals at a junction edge clearly was correlated with junction leakage, and the next step in the procedure was an examination of the EBIC site at high resolution. The site was marked with locally polymerized hydrocarbon resulting from the SEM study (Fig. 102*c*) The sample was thinned by ion milling from the back side until a hole close to the marked site was produced (Fig. 102*c*), and the site was then examined by TEM (Fig. 103).

Figure 103*a* is a low-magnification micrograph of the defect region with a stacking fault intersecting the junction in the approximate position expected from the polymerized hydrocarbon mark. The junction is made visible because of differential etch rates of n and p^+ material (see Chapter 5). The clearer micrographs in Fig. 103*b* and Fig. 103*c* show the stacking fault plus precipitate and the precipitate alone, respectively. A two-beam condition was used to obtain the image shown in Fig. 103*b* and a slight change in sample tilt effectively extinguished the stacking-fault contrast so that the precipitate could be made clearly visible in Fig. 103*c*. Copper was found in the precipitate by X-ray emission spectroscopy, and a dislocation tangle has been created at the precipitate.

Leakage was found to correlate with the presence of a precipitate within a stacking fault. Subsequent processing used a higher temperature for PBr_3 gettering, and leakage decreased to acceptable values. Although these (latter) wafers still contained stacking faults intersecting the p^+n junctions created by an oxidizing step, the faults were, however, free of heavy metal precipitates. Thus, precipitates, not stacking faults, were shown to be the major cause of leakage.

8.3 GATE OXIDE LEAKAGE AND BREAKDOWN

Another type of failure analysis problem was the occurrence of low breakdown fields (4–6 MV/cm) on certain capacitors made with oxides having thicknesses of 90–250 Å; other capacitors on the same wafers had higher breakdown fields (9 MV/cm).[72] Figure 104 shows typical leakage curves for two of these capacitors. EBIC study showed a high density (10^5 cm^{-2}) of high-current EBIC sites on leaky capacitors, while capacitors with low leakage showed few or no such sites. Increasing a capacitor bias to breakdown caused failure at one of these EBIC sites. When the polysilicon field plate and gate oxide were stripped off the discrete capacitors and replaced with thin aluminum films to form Schottky barriers, EBIC signals were found at identical positions, proving that the defects responsible for the leakage were within the substrate (Figs. 105*a* and 105*b*).

TEM study of a number of such sites showed the same type of defect in all sites (Fig. 106): a precipitate mass in a stacking fault. Microanalysis of these precipitates showed varying amounts of copper, nickel, and tin (Fig. 107). Both the stacking faults and decoration of the faults were found to occur at the field-oxidation processing step. These faults were removed by applying a gettering step prior to field oxidation.

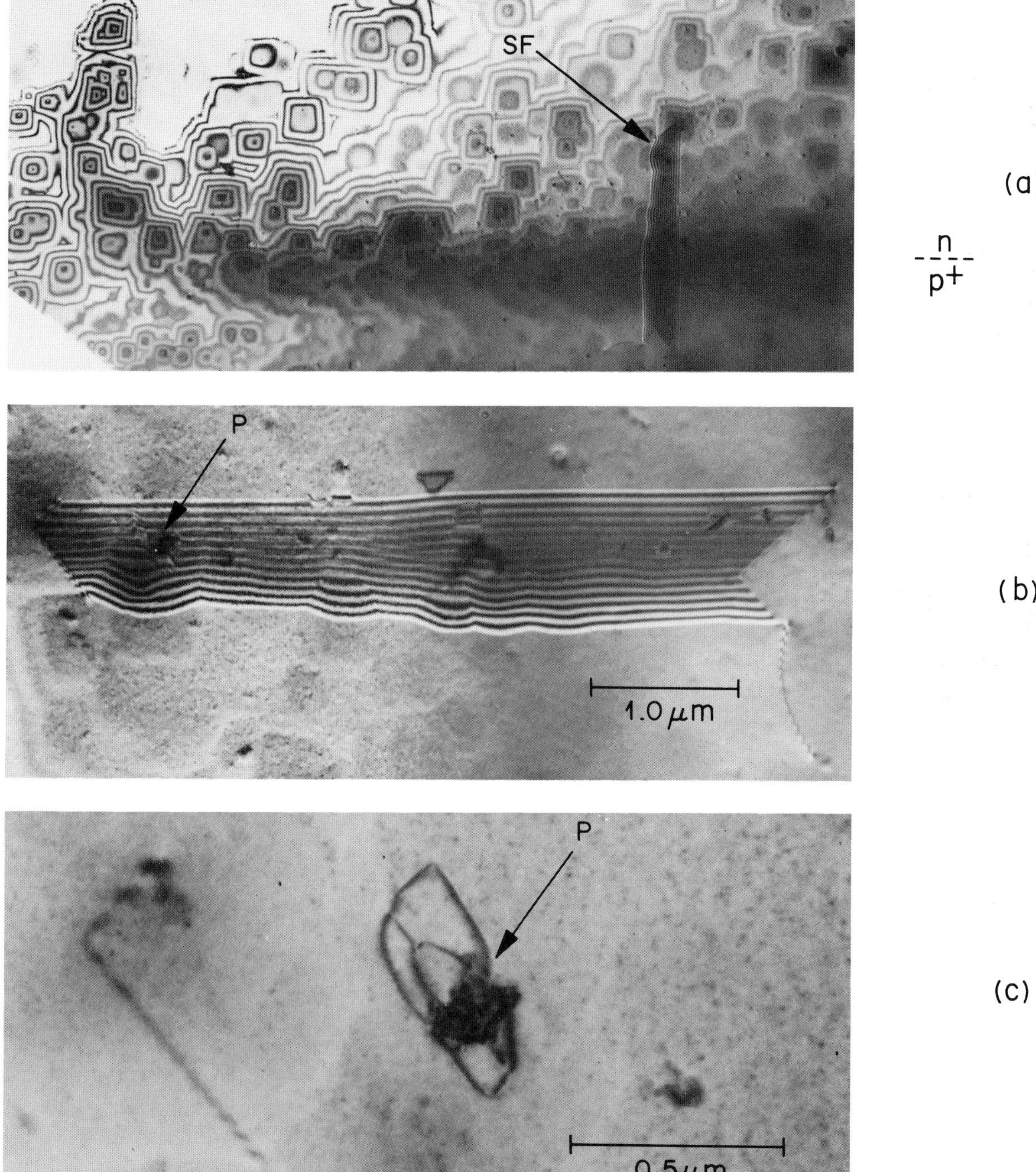

FIGURE 103. TEM photographs of a p^+n junction leakage region (see Fig. 102*b*,*c*). A stacking fault SF intersects the junction at the leakage site (upper). Higher magnification view of the stacking fault with the precipitate P (middle), and higher magnification view of the precipitate alone P (lower).

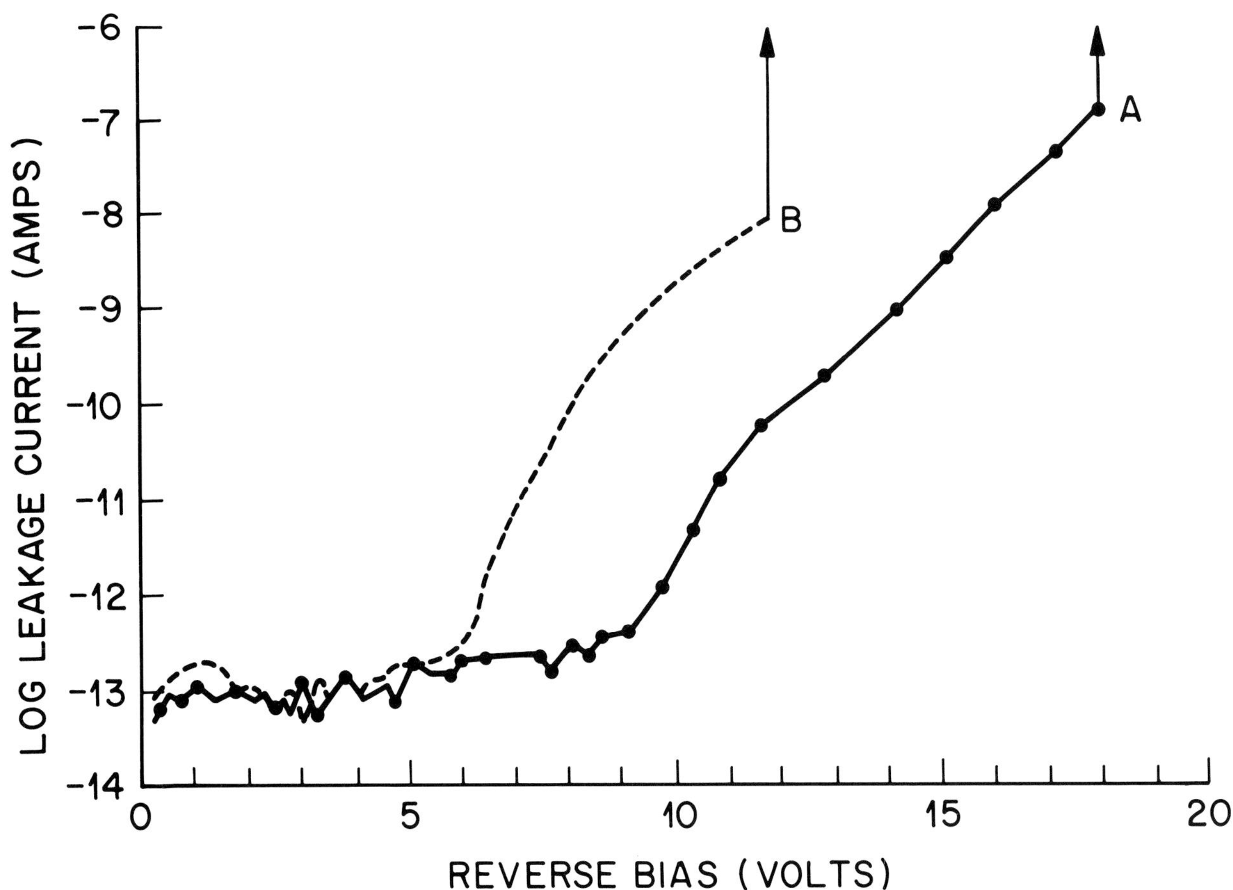

FIGURE 104. Leakage currents for MOS capacitors with 220-Å gate oxide showing a higher leakage from capacitor B.

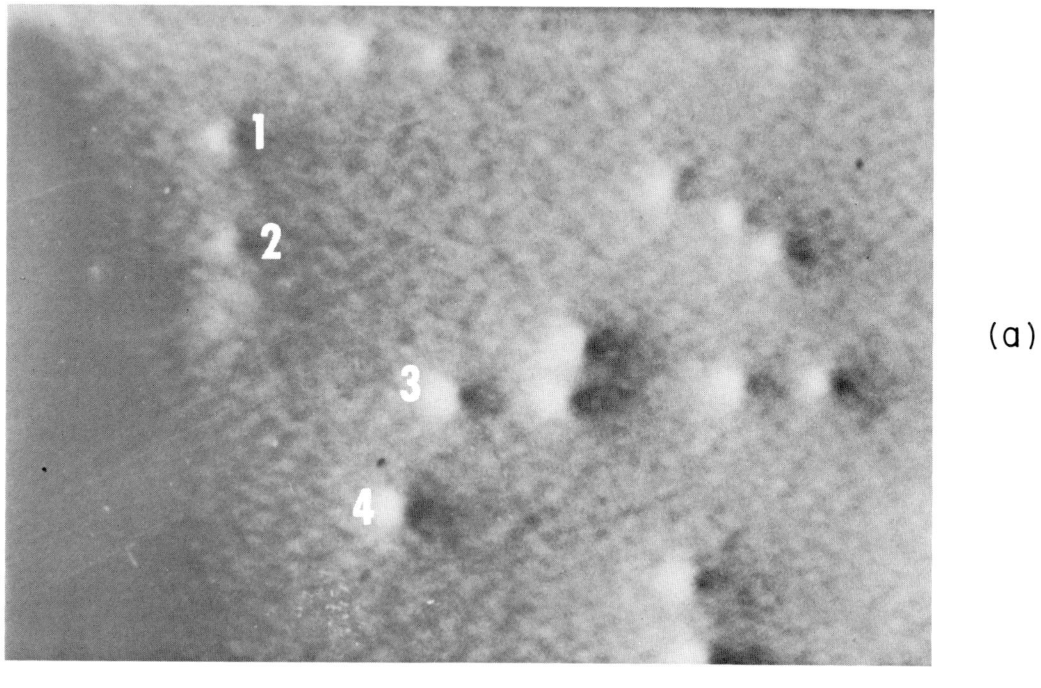

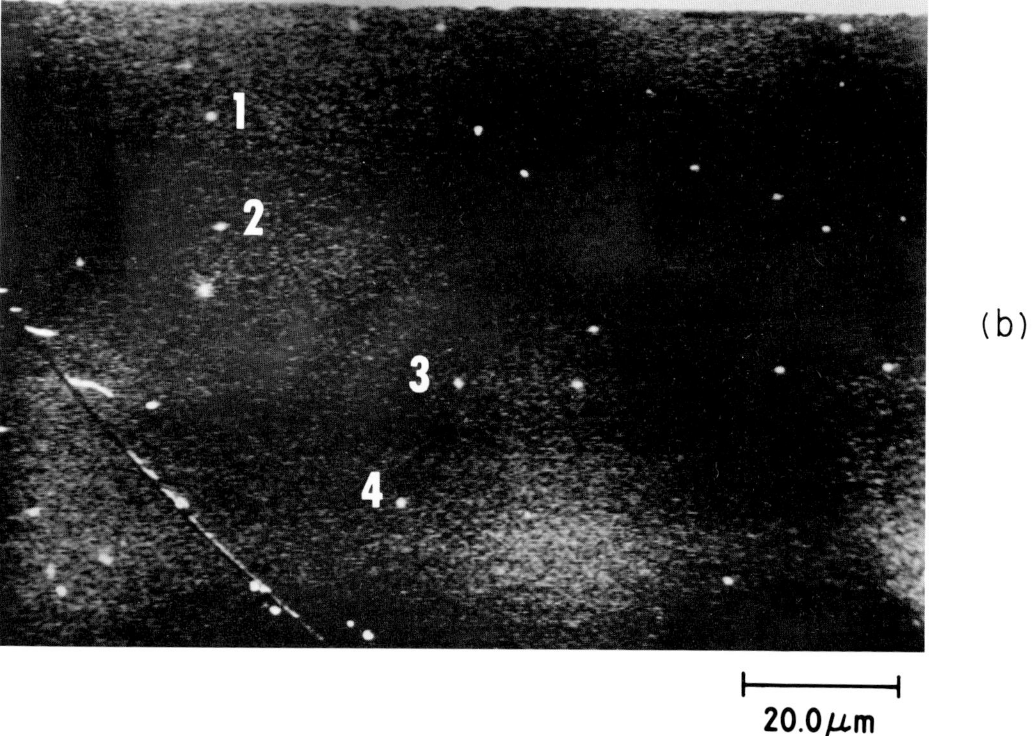

FIGURE 105. EBIC study of MOS capacitors. (*a*) EBIC image of a leaky capacitor. (*b*) EBIC image of a Schottky-barrier junction at the same site. The leakage sites are shown as black–white spots and white spots, respectively. The presence of identical sites in (*a*) and (*b*) (four sites are marked) shows that the defects are in the substrate.

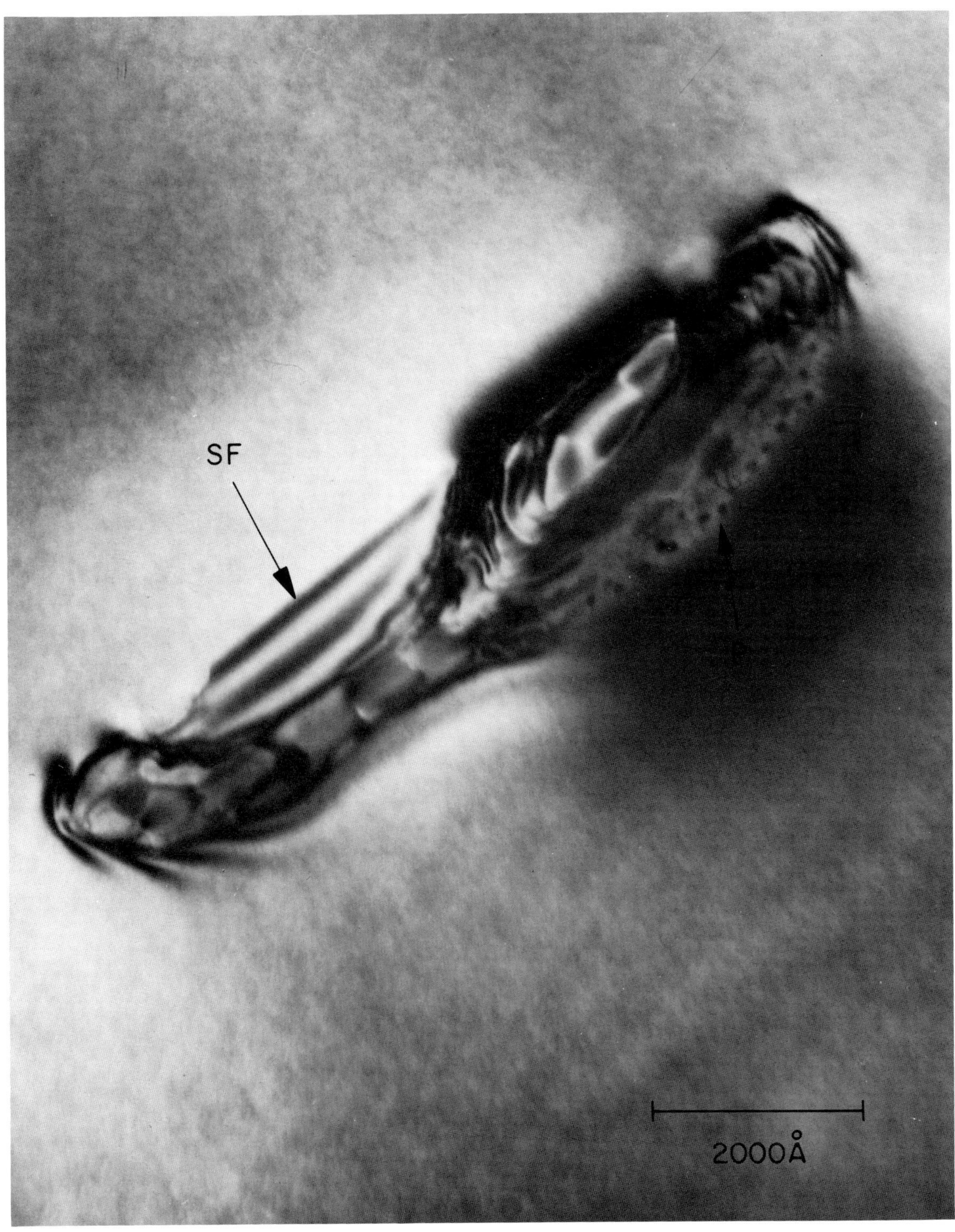

FIGURE 106. TEM photograph of the defect at leakage site 2 (Fig. 105*b*). This defect is typical and shows a precipitate mass P at a stacking fault SF.

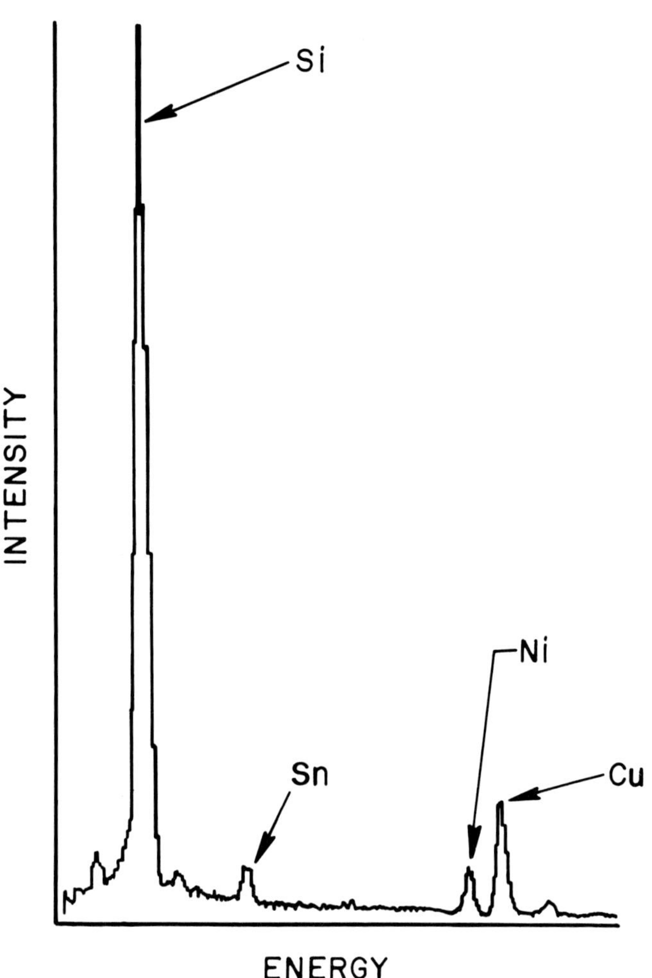

FIGURE 107. X-ray spectrum of the upper part of the precipitate complex in Fig. 106, showing the presence of copper, nickel, and tin.

REFERENCES

1. J. S. Kilby, *IEEE Trans. on Electron Devices,* **ED-23,** 648 (1976).
2. G. C. Dacey and I. M. Ross, *Proc. IRE* **41,** 970 (1953).
3. W. Fichtner, in *Proceedings of the NATO Advanced Research Institute on Microelectronics,* R. Dingle, Ed., Plenum, New York, 1982.
4. G. Nomarski and A. R. Weill, *Bull. Soc. Fr. Mineral. Crystallogr.,* **77,** 840 (1954).
5. J. I. Goldstein, D. E. Newbury, P. Echlin, D. C. Joy, C. Fiori, and E. Lifshin, *Scanning Electron Microscopy and X-Ray Microanalysis,* Plenum, New York, 1981, Chapters 3 and 4.
6. P. B. Hirsch, A. Howie, R. B. Nicholson, D. W. Pashley, and M. J. Whelan, *Electron Microscopy of Thin Crystals,* Plenum, New York, 1965.
7. G. Thomas, *Philos. Mag.,* **17,** 1097 (1968).
8. M. C. King, in *VLSI Electronics,* N. G. Einspruch, Ed., Academic, New York, 1981, Chapter 2.
9. E. Spiller and R. Feder, in *X-Ray Optics,* H. J. Queisser, Ed., Springer-Verlag, New York, 1977, Chapter 3.
10. K. J. O'Conner and R. A. Kushner, to be published in *Proceedings of the International Solid State Circuits Conference,* Feb. 23–25, 1983, New York.
11. R. A. Kushner, in *Proceedings SEMICON-West California,* 1982, p. 7.
12. A. C. Adams and C. D. Capio, *J. Electrochem. Soc.,* **128,** 423 (1981).
13. R. D. Heidenreich, *Fundamentals of Transmission Electron Microscopy,* Wiley, New York, 1964.
14. J. J. Hren, J. I. Goldstein, and D. C. Joy, *Introduction to Analytical Electron Microscopy,* Plenum, New York, 1979.
15. T. T. Sheng and R. B. Marcus, *J. Electrochem. Soc.,* **127,** 737 (1980).
16. T. T. Sheng, to be published in *Rev. Sci. Inst.*
17. G. Wallis, *Electrocomp. Sci. Tech.,* **2,** 44 (1975).
18. T. T. Sheng and R. B. Marcus, *J. Electrochem. Soc.,* **128,** 881 (1981).
19. A. C. Adams, F. B. Alexander, C. D. Capio, and T. E. Smith, *J. Electrochem. Soc.,* **128,** 1545 (1981).
20. E. Bassous, H. N. Yu, and V. Maniscalo, *J. Electrochem. Soc.,* **123,** 1729 (1976).
21. E. Kooi, J. G. van Lierop, and J. A. Appels, *J. Electrochem. Soc.,* **123,** 1117 (1976).

22. T. A. Shankoff, T. T. Sheng, S. E. Haszko, R. B. Marcus, and T. E. Smith, *J. Electrochem. Soc.*, **127**, 216 (1980).
23. L. Katz, Bell Laboratories, private communication.
24. T. T. Sheng and R. B. Marcus, *J. Electrochem. Soc.*, **125**, 432 (1978).
25. L. O. Wilson, *J. Electrochem. Soc.*, **129**, 831 (1982).
26. A. S. Grove, *Physics and Technology of Semiconductor Devices*, Wiley, New York, 1967, pp. 23–31.
27. M. A. Hopper, R. A. Clarke, and L. Young, *J. Electrochem. Soc.*, **122**, 1216 (1975).
28. F. E. Holmes and C. A. T. Salama, *Solid State Electron.*, **17**, 791 (1974).
29. R. B. Marcus and T. T. Sheng, *J. Electrochem. Soc.*, **129**, 1278 (1982).
30. E. P. EerNisse, *Appl. Phys. Lett.*, **35**, 8 (1979).
31. E. A. Irene, E. Tierney, and D. W. Dong, *J. Electrochem. Soc.*, **127**, 705 (1980).
32. D. J. DiMaria and D. R. Kerr, *Appl. Phys. Lett.*, **27**, 507 (1975).
33. R. B. Marcus, T. T. Sheng, and P. Lin, *J. Electrochem. Soc.*, **129**, 1282 (1982).
34. P. A. Heimann, S. P. Murarka, and T. T. Sheng, *J. Appl. Phys.*, **53**, 6240 (1982).
35. R. Edwards et al. in *Semiconductor Silicon 1973*, H. R. Huff and R. R. Burgess, Eds., The Electrochemical Society, Princeton, N.J. 1973, p. 905.
36. S. P. Murarka, Bell Laboratories, private communication.
37. A. K. Sinha and T. T. Sheng, *Thin Solid Films*, **48**, 117 (1978).
38. J. M. Poate, K. N. Tu, and J. W. Mayer, Eds., *Thin Films—Interdiffusion and Reactions*, Wiley, New York, 1978, p. 293.
39. S. Vaidya, D. B. Fraser, and A. K. Sinha, in *Proceedings of the 18th Annual Reliability Physics Symposium*, April 8–10, 1980, Las Vegas, p. 165.
40. I. A. Blech, D. B. Fraser, and S. E. Haszko, *J. Vac. Sci. Technol.*, **15**, 13 (1978).
41. J. T. Clemens and K. Locke, Bell Laboratories, private communication.
42. S. Vaidya, *J. Electron. Mater.*, **10**, (2) 337 (1981).
43. S. P. Murarka, *J. Vac. Sci. Technol.*, **17**, 775 (1980).
44. S. P. Murarka, D. B. Fraser, A. K. Sinha, and H. J. Levinstein, *IEEE Trans. on Electron Devices*, **ED-27**, 1409 (1980).
45. C. J. Koeneke, S. M. Sze, R. M. Levin, and E. Kinsbron, Proc. Int. Electron Device Meeting, Washington, D.C., 1981, p. 367.
46. S. P. Murarka, D. B. Fraser, T. F. Retajczyk, Jr., and T. T. Sheng, *J. Appl. Phys.*, **51**, 5380 (1980).
47. M. von Allmen, S. S. Lau, T. T. Sheng, and M. Wittmer, in *Laser and Electron Beam Processing of Materials*, C. W. White and P. S. Perry, Eds., Academic, New York, 1980, p. 524.
48. W. H. Schroen, G. A. Lee, and F. W. Voltmer in *Semiconductor Measurement Technology: Spreading Resistance Symposium*, J. R. Ehrstein, Ed., National Bureau of Standards Special Publ. 400–10, U.S. Government Printing Office, Washington, D.C., 1974, p. 155.
49. J. Poate, Bell Laboratories, private communication.
50. C. C. Chang, Bell Laboratories, private communication.
51. V. Deline, Charles Evans Associates, San Mateo, Cal., private communication.
52. W. Schockley, U.S. Patent 2,787,564 (1954).
53. P. Sigmund and J. Sanders, in *Proceedings of the Conference on the Applications of Ion Beams to Semiconductor Technology*, P. Clotin, Ed., Grenoble, 1967, ed. Orphyrs, p. 215.
54. G. Dearnaley, J. H. Freeman, R. S. Nelsen, and J. Stephen, *Defects in Crystalline Solids*, North-Holland, Amsterdam, 1973, p. 173.
55. Y. Wada and N. Hashimoto, *J. Electrochem. Soc.*, **127**, 461 (1980).
56. S. Mader and A. E. Michel, *Phys. Status Solidi*, **33**, 793 (1976).
57. T. E. Seidel, Bell Laboratories, private communication.
58. J. F. Gibbons, W. S. Johnson, and S. W. Mylroie, *Projected Range Statistics*, Dowden, Hutchinson and Ross, Stroudsburg, Pa., 1975.

59. W. Fichtner, Bell Laboratories, private communication.

60. R. B. Fair and J. C. C. Tsai, *J. Electrochem. Soc.*, **122,** 1689 (1975).

61. E. Kinsbron, S. P. Murarka, T. T. Sheng, and W. T. Lynch, presented at Electronic Materials Conference '82, June 23–25, 1982, Colorado State University, Fort Collins, paper H-2.

62. W. Fichtner, R. M. Levin, T. T. Sheng, and R. B. Marcus, in *Proceedings of the Electrochemical Society Meeting,* May 9–14, 1982, Montreal, Canada, p. 286.

63. A. K. Sinha, T. T. Sheng, T. A. Shankoff, W. S. Lindenberger, E. N. Fuls, and C. C. Chang, in *Proceedings of the 17th Annual Reliability Physics Symposium,* April 24–26, 1979, San Francisco, p. 35.

64. R. M. Levin, Bell Laboratories, private communication.

65. T. E. Seidel, S. E. Haszko, and D. M. Maher, *J. Appl. Phys.*, **218,** 5038 (1977).

66. H. Shiraki, *Jpn. J. Appl. Phys.*, **13,** 1514 (1974).

67. K. V. Ravi, C. J. Varker, and C. E. Volk, *J. Electrochem. Soc.*, **120,** 533 (1973).

68. R. B. Marcus, in *Silicon VLSI Technology*, S. Sze, Ed., McGraw-Hill, New York, 1983, Chapter 12.

69. J. M. Dishman, S. E. Haszko, R. B. Marcus, S. P. Murarka, and T. T. Sheng, *J. Appl. Phys.*, **50,** 2689 (1979).

70. R. B. Marcus, S. E. Haszko, and J. C. Irwin, *J. Electrochem. Soc.*, **121,** 692 (1974).

71. P. Roitman and W. R. Bottoms, in *Scanning Electron Microscopy 1977,* O. Johari, Ed., IITRI, Chicago, 1977, p. 731.

72. P. S. D. Lin, T. T. Sheng, and R. B. Marcus, to be published in *J. Electrochem. Soc.*

73. C. J. Varker and K. V. Ravi, *J. Appl. Phys.*, **45,** 272 (1974).

INDEX